MATHEMATICS
in a
POSTMODERN AGE

MATHEMATICS *in a* POSTMODERN AGE

A Christian Perspective

Edited by

Russell W. Howell and W. James Bradley

William B. Eerdmans Publishing Company
Grand Rapids, Michigan / Cambridge, U.K.

Wm. B. Eerdmans Publishing Co.
255 Jefferson Ave. S.E., Grand Rapids, Michigan 49503 /
P.O. Box 163, Cambridge CB3 9PU U.K.

Printed in the United States of America

06 05 04 03 02 01   7 6 5 4 3 2 1

Library of Congress Cataloging-in-Publication Data

Mathematics in a postmodern age: a Christian perspective /
edited by Russell W. Howell and W. James Bradley.

p. c.

Includes bibliographical references.

ISBN 0-8028-4910-5 (pbk.: alk. paper)

1. Mathematics — Philosophy. 2. Religion and science.
I. Howell, Russell W. II. Bradley, W. James, 1943-

QA8.6.M385 2001
510′.1 — dc21

2001019360

www.eerdmans.com

Contents

Acknowledgments

This book is not simply a collection of essays independently authored by different individuals. We all have "signed off" on the project as a whole. This does not mean we are in complete agreement with everything that is said throughout this book; indeed, on many points there is considerable disagreement. It does mean, however, that collectively we think the ideas presented are done so in an academically respectable manner, and deserve serious consideration. The principal authors for the various chapters are Paul Zwier of Calvin College (Chapter 1); Glen VanBrummelen of Bennington College (Chapter 2); Christopher Menzel of Texas A&M University (Chapter 3); William Dembski of Baylor University (Chapters 4 and 10); Calvin Jongsma of Dordt College (Chapters 5 and 6); James Bradley of Calvin College (Chapter 7); Michael Veatch of Gordon College (Chapter 8); Russell Howell of Westmont College (Chapter 9); Scott VanderStoep of Hope College (Chapter 11); and David Klanderman of Trinity College (Chapter 11). The authors owe a great debt to their spouses and families for their patience, encouragement, and assistance.

There are two organizations whose help was indispensable in the formation of this book. We are grateful to the Calvin Center for Christian Scholarship (CCCS) for the financial support that made it possible for a team of ten people to be co-authors of this project. We are also grateful to the Association of Christians in the Mathematical Sciences (ACMS), who generously assisted with the publishing costs and whose organization facilitated the formation of our team in the first place. Ronald Wells, James Bratt, and Donna Romanowski of CCCS were invaluable in their support and encouragement, as was Robert Brabenec, executive secretary of ACMS. Several reviewers also provided invaluable input, both formal and informal. We would like to thank especially Jim Turner of Calvin College and Paul Bialek of Trinity International University for the care they took in going through the manuscript, and for the many valuable comments and suggestions they made. And, of course, we thank God, who is the author of all wisdom and knowledge.

Introduction

We can almost hear the hallway whispers in response to the title of this book. "We certainly see the importance of discussing mathematics in a postmodern age, but, come on, what possible relevance could a Christian perspective bring to the issue? Is the Pythagorean Theorem different for Christians?"

Of course, the Pythagorean Theorem is not different for Christians. Even so, we believe there are a number of ways in which a Christian perspective can enrich our understanding of mathematics. Conversely, we think that many ideas in mathematics can enhance our understanding of the Christian faith. But before we elaborate on this thesis, we should clarify what we mean by a *Christian* perspective. The Christian faith has its roots in Judaism, which has existed from about 2000 B.C. Since that time many Jewish and Christian thinkers have reflected carefully and deeply on the human situation. Guided by their faith, they have developed rich and powerful ideas that are applicable to many different areas. In looking at the possibility of forming a Christian perspective on mathematics, then, we seek to do the same. That is, we make a serious attempt to ask whether any ideas that might spring from a Christian faith commitment can enrich our understanding of mathematics, and whether the ideas of mathematics can contribute to and enlarge our understanding of the Christian faith. Thus, we examine mathematics and the Christian faith from an unfamiliar angle in the hope that such an examination will yield new insights. George Marsden calls such an endeavor *Christian-informed scholarship,*[1] and we are comfortable with that description of our work here.

1. George Marsden, *The Outrageous Idea of Christian Scholarship* (New York: Oxford University Press, 1997).

What would Christian-informed scholarship mean in our project? Since the time of Galileo (early 1600s), the notion that mathematical and religious thought belong to different realms having little or nothing in common has gained considerable credibility. It is not hard to understand why this is the case. Mathematicians have worked hard to frame their discipline as abstract and formal. Especially in the twentieth century they intentionally separated mathematics from the scientific and cultural contexts in which mathematical questions originated. Many had laudable purposes in doing this. They were pursuing goals such as gaining generality and laying a foundation for mathematics that would be independent of any particular individual or cultural experience. Their belief was that abstraction would free mathematics from the biases and presuppositions that so easily slip into thought from one's personal experience or perspective. Indeed, much of the global credibility that mathematics has enjoyed is due to its success in attaining these goals. From this perspective it is easy to see how the jibe, "Is the Pythagorean Theorem different for Christians?" arises. Religious presuppositions are precisely the kinds of distracters that mathematical abstraction has sought to avoid.

So, when we speak of a Christian perspective, we are not challenging the value of abstraction or attempting to turn the clock back to a different era. To the contrary, we actively support the notions of abstraction and rigor. But we don't want to stop there. Rather, we want to look at the mathematical endeavor, its place, nature, and contribution within the broader context of human thought and culture. We want to ask, "How can a vision of mathematics that incorporates its cultural role, its relationship to other areas of thought, its philosophical foundations, and its technical content enrich our understanding of it?" Conversely, we ask whether mathematics itself might contribute to a Christian worldview. Specifically, we are concerned with issues like the following:

- Many claim that Western culture is in a major paradigmatic shift from the "modern" to "postmodern" eras. The intellectual shape of the modern age was, however, highly influenced by mathematical ideals. What exactly were those ideals and how did modernism instantiate them? Does the postmodern spirit affirm or deny them? What are the consequences of this shift for mathematics and for its role in the larger culture?
- Mathematics has often been viewed as the epitome of the application of "pure reason." The issue of the proper relationship between faith

and reason, however, has occupied the thought of many outstanding thinkers throughout the Christian era. If the view of mathematics as an application of reason is correct, any limitations of reason are also limitations of mathematics; any capabilities mathematics possesses are also capabilities of reason. Thus, consideration of the nature of mathematics may help clarify this relationship.

- For the past three hundred years or so "foundationalism," an approach modeled on the axiomatic method of mathematics, has influenced philosophic thinking. Many scholars today, however, regard foundationalism as misguided. Does this change in the status of foundationalism have any implications for the practice of mathematics? What does it portend for the role of mathematics in Western culture?
- Mathematicians tend to be Platonists. That is, they view mathematical objects such as groups, vector spaces, or the axioms of plane geometry as having an existence independent of human minds, although it is unclear where this existence can be located. Many Christians would locate mathematical objects "in the mind of God." Some thinkers, however, have been nominalists and deny such an independent existence. Furthermore, postmodern thinkers have tended to move away from Platonism. What, then is the nature of mathematical objects, and of mathematical thinking?
- Religious concepts use terms such as 'infinity' or 'paradox' that are also common in mathematics. Is this just a coincidence, or could these mathematical ideas give us insights into aspects of our religious beliefs? What other specific mathematical ideas might say things about the Christian faith?
- Mathematics has had a profound impact on human cultures, especially contemporary Western culture. Can we articulate what that impact has been? Can we provide a framework of norms and values for assessing that impact? Is it desirable, or even possible, to shape or predict the future form of that impact?
- Most contemporary societies invest an enormous amount of their wealth and their children's time in mathematics education. Substantial investments are also made in mathematical research. That is, contemporary cultures have placed an enormous amount of trust in their mathematical communities; these communities in turn have an enormous responsibility to their respective cultures. How can mathematical communities best meet this responsibility?

- David Hilbert, one of the greatest mathematicians of the twentieth century, has claimed that mathematics is a "presuppositionless science." Is this the case, or does mathematics have its own values and presuppositions? If so, can we clearly articulate those? Can we constructively critique them?

It seems to us that consideration of issues like these is important for a broader understanding of mathematics. As Christian thinkers have thought deeply about many of these matters, we believe that Christian concepts have a great deal to contribute to the discussion of such issues. We hope that this book will stimulate further conversation about them.

We also want to be straightforward concerning our own presuppositions. While we will try throughout the book to acknowledge our assumptions wherever they are used, they provide a general overarching framework that shapes even the way we ask the questions we seek to address, so it is impossible to point out every instance of their application. Here, then, is a brief summary of the main features of our presuppositions. First, we believe in the God of the Old and New Testaments, and believe that he is the creator of the universe. In particular, he is our creator, and for reasons we don't fully understand he has created us with the capacity to engage in mathematical inquiry. Thus, we reject the notion that this capacity is neutral, a tool to be used for good or ill. Rather we affirm that such capacity is inherently good and hence has intrinsic value. Second, we believe that humans are made in God's image, but they are "fallen" — that is, they have rejected God's authority over them and have replaced it with a claim of autonomy. This attempt at autonomy is a source of human evil. Hence, although mathematical capacity is good, humans can engage in mathematical exploration for evil ends. Third, we believe it is the responsibility of human beings to work for "redemption" — that is, to work to overcome the consequences of the fall. In a fundamental sense, of course, true redemption is possible only by the work that Christ did on the cross for us. But out of gratitude for this gift of God, we as believers desire to help fulfill God's purposes for creation, and that is the sense in which we work for redemption. For Christians interested in mathematics, this entails endeavoring to discern God's purposes in giving human beings the capacity to engage in mathematical activity, and seeking to help the mathematics community fulfill those intentions.

What exactly were God's intentions in giving us the capacity to engage in mathematical inquiry? Scripture does not provide a direct answer to this question, so the answer must be inferred from broader purposes that have been revealed. If we go back to Genesis 1 and 2, we see that God's original purpose was that we be co-creators with him in two ways: that we be stewards of this world, walking closely with him in using his creation to build cultures and to care for this world, and that we ourselves would be built into "sons of God." But carefully studying anything in this world often involves forming precise definitions, measuring, and thinking deductively about the way things are and the way they might be. Thus mathematics is an essential component of co-creating. While such a vision does not give precise answers to every value question we might ask, it gives us a framework from which to start. If, as we affirm, the capacity to do mathematics is a good gift of God, it reveals something about his nature, for example, his subtlety, order, beauty, and variety. When people respond to these qualities of mathematics with awe and joy and turn to God with reverence and thankfulness, they are fulfilling this purpose. Thus, we immediately see one consequence of a Christian perspective — mathematics does not need to be "applicable" to be of value. However, applicability is also important, as a second purpose of mathematics is its use in helping to build human cultures, to serve people, and to care for the earth. Thus a second consequence of a Christian perspective is that the mathematics community cannot stop at considering the abstract, formal aspects of mathematics, but must consider the consequences of these abstractions when they are reintroduced into the human community.

This concludes our brief sketch of our underlying assumptions. We do not expect every reader to agree with all of these — some will reject them *in toto*. Some will disagree with certain points or with emphases. Furthermore, we do not expect that those readers who do agree with our assumptions will necessarily agree with every position put forth in this book. Indeed, the individual authors of various chapters do not uniformly agree with each other about what they have said. We believe, however, that the ideas we present here are credible, and we hope that this book succeeds in demonstrating that the kind of thinking we are seeking to do can enrich our understanding of both mathematics and the Christian faith.

Returning to the title of this book, what exactly do we explore? Painting with broad strokes, we ask if our Christian faith has any bearing on how we look at the nature of mathematics, from our views of its truth claims to the status we ascribe to mathematical objects. Some of our thinking will be based on a comparative study of modern and postmodern thinkers, as well as a historical cultural study relating to how mathematics progressed in various settings. We also ask if our Christian faith has any bearing on how we might view the place of mathematics in culture today. We propose to review, from the stance of the Christian faith operating within the mathematics profession, some of the key historical epochs in mathematical thinking, including the philosophical assumptions on which that thinking is based. We look at how mathematical ideas can directly interact with our Christian faith, and finally give an appreciative critique of mathematics, and suggest to the contemporary mathematics community that it needs to cast its performance net more broadly, and its self-critical gaze more intently, if mathematics is to do all it ought to do.

Specifically, we begin in Chapter 1 with a discussion of what we mean by modernism and postmodernism and then look at representatives of mathematical thinking from those traditions. In particular, we explore the ideas of Gottlob Frege, a prominent logician, and Paul Ernest, a contemporary philosopher of mathematics. Frege (1848-1925) predates Ernest (1944-), and Frege's desire to put all of mathematics on an indisputably firm logical foundation did not succeed. It would be a mistake to infer, however, that because of this we somehow think Frege's views are no longer credible. Today, Frege's work is still recognized as very perceptive and influential.[2] We chose him because he is a good example of how a contemporary thinker in the modern school of thought might view mathematics. What, then, were Frege's views, what caused part of his program to fail, and what benefits can we glean for ourselves today by studying his ideas? Questions such as these, of course, must be answered in context, and Paul Ernest provides an interesting contrast with Frege. Whereas Frege views mathematical propositions as true independent of experience, Ernest emphasizes the importance of social agreement in the acceptance of mathematical theories. On the one hand, Frege would hold that a mathematical theory is true for all time. On the other hand, Ernest views mathematical constructs as fallible, reminding us that even published proofs are unreliable guides to any sort of universal truth as they are very commonly flawed.

How should we adjudicate these claims? We do not intend to present Frege and Ernest as polar opposites, forcing our readers to choose between them. Honest thinkers may side completely with one or the other, of course,

2. See Mark Steiner, *The Applicability of Mathematics as a Philosophical Problem* (Cambridge, Mass.: Harvard University Press, 1998). Steiner argues that Frege completely solved the *semantical* problem of mathematics' applicability.

but they may also favor some middle ground. Or, they may opt for an entirely different approach. But the position a person finally adopts must come from an informed historical perspective. With this in mind, Chapter 2 traces the development of mathematical ideas in three cultures, ancient Greece, medieval Islam, and pre-modern China. Favoring Ernest, we present differences between Greek and Islamic mathematics that appear to depend on world and religious views, even though these two cultures commingled. Favoring Frege, we show that, perhaps surprisingly, the isolated Chinese mathematicians developed many of the same classic theorems (such as the Pythagorean Theorem) common to other cultures. Their methods of proof were quite different from those of Western mathematicians, but this cross-cultural commonality makes it hard to hold a position that mathematical knowledge is merely a social or linguistic convention. Such knowledge may not be universal in a Platonic sense, as the commonality we noted may speak more about the structure of our brains than anything beyond them. Nevertheless, this observation seems to counter the complete relativism that often accompanies an extreme form of postmodern thinking.

But how would we account for this observed universality of mathematical ideas? In particular, what do we think about the status of mathematical objects such as propositions, relations, and properties? Are they simply a construct of human thought (however universal that might be) that would cease to exist if humanity perished? If not, are they located within the mind of God, or do they exist independently of him? We take up these questions in Chapter 3, where we discuss the ontology of mathematical objects. We tentatively argue for a certain form of mathematical realism. This position creates problems, which we discuss. The position also raises numerous interesting paradoxes relating to the nature of God, specifically to the meaning of terms such as 'omniscient' and 'omnipotent.'

Swinging back to a more postmodern stance, Chapter 4 argues that raw logic by itself is an insufficient model for explaining the process by which mathematics progresses. For example, how do we account for the conviction in the mathematical community that certain well-established portions of mathematics, like Euclidean geometry and number theory, are consistent? Mathematicians may leave open the possibility that Euclidean geometry and number theory are inconsistent, but their confidence that these theories won't sprout contradictions is analogous to the layperson's confidence that the sun will rise tomorrow. The layperson's confidence rests on an induction from past experience (supplemented perhaps by theoretical support from the layperson's physical understanding of the world). Similarly, we argue that the mathematician's confidence in consistency rests on an induction from mathematical experience. We discuss the partial failures of Hilbert's program and propose a way of looking at mathematics that fits well in both the modern and postmodern contexts. We explore the role that pragmatism plays in mathematical inquiry to include things such as effort and strategies for tackling certain problems.

Taken together, Chapters 1-4 form our first section, which broadly concerns itself with how we might view the nature of mathematics from a Christian perspective. As important as thinking in this area is, no one would be in a position to critique the role of mathematics in culture without a good understanding of how it has influenced culture, and this is the task of Section II, which consists of Chapters 5-7.

For the purposes of our discussion, we date the modern era from the time of Descartes (1596-1650) to about 1962, the year Thomas Kuhn published his controversial book, *The Structure of Scientific Revolutions.* Thinkers of this modern era shared the belief that it is possible to construct a coherent and true view of reality, though they certainly disagreed as to the methods by which such a view would be constructed. In either case, however, they saw mathematics as capable of providing absolutely certain knowledge about the natural world without the need to appeal to any form of authoritative revelation. For some, mathematics was in the mind of God at creation. Thus, mathematics enabled them to think thoughts that are part of the very essence of God. However, from the typical modernist point of view, revelation, intuition, and other epistemologies ought to be excluded from the public sphere and should (at best) be regarded as private matters. This perspective has had enormous consequences in shaping Western science; furthermore, its influence has extended well beyond science into culture more broadly. Chapter 5 explores the historical roots of this perspective beginning with Pythagoras, Plato, Euclid, and Aristotle and traces its development up to the end of the sixteenth century. Chapter 6 focuses in depth on the seventeenth century, as this was a period in which these ideas particularly flourished. Both chapters focus primarily on what we call the *mathematization* of science.

Mathematics played a central role in the thinking of the modern era. It was seen as the language of science. That is, careful scientific observations frequently involved quantitative measurements; these were organized, studied, and served as the basis for inferring laws of nature, such as Newton's law of gravitation. These laws also had mathematical expressions. Furthermore, mathematics provided a means to apply reason more systematically and more powerfully to infer new truths from the newly discovered natural laws. Whereas theology had been seen as the queen of the sciences before the Enlightenment period, that role was now assumed by mathematics. Moreover, Enlightenment thinkers extended the influence of mathematics well beyond natural science to economics, commerce, politics, ethics, literature, and popular thought. Chapter 7 extends our study of the influence of mathematics beyond science to culture more broadly; in it we evaluate this "mathematization of culture" from a Christian perspective and make some recommendations as to how the mathematics community can address the excesses of the modern period.

Whereas Section II focuses on the process of mathematization that occurred over an extended period of time, Section III (Chapters 8-12) comes full circle to the question posed in our opening paragraph. While the Pythagorean theorem may not be different for Christians, we think there are aspects about our faith that at least shape how we view the practice of mathematics. Conversely, we think there are aspects of mathematics that speak to our Christian faith.

Chapter 8 begins with the recognition that mathematics possesses a collection of professional values that give direction to the discipline. What problems are important? Which approaches are most likely to be fruitful? In what style should papers, and particularly proofs, be written? What constitutes good mathematics? The answers to these questions have changed over time; they are governed in part by the values of the mathematics community. Furthermore, additional questions arise in many contexts about how — and if — mathematics should be used. While not solely the concern of mathematicians, they are part of a broad understanding of the meaning of mathematics. The remainder of this chapter discusses how to assess the value of mathematical work, the contribution mathematics has made to culture, its role as a conceptual framework, and its aesthetic value. It also critically examines the future use argument, argues that mathematics has intrinsic value, and provides a framework, based on Christian principles, that can help an individual assess the value of particular mathematical projects.

A sweeping mathematical project that has come with the computer age is popularly known as artificial intelligence (AI). Proponents of its strong form illustrate an epitome of mathematical reductionism by claiming that humans are, at bottom, sophisticated automata. Chapter 9 looks at some results in theoretical computer science and explores whether they have anything to say about an old question dealing with the possibility of computer intelligence: Can a computer simulate human thought in its fullest form? Many experts have answered and continue to answer this question strongly in the affirmative, despite the continued failure of computers to come even close to passing Turing's famous "test" for general intelligence. We attempt to illustrate that the problem of simulating human thought in its fullest form is at least much more difficult than what some proponents of "strong AI" indicate in popular literature,[3] and suggest that it is not at all clear that the case for strong AI is a foregone conclusion. That is, we seek to cast doubt on claims that an algorithm can duplicate human activity in its fullest form. We also ask whether what we call creativity involves being able to jump out of a rule-governed system, thus giving an example of an orderly but non-algorithmic aspect of human capability.

There is an informal algorithm people tend to use in everyday life that has interesting implications. The central question of Chapter 10 asks whether it is possible to construct a rational method or algorithm by which we may detect design. We argue that people, in fact, infer design in an entity when that entity cannot be explained on the basis of known physical laws, when the probability of its occurrence is sufficiently low, and when the entity itself can be specified in a way that is independent of our having observed it. Thus, being dealt an "arbitrary" bridge hand is not cause for suspicion, but getting all spades would be, since the latter conforms to a pattern that is possible to construct independently of our observing that hand. This chapter seeks to construct a rational basis for such inferences, and to unpack in more rigorous terms the rather loose language of the preceding sentence. We conclude by asking whether we can apply this theory to certain biological systems.

Our motivating thesis for Chapter 11 is that understanding mathematics in a postmodern world requires an awareness of the human side of mathematics. We thus provide a psychological analysis of the nature of mathematical learning and thinking. In other words, we discuss how people do mathematics and understand mathematics. We begin by examining human thinking in general, and the extent to which people's thinking in one context can be used in another context. Then we describe the process of mathematical thinking more specifically, with a psychological analysis of how people

3. See, for instance, Ray Kurzweil, *The Age of Spiritual Machines* (New York: Viking Press, 1999).

solve mathematical problems. This leads to an exploration of the social aspects of mathematical thinking, and how culture and schooling affect mathematical thinking. We then review research on students' beliefs about mathematics learning in relation to learning in other academic disciplines. Finally, we attempt to map out a Christian perspective on mathematical learning and thinking and indicate the issues Christian mathematicians might consider when trying to understand the psychology of learning mathematics.

Christians who have chosen careers that involve the teaching and learning of mathematics are often asked if their Christian view of the world makes any difference in how they approach mathematics and mathematics education. The typical assumption of the questioner is that it does not make any difference. In Chapter 12 we argue that in several areas critical to the teaching and learning of mathematics, our Christian perspective affects actual choices that are made dealing with pedagogy and curriculum. We take a closer look at a theory of learning and teaching called constructivism. We begin with a careful delineation of several important versions of constructivism. In the views of some constructivists, the adoption of a constructivist viewpoint necessitates significant changes in how one approaches teaching, learning, mathematics itself, and even research methodologies within mathematics education. Next, we document the wide-ranging impact that these related theories have had on the teaching and learning of mathematics in the decade following the release of the first of several documents known as the *Standards.* Finally, we analyze both the nature and effects of constructivism from a Christian perspective. This analysis identifies positive aspects of constructivism that Christians may well embrace and attempt to implement, and it highlights a few areas where Christians may draw some careful distinctions, particularly with those proponents of radical constructivism.

Our concluding chapter discusses several issues that are important to Christians concerned about mathematics. These include the role of beliefs in mathematics, the implications of formalism, what God's purposes for mathematics might be, the "unreasonable effectiveness of mathematics," and the limits of mathematics. We recognize, of course, that much more can be said and that we have barely scratched the surface. Thus, we hope those reading our book will be persuaded to contribute their own thoughts to the important task of articulating a proper role of mathematics in our culture today.

SECTION I

THE NATURE OF MATHEMATICS

CHAPTER 1

Mathematical Truth: Static or Changing?

Introduction

It is rare for someone to pose a mathematical question that elicits a variety of answers, but Philip J. Davis has done just that, commenting wryly, "There are probably more answers to this question than there are people who have thought deeply about it."

His question is this: "Why are the theorems of mathematics true?"

The following list is a synopsis of some of the common answers that Davis states may be given to this question. In this list, the word *mathematics* may have more than one meaning. It may mean the collection of all the propositions submitted to the community of mathematicians that has been certified and is now recognized as part of the corpus of mathematics. On the other hand, it could refer to the set of all true mathematical propositions, independent of cultural effort.[1] The list given below is inextricably bound with what is meant by *true* in Davis's original question. Its meaning may vary from "saying of what is that it is" to "socially agreed upon." As you read through the list try to discern the various perspectives represented. Some exude a modern spirit; others a postmodern spirit (we will define these terms later). Some use metaphysical and theological categories; others are secular and naturalistic.

1. Mathematics is true because it is God-given.

1. "When a Mathematician Says No," *Mathematics Magazine,* April 1986, p. 70.

2. Mathematics is true because humans have carefully constructed it; its fabric is knit from its axioms as a sweater is knit from a length of yarn.
3. Mathematics is true because it is nothing but logic and what is logical must be true.
4. Mathematics is true because it is tautological.
5. It is true because it is proved.
6. It is true in the way that the rules and the subsequent moves of a game are true.
7. Mathematics is true because it is useful (or because it is beautiful, or because it is coherent).
8. Mathematics is true because it has been elicited in a way that reflects accurately the phenomena of the real world.
9. Mathematics is true by agreement. It is true because we want it to be true, and whenever an offending instance is found, the mathematical community rises up, extirpates that instance and rearranges its thinking.
10. Mathematics is true because it is an accurate expression of a primal, intuitive knowledge.
11. Mathematics is not true at all in the rock-bottom sense. It is true only in the probabilistic sense.
12. Truth is an idle notion, to mathematics as to all else. Walk away from it with Pilate.

In this chapter we investigate how answers to the question regarding the truth status of mathematics have shifted over time. Rather than conduct an exhaustive survey, we focus on two representatives, one from the modern era (Gottlob Frege), the other from what we call the postmodern school (Paul Ernest). We judge that there is some urgency for the mathematics community to better understand these two approaches. In fact, the view you finally adopt on this question will have important practical consequences. For instance, we argue in Chapter 12 that the modern spirit we describe gave rise to a position about the nature of mathematics and the role of mathematical proof that was implemented in the "new math" curriculum of the sixties and seventies. Likewise, a philosophy of mathematics education associated with the postmodern spirit has arisen. It is seeking recognition and implementation in the mathematics classrooms of today. Because the purpose of this chapter is to uncover the underlying assumptions of the modern and postmodern camps, we do not take sides as though these views are contradictory. Rather, we seek understanding with a view to evaluating the consequences of these assumptions. At the end of the day, you might not be comfortable with either of these alternatives in full-blown form.

Before we begin, we must clarify an important point, namely, what we mean when we claim we know something to be true. A cadre of philosophers has made a concerted effort to specify necessary and sufficient conditions for an individual to have knowledge of a proposition such as "In a right triangle, the square of the hypotenuse equals the sum of the squares of the other two sides." If such conditions are met, philosophers say that a person has sufficient *warrant* for that proposition. This is a term we will use frequently in this book, and Alvin Plantinga has proposed a succinct, preliminary description of it. Briefly, he says that for a person to *know* (or to have sufficient warrant for) a proposition, (1) the proposition must be true, (2) the person must believe the proposition, and (3) the person must be justified in believing it. To do justice to these ideas would require several volumes, so we will leave our discussion of warrant at this intuitive level.[2]

We are now equipped to proceed with the task we have laid out for this chapter. We begin with a discussion of what we mean by the terms *modern* and *postmodern.*

Modernism and Postmodernism

In a myriad of different contexts varying across the standard disciplines, we hear that a new worldview or zeitgeist called postmodernism is emerging as distinguished from a former worldview called modernism.

Many postmodernists would take a dim view of any attempt to "define" the movement, as such a process, they would claim, is part of the modern agenda. Thus, in describing these two camps, we will use lists of words (note the capitalization) that are rich in connotations and that are designed to communicate the spirit of a time by images and suggestions. Postmodernism is after all a zeitgeist, the supposed spirit of our time since the tumultuous sixties. That spirit is different from the spirit of the modern era, which many say began with the French mathematician and philosopher, René Descartes (1596-1650). In "modern" times the concern was with THE

2. For a fuller discussion, see Plantinga's book, *Warrant and Proper Function* (Oxford: Oxford University Press, 1993).

TRUE, THE JUST, THE BEAUTIFUL, THE RATIONAL, THE ETHICAL, THE TRANSCENDENT, THE LAWFUL, THE ESSENCE OF THINGS, THE CERTAIN, THE UNIVERSAL, and THE CORRECT INTERPRETATION. Modern people view themselves as inherently in tune with the world around them and capable of discerning what its STRUCTURE is. As time passes by, PROGRESS will be made discerning this structure. Thus we are able to form LAWS OF PHYSICS, LAWS OF CHEMISTRY, LAWS OF LOGIC, ETHICAL PRINCIPLES, THEOLOGY, AESTHETIC PRINCIPLES, LAWS OF GRAMMAR and LINGUISTICS, LAWS OF LANGUAGE and RHETORIC. According to the modern stance, school instruction should lead students to the law-like structure of the universe — to THE GOOD, THE TRUE, THE JUST, and THE BEAUTIFUL, which is the same for all.

A postmodernist would tend to focus on a view in which the earth is inhabited with many different kinds of peoples with different histories, cultures, and zeitgeists. Furthermore, the considerable effort put forth by various communities and cultures in trying to achieve the modern agenda has not resulted in universal agreement. While a great deal of progress may have been made, often the net result has been OPPRESSION and EXPLOITATION by those in the upper echelons of a hierarchical social order. We should change our aspirations to more realistic ones and respect the DIVERSITY that we see around us, concentrating on the PRAGMATIC, the IMMANENT, the CORRIGIBLE, the FALLIBLE, the METAPHORICAL, the HETEROGENEOUS, the UNCERTAIN, the AGREED UPON, the LOCAL, the PERSONAL, ONE'S INDIVIDUAL FREEDOM, HAPPINESS, COMMUNITY DISCUSSION, ARBITRATION, and MUTUAL UNDERSTANDING. At least, we should take into account the social and cultural influences that are bound to be part of any thinking process.

From this perspective, the commentary that we make about our state of affairs and the content of what we teach our children should be guided more by how we actually conduct our cultural practices and not by idealized constructs that may not exist.

Gottlob Frege: A Modern Spirit

As noted in the introduction, we choose Gottlob Frege as a representative of the modern era from among many candidates because he is a good example of how a thinker in the modern school of thought might view mathematics.

Frege was born on November 8, 1848, in Wismar, Mecklenberg-Schwerin (now Germany) and died on July 26, 1925, in Bad Kleinen, Germany. He studied mathematics, physics, and chemistry at the universities of Jena and Göttingen, where he obtained the doctorate degree in 1873. He taught at Jena from 1874 until his retirement in 1918. Although he lectured on all branches of mathematics, his most influential writings concerned the philosophies of logic, mathematics, and language. Sadly, his ideas were mostly ignored at the time they were written, though his scholarship was voluminous. Today, Frege's work is recognized as very perceptive and influential. In fact, in his contribution to *The Oxford Companion to Philosophy,* Sir Anthony Kenny refers to Gottlob Frege as the founder of modern mathematical logic: "As a logician and philosopher of logic he ranks with Aristotle; as a philosopher of mathematics he has no peer throughout the history of the subject." Today, many agree that Frege's mathematical contributions in the foundations of mathematics and the clarity and lucidity of his philosophical work qualify him for first rank as a philosopher of mathematics. What were some of his ideas, and why did they have such influence? In the next few pages we will survey Frege's basic philosophic framework and show how this tied in to an agenda he developed for advancing the field of mathematics.

In *Logical Investigations,* Frege makes a distinction between the terms *propositions* (which he also calls *thoughts*) and *ideas.* For Frege, a proposition would be an assertion like, "In a plane triangle, the sum of the degrees of its angles is 180." This contrasts with ideas, which are psychological entities in the mind of the individual thinker. Propositions, on the other hand, are entities that truly exist independently of our mind. We may *grasp* propositions as thinking persons, but we do not create them.

> In thinking we do not produce propositions, we grasp them. For what I have called propositions stands in closest connection with truth. What I acknowledge as true, I judge to be true apart from my acknowledging its truth or even thinking about it. That someone thinks it has nothing to do with the truth of a proposition. . . . "Facts, facts, facts" cries the scientist if he wants to bring home the necessity of a firm foundation for science. What is a fact? A fact is a proposition that is true. But the scientist will surely not acknowledge something to be the firm foundation of science if it depends on man's varying states of consciousness. The work of science does not consist in creation, but in the discovery of true propositions. The astronomer can apply a mathematical proof in the investiga-

> tion of long past events which took place when — on Earth at least — no one had yet recognized that truth. He can do this because the truth of a proposition is timeless. Therefore that truth cannot have come to be only upon its discovery.[3]

According to Frege, a proposition belongs neither to our personal inner world as an idea nor to the objective external world, the world perceptible by the senses. Although propositions are the entities we know as truth-seeking persons, this knowledge is not based on sense perception. But this view creates problems. It seems incongruous to believe that people can obtain information about something not belonging to the inner world without recourse to sense perception. Frege would claim that his notion is no more incongruous than that of sense perception itself. It too has constituents not part of our inner world. Upon viewing a tree, for example, a separate image is made upon our two retinas, and two different people viewing this tree would each have two images of it. Since they are able to share in this access to the external world, something that does not pertain to the senses must be involved in this phenomenon.

For Frege, the claim that a proposition is true is not relative to a particular instant in time. True propositions are timeless. Even the timeless, however, if it is to mean anything to us, must somehow be connected with the temporal world in which we live. Frege hints that the changes a proposition undergoes may be only in its inessential properties, that is, properties that it has as a result of its being grasped by a thinker.

Propositions *act* by being grasped and taken to be true. We grasp the proposition of the Pythagorean Theorem and recognize it to be true. According to Frege, "The fundamental propositions of arithmetic should be proven, if in any way possible, with utmost rigor. Only if every gap in the chain of deductions is eliminated with the greatest care can we identify with certainty those primitive truths on which the proof depends; and only when these are known shall we be able to answer our original questions."[4]

Indeed, in his 1879 pamphlet *Begriffsschrift* (translated *Concept Script*), the subtitle for the book reads, *A formula language, modeled upon*

3. Gottlob Frege, *Logical Investigations,* a translation of three articles by Frege that appeared in *Beitrage zur des deutschen Idealismus* in 1918 and 1923 (New Haven: Yale University Press, 1997). We have substituted here the word *proposition* for *thought* throughout.

4. Gottlob Frege, *The Foundations of Arithmetic,* a translation of *Die Grundlagen der Arithmetic,* which appeared in 1884 (Oxford: Basil Blackwell, 1950), p. 4.

that of arithmetic, for pure thought. Within some eighty-eight pages, he set forth a new calculus (the propositional calculus) that set the direction maintained in mathematical logic today. More importantly, for the first time Frege identified the fundamental principles underlying valid arguments using sentences that involve multiple quantification. In short, he founded quantification logic, including sentences of first order logic. He also included the second order quantification where the quantification is done over sets of objects.

Today most students of logic unwittingly use notions that Frege developed. Consider the following mathematical sentence given in plain English: There are at least two perfect numbers that are larger than 6. (Recall that a perfect number is one that equals the sum of its proper divisors.) In the symbolism of the mathematical logic of today, it would be written as follows: $\exists x \exists y (P(x) \wedge P(y) \wedge (x \neq y) \wedge L(x,6) \wedge L(y,6))$

Here $\exists$ is the existential quantifier, x and y are variables, P is the predicate *is perfect,* $P(x)$ means that x is perfect, $\wedge$ is the conjunct "and," and L is the binary predicate "larger than." Thus, $L(x, 6)$ means x is larger than 6. All these constructs (though not all the symbols) are due to Frege.

By creating the intricate symbolism found in the *Begriffsschrift,* Frege could envision a perfectly precise language for arithmetic in which all the assertions of arithmetic could be written. But would such assertions rest upon pure logic (in that they are based upon general laws operative in every sphere of knowledge), or would they need support from empirical facts? From what we have said, it should be no surprise to learn that Frege argued they rest on pure logic, and that arithmetic can be formally presented without the use of any "non-logical" axioms. Frege believed that if he could put arithmetic (this most fundamental part of mathematics) on this firm foundation, then he would be able to build upon it to study the rational numbers and, eventually, the real numbers. In fact, Frege held that arithmetic is nothing more than a branch of logic. Thus, Frege would give the third answer we listed as options to Davis's question, "Why are the theorems of mathematics true?" This view has come to be known as *logicism.* Frege set out to defend this thesis in his second book, *Grundlagen der Arithmetic,* written in 1884.

In formulating his thesis of logicism, Frege sought to improve upon the work of Immanuel Kant, and argued strongly against J. S. Mill. For Kant, an analytic statement is one where the concept expressed by the predicate is implied by that of the subject. For Example, the statement *all bachelors are unmarried males* has *bachelor* as the subject and *unmarried males* as the predicate.

It is an analytic statement because the concept of being an unmarried male is implied by the term *bachelor.* A synthetic statement, by contrast, is one that extends knowledge of the concept involved beyond what we can obtain by merely analytic means. For example, the statement *Sacramento is the capital of California* is synthetic because there is no way that the concept of the predicate (i.e., being the capital of California) is implied by the subject (the city of Sacramento). Now, Kant held that the truths of mathematics are "synthetic a priori." They are a priori in that they are known independently of experience, yet they are also synthetic (rather than purely analytic) in that they extend the knowledge of the concept involved beyond what we can obtain by merely analytic means. For example, the statement *two plus two equals four* is, for Kant at least, a synthetic a priori statement. Kant proposed an intricate theory for human intuition that accounts for our a priori knowledge. He thought our sense of time and space induces an order on our perceptions, which precedes and affects the judgments that we make. In particular, this structuring of our perceptions determines the geometric axioms we adopt as we formalize geometry.

On the other hand, J. S. Mill saw mathematical truths as *a posteriori,* that is, as empirical generalizations of the manipulations we all make on small collections of physical objects in our early training. Frege disagreed. He held that the truths of arithmetic are analytic in that they can be defined in purely logical terms. Therefore, the objects under consideration in arithmetic (numbers) must be carefully defined and identified as clearly logical in character. Furthermore, Frege held that the natural numbers are not only purely logical entities, but that, as logical objects, numbers exist independently of humans. People who share Frege's view on the independent existence (or *being*) of numbers are known as mathematical realists, although they may disagree as to the particulars of what, exactly, this notion of being entails. For example, one might say that numbers exist independently of humans, and are abstract objects located within the mind of God. The word *ontology* is used to describe the study of *being.* It comes from the Greek word *ontos,* which literally means *being.* In Chapter 3 we will take up the question of the ontology of mathematical objects.

In order to complete his argument for logicism, Frege saw the necessity of establishing the logical character of numbers themselves. He laments the fact that no acceptable definition of a number had heretofore been given:

> Yet, is it not a scandal that our science should be so unclear about the first and foremost among its objects, and one which is apparently so

> simple? Small hope, then, that we shall be able to say what number is. The fact is, surely, that if a concept fundamental to a mighty science gives rise to difficulties, it is an imperative task to investigate it more closely until those difficulties are overcome, especially as we shall hardly succeed clearing up negative numbers, or fractional numbers or complex numbers, so long as our insight into the foundation of the whole structure of arithmetic is still defective.

His frustration reaches a peak in his *Foundations of Arithmetic.* Note the famous, and scathing, indictment of J. S. Mill at the end:

> Never again let us take a description of the origin of an idea for a definition, or an account of the mental and physical conditions on which we become conscious of a proposition for a proof of it. A proposition may be thought, and, again, it may be true; never again let us confuse these two things. We must remind ourselves, it seems, that a proposition no more ceases to be true when I cease to think of it than the sun ceases to exist when I shut my eyes. Otherwise, in proving Pythagoras' theorem we should be reduced to allowing for the phosphorous content of the human brain; and astronomers would hesitate to draw any conclusions about the distant past, for fear of being charged with anachronism — with reckoning twice two as four regardless of the fact that our idea of number is a product of evolution and has a history behind it. How could they profess to know that the proposition $2 \times 2 = 4$ was already in existence in that remote epoch? Might not creatures then extant have held the proposition $2 \times 2 = 5$, from which the proposition $2 \times 2 = 4$ was only evolved through a process of natural selection in the struggle for existence? . . . The historical approach, with its aim of detecting how things begin and of arriving from these origins at a knowledge of their nature, is certainly legitimate; but it also has its limitations. If everything is in continual flux, and nothing maintained itself fixed for all time, there would no longer be any possibility of getting to know anything about the world and everything would be plunged into confusion. We suppose, it would seem, that concepts grow in the individual mind like leaves on a tree, and we think to discover their nature by studying their growth: we seek to study them psychologically, in terms of the human mind. But this account makes everything subjective, and if we follow it through to the end, does away with truth. What is known as the history of concepts

> is really a history of our knowledge of concepts or of the meaning of words. Often it is only after immense intellectual effort, which may have continued over centuries, that humanity at last succeeds in achieving knowledge of a concept in its pure form, in stripping off the irrelevant accretions which veil it from the eyes of our mind. What, then, are we to say of those who, instead of advancing this work where it is not yet completed, despise it, and betake themselves to the nursery, or bury themselves in the remotest conceivable periods of human evolution, there to discover, like John Stuart Mill, some gingerbread or pebble arithmetic?[5]

In writing this paragraph near the later stages of the modern era, Frege shows how firmly he believed that by dint of hard work we can succeed in establishing the certainty of arithmetic and the mathematics that rests upon it.

In summary, it is no surprise to find that Frege undertook the daunting task of defining the natural numbers in terms of notions that he thought were purely logical in character. Assuming that could be accomplished, he envisioned a purely logical construction of the integers, rational numbers, and real numbers in terms of natural numbers. With the grounding of these fundamental number systems in logic, he believed it would be possible to account for the confidence that we have in our beliefs concerning the properties of propositions. Note how this posture fits in with our earlier description of the modern spirit.

With his view of logicism guiding him, Frege worked tirelessly on his magnum opus, *Die Grundgesetz der Arithmetic,* a two-volume work that he developed from 1893 to 1903. Tragically, this groundbreaking effort had to be aborted before it was completed. While it was in press, Frege received a letter from Bertrand Russell pointing out that the fifth axiom in Frege's presentation made the entire work inconsistent. Using this axiom Russell was able to form "the class of all classes that are not members of themselves," which was the basis for that devastating nemesis known today as Russell's Paradox.

Bertrand Russell and A. N. Whitehead set out to repair Frege's work and complete his agenda of logicism. Their prodigious effort came to fruition in *Principia Mathematica.* This monumental treatise went a long way toward establishing Frege's program. Since its publication, however, two barriers have emerged in the road this program took in attempting to

5. Frege, *The Foundations of Arithmetic,* pp. ii, iii.

move from pure logic to full-blown arithmetic. First, the additional axioms created by Russell and Whitehead that finally enabled them to circumvent Russell's paradox are widely held not to be purely logical in character. That is, they do not appear to be self-evident axiomatic propositions, their formulations being guided instead by other principles than those Frege sought. This observation seems to provide grounds for the belief that mathematics cannot be reduced to mere logic. Second, Gödel's famous theorem (proved in 1931) shows that there can be no complete and consistent axiomatization of arithmetic.[6] Therefore, Frege had set an unattainable goal so far as it involved producing a *complete* and grounded knowledge of arithmetic.

The Road from Modernism to Postmodernism in Mathematics

What is the nature of the road that led from a modern view of the nature of mathematics to one that is postmodern? When we apply the categories of modernism and postmodernism to mathematics, there is a danger that we will transfer all the various and sundry connotations associated with these words to the mathematical situation. In order to avoid this, it may be helpful to use an especially perceptive comparison of modernism and postmodernism in general terms as sketched by Lawrence Cahoone. By applying this general comparison to mathematics we shall have a more tractable description of what we mean by a *postmodern* view of mathematics.

In his analysis, Cahoone identifies five important themes characterized by single words. Four are negative and one positive, and we will discuss what he means after we review the list:

- Postmodernism criticizes presence and presentation versus representation and construction.
- Postmodernism criticizes origin versus phenomena.
- Postmodernism criticizes unity versus plurality.

6. Gödel's theorem says that any formal system of arithmetic that has terms, formulas, and proof rules recognizable by a computer and that has a language rich enough to describe the fundamentals of arithmetic will either be inconsistent or will have a statement such that neither it nor its negation can be proved in the system.

- Postmodernism criticizes transcendence of norms versus their immanence.
- Postmodernism advocates an analysis of social entities using constitutive other-ness.

If we apply this general description to mathematics, we get an idea of what the modern-postmodern distinction means in that field.

Presence

From the modern point of view, the fundamental entities under consideration in mathematics are clearly given to us in our mathematical intuition using the processes of abstraction and idealization. These entities are to be distinguished from the language and symbols we use to represent them and the relationships between these entities. Although the language we use and the symbolism we adopt are of human construction, they are nevertheless effective in uniquely capturing that intuition. However, from the postmodern view nothing is "immediately present," not even mathematical entities.

Origin

From the modern view, the philosopher of mathematics is trying to penetrate beyond the cultural phenomenon of mathematical activity to the ultimate origin or foundation of this activity. From the postmodern viewpoint there is no such origin or deeper reality behind mathematical activity. All we have is how mathematics has been practiced in historical context.

Unity

Again, from the modern view, mathematics as independent of cultural elements has a unity, universality, and coherence of its own that we humans seek to plumb. The postmodern thinker says that this is nonsense. There is nothing at all that is represented in culture that is not shaped and fashioned by this culture — and in many different ways from one culture to the next and from one era to the next.

Transcendence

From the modern point of view, there is a transcendental norm of truth that mathematical propositions have independent of the social processes that are at work to ascertain it. The way to show that something is true is to present a proof for it that is a faithful representation of a transcendental proof standing behind it. On the other hand, the postmodern view is that the only way to account for the designation of a proposition as true is by displaying the heuristic processes by which individuals arrived at it, by writing down a proof for it for all to see, and thereafter to negotiate with others as to whether it is socially acceptable.

Constitutive Other-ness

Finally, the postmodern application of the first four themes is to assert that the present mathematics community maintains itself as a single monolithic unity by exclusion, opposition, and hierarchization. Thus, in its attempt to maintain itself as one universe-wide entity of mathematics, it must devalue the role of other past or present mathematical cultures. It sets one universal standard for warranting mathematical knowledge and undervalues conjecture and other forms of tentative contributions. In mathematics education it applies universal standards that pay no attention to gender, to ethnicity, to social status and the like.

The postmodern attack on the modern view of the nature of mathematics is concentrated on the following aspects. The first ones are by way of denial. The last two are affirmative in nature.

- Deny the long-standing view called mathematical realism (or Platonism) that holds the mathematical entities like numbers, functions, structures and the like exist independently of us. We will take a closer look at this ontological issue later in the book.
- Deny the correspondence theory for the truth of propositions. The correspondence theory says that the way one describes the truth of everyday propositions applies to mathematical propositions. Thus a mathematical proposition is true if and only if it says of *what* is the case *that* it is the case. The mathematical realist would be inclined to affirm a correspondence theory while anti-realists would tend to

see propositions as having a validity that depends on cultural context.

- Deny the a priori character of mathematical propositions. Some who say that a mathematical proposition is a priori mean that once we grasp it, the methodology that we use to provide warrant for it is independent of experience.
- Deny that mathematical knowledge is certain, incorrigible, infallible, and absolute. To say that mathematical knowledge is certain is to say that it is necessary; that is, a proposition is certain if it could not be false or that it is true in all possible worlds. To say that mathematical knowledge is incorrigible means that once I know a mathematical proposition, there is no future time at which I can sanely deny it. To say that mathematical knowledge is infallible means that the strategy that we adopt — namely that of proving it — is totally reliable. Thus, the method of using proof is sound; it delivers true propositions from true ones. Absolute knowledge is incorrigible and infallible. In Chapter 3 we take up the question of how this relates to God, and in Chapter 4 we will suggest that there very well may be a pragmatic element in mathematical inquiry.
- Deny that mathematics is a single, unified, coherent body of knowledge that can be organized in the Euclidean way, that is, by identifying certain undefined terms, designating certain propositions as axioms, making careful definitions, and then, using logical deductive processes, proving theorems.
- Deny that the defining process is efficacious in portraying independently existing entities, either as irreducible entities or as described in terms of them. Also, deny that the definitions in mathematics are stable over time.
- Deny that we human beings have any way at all of finding out the truth about abstract, non-material entities.
- Deny that the locus of mathematical practice is with the individual mathematician.
- Deny that formal proof is the only vehicle to provide enough warrant for a true mathematical proposition to be known and to be accepted by the mathematical community.[7]

7. See Israel Kleiner and Nitsa Movshovitz-Hadar, "Proof: A Many-Splendored Thing," *The Mathematical Intelligencer* 19, no. 3 (Summer 1997).

- Deny that the foundational enterprises such as logicism, formalism, and intuitionism are the only real options as we tell the story of what mathematics is really like. In fact, we have strong reasons for questioning the fundamental theses that were posed for each. It is said that logicism has been inadequate since several of the fundamental concepts and axioms as proposed in Russell, for example, are not purely logical.[8] If we opt for a formalist position and say that mathematics is merely a game using symbols according to well-defined rules, then the limitation theorems of metamathematics, especially those of Gödel, show that the formalist methodology is inadequate. Furthermore, some regard the isolation from mainstream mathematics and the tendency toward solipsism as signs that intuitionism is also not adequate.
- Deny the assertion that the agenda and the guidelines for framing a philosophy of mathematics should come from the philosophers. Philosophers of mathematics should not necessarily take the general questions and methodology of the philosophers as the ones to consider in describing the nature of mathematics. Instead, the better way is to take our cues from the historian of mathematics and the sociologist of mathematics using their findings as the basis for formulating our descriptions of the nature of mathematics.
- Instead, affirm again and again, that our philosophies of mathematics must take into account the actual practice of the individual mathematician and of the mathematical community. Furthermore, these descriptions of the nature of mathematics must include the techniques of problem solving and must include conjecture formulation and counterexamples. Also, the fact that there are various different levels of justifying and warranting mathematical beliefs is very important.
- Again, affirm that the best descriptions of what mathematics is should have applications in the area of pedagogy and curricula. Thus, the road from the philosophy of mathematics should lead directly to the classroom. Furthermore, it should include the historical development of mathematics and should include not only Western culture, but also

8. Today, the favored formal system for presenting a large part of current mathematics is Zermelo-Fraenkel Set Theory (ZFC). But there are axioms in that proposed system generally acknowledged as not purely logical. One of these is an Axiom of Infinity, which admits the existence of a set that is infinite.

other cultures as equally deserving of our attention. Furthermore, affirm that it is essential that we include a wide variety of cultures in all of our descriptions of the nature of mathematics.

To summarize a postmodern view, we present a list that historian Michael J. Crowe calls *misconceptions* about mathematics and its history:[9]

- The method of mathematics is deduction.
- Mathematics provides certain knowledge.
- Mathematical statements are invariably correct.
- The structure of mathematics accurately reflects its history.
- Mathematical proof is unproblematic.
- Standards of rigor are unchanging.
- Mathematical claims admit of decisive falsification.
- In specifying methodology used in mathematics, the choices are logicism, formalism, intuitionism, and Platonism.

We now proceed to describe the work of Paul Ernest.

Paul Ernest and the Warranting of Mathematical Knowledge

Paul Ernest adds a new dimension to the activity of philosophizing about mathematics. His main academic concern is with the learner of mathematics and with classroom math activities. He is a graduate of the Universities of Sussex and London, where he studied mathematics, logic, and philosophy. Currently he is Professor of the Philosophy of Mathematics Education at the University of Exeter in the United Kingdom. Professor Ernest is widely recognized for his philosophical work in the philosophy of mathematics education and also for conducting and supervising research in the teaching of mathematics. His philosophical work on mathematical epistemology relies heavily on the writings of Ludwig Wittgenstein, Imre Lakatos, and the sociologist David Bloor. His theories recognize the important role that historical and social forces play not only in the practice of the mathematics researcher

9. See Philip Kitcher and William Aspray, eds., *History and Philosophy of Modern Mathematics* (Minneapolis: University of Minnesota Press, 1988), pp. 260-77.

but also in an individual's learning of mathematics. We begin with a summary of Ernest's philosophy of mathematics.[10]

Ernest criticizes what he calls "absolutism" in the philosophy of mathematics. He uses this word to describe the view that all mathematical propositions that have been certified as true in the mathematics community by virtue of a proof *are* true and, in fact, are *necessarily* so. Rather, he says that mathematical propositions are fallible and corrigible and, "should never be regarded as above revision and correction."[11] He reminds us that published proofs are unreliable as they are very commonly flawed.[12] Furthermore, even if we granted Frege the certainty that would come from following his logistic approach (something that Ernest would *not* do), published proofs would still be at least suspect because it is not at all clear that they can be transformed into the formal proofs that Frege envisioned. Ernest judges it highly improbable that the older theories for which axioms have been proposed (and which appear to be stable at the present time) will not undergo extensive revision at some future point. Neither can we regard our attempts to produce more rigorous mathematics as having the result of progressing toward more certain mathematical findings. We will have more to say about this in Chapter 4.

He complains that, until now, philosophers of mathematics have allowed the more general philosophical questions to dominate their thinking about the nature of mathematics. They have used mathematics as an area to apply and test their theories. Thus, in ontology, philosophers have considered the nature of the abstract entities studied in mathematics. Again, in epistemology, philosophers have tested their views about the nature of knowledge — its possibility, scope, and warrants — using the realm of mathematics. The same can be said about the philosophy of mind, which investigates the nature of mental states. Finally, the philosophical theories of meaning have been applied to the mathematical setting. In each of these cases, while mathematics has provided important examples to keep in mind as well as some of the greatest challenges to philosophical thinking, Ernest claims it has been the philosophical community that has posed the questions and determined the direction of the inquiry.

10. See Paul Ernest, *Social Constructivism as a Philosophy of Mathematics* (Albany: State University of New York Press, 1998).

11. Ernest, *Social Constructivism as a Philosophy of Mathematics,* pp. 31-32.

12. For example, Kempe's 1876 proof of the four-color problem was flawed, but the error was not discovered until ten years after the result was originally published.

Thus, with philosophers at the forefront, Ernest thinks mathematics has been described as it is ideally, stripped of its historical and cultural setting, and not as it really is. One can see this clearly by looking at the philosophical schools of logicism, formalism, and intuitionism and their research programs. Ernest claims that none of these schools has been capable of furnishing an acceptable philosophy of mathematics. Therefore, the philosophy of mathematics needs a complete reformulation. It needs to be loosened from the grip of the philosophical outlook that has been concerned with the foundations of mathematics and must be enlarged to consider the history and practice of mathematics. In short, it must deal with the cultural setting of mathematical activity and must seek to describe the nature of mathematics in those terms.

Ernest also proposes an alternative way for warranting mathematical knowledge. Recall that Frege had identified the proposition as the fundamental unit of knowledge. He also included in his analysis the relations between the proposition and person-dependent concepts like judgment, assertion, ideas, belief, and sentences. Proofs, according to Frege, provide warrant for arithmetic sentences. This is done by creating a formal, symbolic language in which formal proofs can be constructed for the propositions asserted.

Ernest disagrees strongly with Frege, borrowing from the ideas of R. G. Collingwood, who proposed that we replace the logic of propositions by a conversational or dialectical logic of questions and answers. There are rules governing such conversations. Among them the questions raised must be relevant and appropriate, must "belong to the whole," and must arise as a natural part of the conversation. Answers must be "right" in that they respond directly to the question and "move the conversation forward." Against Frege, Ernest argues:

> Logicians have almost always tried to conceive the "unit of thought," or that which is either true or false, as a kind of logical "soul" whose linguistic body is the indicative sentence. There have always been people who saw the true "unit of thought" was not the proposition but something more complex in which the propositions served as an answer to a question. Not only Bacon and Descartes, but Plato and Kant come to mind as examples.[13]

13. Ernest, *Social Constructivism as a Philosophy of Mathematics,* pp. 165-66.

From this perspective, a conversation is a sequence of linguistic utterances or texts in a common language made by a number of speakers or authors, who speak and who later respond with further relevant contributions to the conversation.

Ernest opts for this controlled, question-and-answer, conversation-privileged epistemological status (along with propositional proof). He believes that in doing this we do not overly abstract the warranting process. Instead, we see how much it depends upon human activity. Without conversation as a feedback mechanism, knowledge cannot be validated. In so doing Ernest denies that such conversation functions only to generate mathematical conjectures and not to provide warrant for them. He argues that even the warranting of mathematical knowledge is a sociological act and thus the supposed separation between the discovery process and the warranting process cannot be sustained.

In addition, the production of mathematical knowledge requires a constructive act in the form of a text embodying the proposal of a conjecture, proof, problem solution, or theory. This would-be-knowledge proposal is subjected to the social acceptance mechanism that requires a formal conversation response from the representatives of the mathematical community. The response could be critical (a counterexample) or an extended new proposal. Or the outcome might be more dramatic and involve changes in the cultural, epistemological context. The full range of discursive products of mathematics — conjectures, proofs, problems, solutions, concepts, methods, and informal theories — are involved and are an integral part of the warranting process. Putting the matter more directly, Ernest approvingly quotes the accomplished mathematician Y. I. Manin, who argues, "A proof becomes a proof after the social act of 'accepting' it as proof. This is as true of mathematics as it is of physics, linguistics, and biology."[14] This is similar to the ideas of the prominent spokesman for postmodern thinking, Richard Rorty:

> If, however, we think of "rational certainty" as a matter of victory in argument rather than of relation to an object known, we shall look toward our interlocutors rather than to our faculties for the explanation of the phenomenon. If we think of our certainty about the Pythagorean Theorem as our confidence, based on experience with arguments on such matters, that nobody will find an objection to the premises from which

14. Ernest, *Social Constructivism as a Philosophy of Mathematics*, p. 183.

> we infer it, then we shall not seek to explain it by the relation of reason to triangularity. Our certainty will be a matter of conversation between persons, rather than an interaction with a nonhuman reality.[15]

Recall that Ernest borrowed some ideas from Ludwig Wittgenstein, who viewed mathematics as consisting of a large collection of language games. He held that meaning is located in the function that words have as signals passed back and forth between people in the course of purposeful and shared activity, and that mathematical activity is essentially the learning of a specialized language game with progressing rules. This game does not describe any actually existing entities, but is only a pragmatic way to accomplish some immediate goal. In this context, a mathematical proof is also part of this game. It becomes a rehearsal of the game rules, which leads to a social agreement to accept or reject. Thus, contributions to mathematical knowledge are fundamentally textual utterances within the context of a useful game.

In a scathing criticism of the current detached way that new mathematical results are announced, Ernest states, "Mathematics is the discipline that hides its dialogical nature under its monological appearance and that has expunged the traces of multiple voices and of human authorship behind a rhetoric of objectivity and impersonality."[16]

Let us try to summarize the views of Ernest in a way that is analogous to Plantinga's formulation of the conditions necessary for a person to know a proposition: A mathematical community in a culture learns a proposition or procedure that is part of a language game by social construction when the community decides to accept it on the basis of cultural conversations, questioning sessions, and arbitration using a variety of heuristics and computational devices and any other methods it decides are important. The conversation and arbitration are done in terms of the rules of the language game that the individual has learned or by reformulating new rules that seem appropriate.

How, then, does Ernest handle the question of subjective vs. objective mathematical knowledge? In the introduction to his book *The Philosophy of Mathematics Education,* Ernest situates his ideas within the current postmodern movement:

15. Richard Rorty, *Philosophy and the Mirror of Nature* (Princeton: Princeton University Press, 1979), pp. 156-57.

16. Ernest, *Social Constructivism as a Philosophy of Mathematics,* p. 173.

> The philosophy of mathematics is in the midst of a Kuhnian revolution. For over two thousand years, mathematics has been dominated by an absolutist paradigm, which views it as a body of infallible and objective truth, far removed from the affairs and values of humanity. Instead, [postmoderns] are affirming that mathematics is fallible, changing, and like any body of knowledge, the product of human inventiveness.
>
> This philosophical shift has a significance that goes far beyond mathematics. For mathematics is understood to be the most certain part of human knowledge, its cornerstone. If its certainty is questioned, the outcome may be that human beings have no certain knowledge at all. This would leave the human race spinning on their planet, in an obscure corner of the universe, with nothing but a few local myths for consolation. The vision of human insignificance may be too much, or rather too little for some to bear. Does this last bastion of certainty have to be relinquished? In the modern age uncertainty has been sweeping through the humanities, ethics, the empirical sciences: is it now to overwhelm all of our knowledge?
>
> However, in relinquishing the certainty of mathematics it may be that we are giving up the false security of the womb. It may be time to give up this protective myth. Perhaps human beings, like all creatures, are born into a world of wonders, an inexhaustible source of delight, which we will never fathom completely. These include the crystal worlds and rich and ornate webs which the human imagination weaves into mathematical thought. In these are infinite worlds beyond the (in)finite, and wondrous long and tight chains of reasoning. But could it be that such imaginings are part of what it means to be human, and not the certain truths that we took them to be? Perhaps facing up to uncertainty is the next stage of maturity for the human race. Relinquishing myths of certainty may be the next act of decentration that human development requires.[17]

Notice, also, that the distinction between the objective and subjective is crucial to his entire argument. Ernest believes that subjective, individual knowledge and objective knowledge are mutually interdependent. Objective mathematical knowledge is reconstructed for assimilation as subjective knowledge through interaction with textbooks, teachers, fellow students,

17. Paul Ernest, *The Philosophy of Mathematics Education* (Bristol, Pa.: Falmer Press, Taylor and Francis, Inc., 1991), p. 11.

teaching aids and charts, computers, and the like. Social interaction with the physical environment and with other persons is the means by which the individuals develop a fit between their subjective knowledge and the socially accepted, objective mathematics. This process is what Ernest calls the *learning* of mathematics.

Conversely, one can start with individuals who have formulated new questions originating in such learned knowledge to produce new subjective knowledge. They produce a written presentation of this new knowledge and submit this writing to the community for criticism and certification. Social conversation takes place and often the result gains the certification of these findings as objective mathematics. If it is deemed not acceptable it is returned with suggestions for improvement.

Objective mathematics resides not in texts but in the shared rules, conventions, understandings, and meanings of individuals in the society and in their social institutions. It survives through an enduring, yet evolving social action, as depicted in the following diagram.

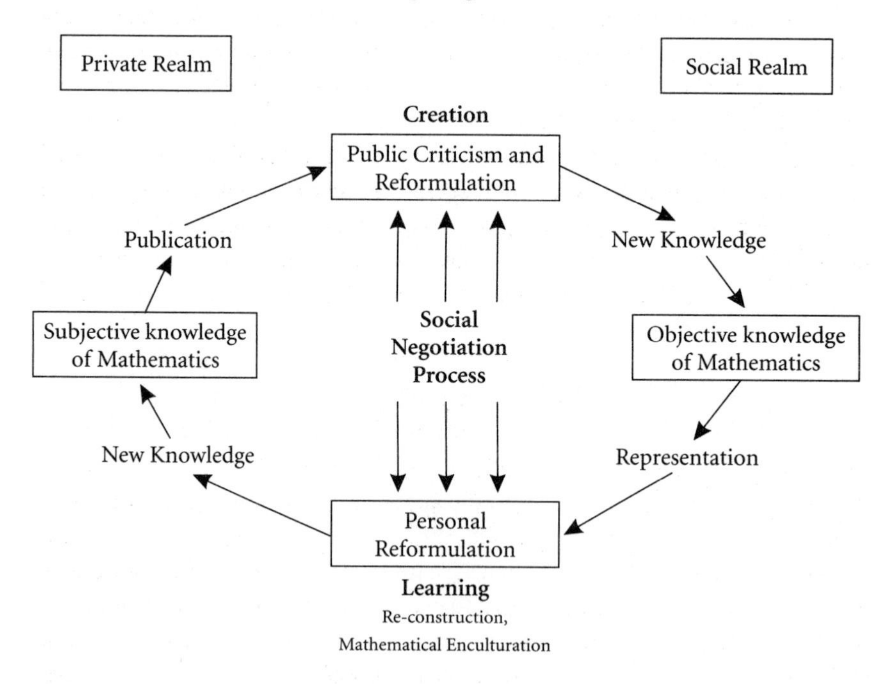

Ernest's diagram showing the relation between objective and subjective mathematical knowledge (from *The Philosophy of Mathematics Education*, p. 85).

As an application of the above diagram, consider the Pythagorean Theorem. A cadre of Babylonians knew a prototype of it. In fact, there is a tablet in the British Museum that contains a problem which, in effect, asks for one leg of a right triangle whose hypotenuse is 5 and that has another leg 4. Egyptians used the (3; 4; 5) right triangle for constructing right angles. Euclid learned the theorem (subjective knowledge) and also described the theorem with a geometric proof in his book *The Elements,* which is a written version of the language game "plane geometry." As such it has become objective mathematical knowledge. For centuries it has been represented in textbooks, each containing a statement and proof for it. School children have learned it over the centuries. All students have reconstructed and processed it for inclusion in their individual knowledge, although for some it has not reached the level of knowledge but only of acquaintance! Since then the theorem has been generalized many times. One use might be to find the diagonal of a rectangular parallelepiped, given the sides. A form of it may appear in defining a metric on a Hilbert space, or a form of it may appear in using the least square error function in statistics. In each case, an individual mathematician reformulates a setting, or theory, which has an extension of the Pythagorean Theorem, and shows how it can be generalized in another way. New individual mathematical knowledge is created. The new theory or cache of theorems is submitted for publication. A public "conversation" takes place and a decision is made whether the new theorem is acceptable as part of the expanded language game. If it is not it is sent back to the individual for reformulation. If it is, then it gets represented in journals and/or texts and the cycle is repeated.

Evaluation and Conclusion

In several core areas Frege and Ernest are polar opposites. Frege was a mathematical realist, while Ernest does not want to allow for any independent objective status of mathematical objects. On the other hand, Frege's approach grows out of certain *ontological* convictions, while Ernest focuses primarily on the *learning* of mathematics, and draws his epistemological conclusions from that perspective. When this is taken into account, it may be that on some issues these two thinkers are not as opposed to each other as it might initially appear.

To illustrate, consider the following imaginary conversation involving three persons. The participants are two students, Alvin and Kathy, who are senior mathematics majors, and their teacher, Professor John Thomas. Thomas wants to know how students will react when they are asked to make a judgment as to whether a given mathematical proposition is true or not and, more importantly, how they will give a justification of their decision. Alvin and Kathy have agreed to participate in the experiment. Because Professor Thomas wants the conversation to have a learning component, he agrees also to participate by including justifications from his own perspective.

Professor Thomas asks Alvin whether the proposition $4 + 3 = 7$ is true, perhaps with tongue in cheek. Alvin replies, "Are you kidding? Of course, I believe it is true and for several reasons. For one, thing, I memorized my addition tables some years ago. Besides, people have known this for centuries! Furthermore, look at these four fingers on my left hand and these three fingers on my right. Push them together and count the resulting collection of fingers and you get seven fingers." Professor Thomas agrees that the justification is a pretty good one. He adds that Georg Cantor would see the justification in terms of sets. He might have said something like this: Let A be a set with cardinality 4 and let B be a set disjoint from A having cardinality 3. Form the set $A \cup B$. The cardinality of $A \cup B$ is 7 since it can be placed into one-to-one correspondence with the set of the days in a week. Therefore $4 + 3 = 7$. Being long-winded, as professors are, he might add that the Italian Giuseppe Peano might have proved the assertion in an axiomatic setting using axioms involving the successor function S, and the definition of addition that involves claiming $S(a + b) = a + S(b)$. Then

$$\begin{aligned}4 + 3 &= 4 + S(2)\\ &= S(4 + 2)\\ &= S(4 + S(1))\\ &= S(S(4 + 1))\\ &= S(S(4 + S(0)))\\ &= S\ S(S(4 + 0)))\\ &= S(S(S(4)))\\ &= 7.\end{aligned}$$

Kathy remarks that Peano's justification is unnecessarily long and complicated.

Indeed, in looking at this last proof, we can see the force of Ernest's comment that most mathematical proofs today do not, in fact, conform to the strict logicism that Frege envisioned. If the proof of the simple proposition $4 + 3 = 7$ gets this convoluted, how would a complicated proof appear if reduced to the form Frege required? To say that all proofs could be put into this form perhaps places too much optimism on the limits of one's endurance! On the other hand, it seems hard to argue that other cultures might come up with a different conclusion using Alvin's finger counting method. Their ways of describing this conclusion would differ, of course, but it does seem that Frege has a point for some form of ontological objectivity here.

But how far can we press this? Professor Thomas now asks Kathy to judge the truth-value of the following proposition:

$$5954682 \times 3493026 = 20799839047732.$$

Kathy seems flustered in her response. "You surely don't want me to multiply these two 7-digit numbers, do you? May I use my calculator?" She enters the numbers and she reads the answer 2.07998590477 E13. She exclaims, "This answer, being in scientific notation, is puzzling to me. It is saying that the product is 20799859047700 and this is only an approximation. But notice that there is a '5' in the 7th decimal place where there is a '3' in the supposed correct answer. Is the proposition false or did the calculator make a mistake?" Kathy has an idea. She notices that 3493026 is divisible by 3, since the sum of its digits is 27. But the sum of the digits of the product given above is 70, which is not divisible by three. Thus, she judges the proposition to be false. They decide to use Mathematica to make the calculation and it finds the product to be 20,799,859,047,732.

Are we relying on an algorithm here for our justification, or on a computer? If the former, how do we know the algorithm is correct, and if it is, whether we followed the algorithm carefully enough, especially if the computation involved is quite lengthy? In the latter case, what assurance do we have that the hardware and software used in the computer are reliable? This, after all, was part of the controversy in Appel, Haken, and Koch's proof of the four-color theorem. What effect does using computers have on the warranting of results? While Frege might be correct in there being only one possible answer, Ernest is certainly right if he points to the dependence we have on our cultural community (those involved in writing software, for example) in identifying what the answer is. That is, there does seem to be some form of social agreement going on here. Notice, also, that it appears easier to refute some propositions than it does to verify them directly.

Professor Thomas puts forward, "The sum of the angles of a triangle in the plane is 180 degrees." Alvin recognizes that he learned this statement as a theorem in a high school geometry class. To the amazement of Professor Thomas he reconstructs the proof: "Let triangle ABC be given. Let EF be a line through C that is parallel to line AB, with C between E and F and E on the same side of BC as A and F on the same side of AC as B. Then the measure of ∠ECA is equal to the measure of ∠A, while the measure of ∠FCB is equal to the measure of ∠B. But the measure of ∠ECA plus the measure of ∠C plus the measure of ∠FCB is 180 degrees. Substituting, we find that the sum of the measures of the angles of the triangle is 180 degrees." Alvin decides that the proposition is true and submits the above proof as justification, but agrees with Professor Thomas that the context of the statement as well as the argument are essential ingredients for such justification. Kathy points out that she learned in her non-Euclidean class that, in hyperbolic geometry, the sum of the angles of a triangle is less than 180 degrees.

Of course, the above proposition is an old saw. All of us who have learned it in high school would be inclined to offer a proof as a justification for it. Many of us would try to reconstruct the proof from memory should we be asked our reasons for believing it. The key step in the proof is to use the proposition that if two parallel lines are cut by a transversal, then the alternate interior angles formed are congruent. To prove this we need the axiom that through a point not on a given line there is exactly one line parallel to the given line. This distinguishes Euclidean Geometry from Hyperbolic Geometry, where it is postulated that there are at least two such lines. All of this reminds us that mathematical propositions are given in a context and that we need to know what the context is in order to decide about its truth or falsehood.

Our hypothetical conversation finishes with an excursion into deeper waters. Thomas proposes the following proposition: "The Riemann Hypothesis is true." Bernhard Riemann framed this hypothesis at the beginning of the nineteenth century. Both Kathy and Alvin admit that they have never heard of the Riemann Hypothesis. Kathy asks, "Has this Riemann Hypothesis even been settled?"

In this case we are no longer in the realm of school mathematics but at the cutting edge of current research. Students are stretched in even trying to understand what the Riemann Hypothesis (RH) says. (Briefly, it is a conjecture about the location of points where a very complicated function in the complex domain equals zero.) This hypothesis is currently not proven, nor has it been refuted. A great deal of numerical evidence has been produced that is consistent with RH but, without a proof or definitive refutation, it continues on the list of open mathematical propositions. Frege, being a mathematical realist, would certainly think that the proposition is either true or false, even if a proof or counterexample could not be found. Ernest would probably say that such a judgment would evolve over time, according to the conversation generated by the mathematical community.

How do you react to this? Perhaps you side with neither Frege nor Ernest. You may say you are not in a position to judge whether the proposition is true or false. If a proof is eventually found, of course, you would deem the proposition to be true. Likewise, if a counterexample were found you would say the proposition is false. But you may think that the proposition cannot be proved from the axioms we now have, and that it is independent of those axioms (as was the case with the continuum hypothesis). If you were to judge the proposition as neither true nor false under those circumstances, you would be leaning towards mathematical formalism. You might *wish* that all propositions could be proven to be true or false, but you would have to recognize that Gödel's incompleteness theorem rules out that possibility. And as far as the existence of mathematical objects is concerned, if there is an independent ontology to them, how do they become grasped if they are conceived by developed languages that are non-unique and varying? On the other hand, how is it that many mathematical propositions can be conceived utilizing different languages across disjointed cultures?

Gödel's incompleteness theorem illustrates that on some level the vacuum left by the partial failures of the programs associated with logicism, formalism, and intuitionism and the threat of inconsistency engendered by the paradoxes of set theory and language have opened the way for the use of postmodern categories and the postmodern spirit in mathematics today. Although the transition has had little effect on the community of practicing mathematicians, postmodern ideas are quite evident among those interested in making changes in the classroom and in the mathematical curriculum.

We have mentioned only Paul Ernest in our sketch of the postmodern spirit in mathematics; there appears to be today a steady progression in the promulgation of postmodern ideas, such as those espoused by the contemporary philosopher of mathematics, Reuben Hersh. Of course, we must be careful in labeling persons as modern or postmodern, but it seems that the transition is becoming more pronounced and self-conscious.

Our discussion concerned "polar opposites" in the way mathematical knowledge is warranted. At one end of the spectrum, Frege represented a modern spirit in mathematics. His work occurred at just the time when there was a great deal of optimism that the venerable discipline of mathematics could be properly founded and freed of error once and for all. Frege's great contribution was his creative mathematical work in exhibiting the possibility of a formal language in which to present mathematical findings. Reflecting upon his mathematical insights now, one can imagine how thrilled he was to discover deep properties of logic and to see how these properties could be used to formalize the mathematics undergirding its upper reaches. Even today an exposure to mathematical logic will engender the same feeling of wonder.

The idea of a formal language with a clearly defined recipe for forming mathematical sentences and also clearly identified proof techniques opened the way for the mathematics of mathematics — metamathematics. For Frege the formal proof was essential in the warranting of mathematical knowledge, and to the extent that practicing mathematicians produced proofs that could be presented in a formal setting, their work should be accepted into the corpus of mathematics. Mathematical rigor, though idealized, is the standard for acceptance in mathematics. The standards are to be set high so as to insure that error cannot creep in. In addition, there should be a cadre of mathematicians whose business it is to pay attention to the foundations of mathematics. Although Frege does engage in conversation with contemporaries, his stance is polemical and is designed to settle the difficulties of warranting mathematical knowledge once and for all and to identify an excellence to which practicing mathematicians should aspire. The formal proof is the ideal way to present a sufficient warrant for mathematical knowledge.

On the other hand, Paul Ernest represents the postmodern spirit and is strongly affected by it. Admittedly, Paul Ernest is not a mathematician and one might criticize him for taking views that may be contrary to many in the mathematics community. Yet he is a theorist on the philosophy of mathematics as applied to mathematics education. His philosophy of mathematics education is closely coordinated with his general views on the philosophy of mathematics. To the extent that an adequate philosophy of mathematics should include an account of the learning process, this is a valuable contribution. Furthermore, his philosophy of mathematics is not an idealized account describing how mathematics should be conducted. It attempts, rather, to describe mathematics as it is currently being practiced. Perhaps paying attention to the history of mathematics will restore some balance to our account and counteract the view that only a formal mathematical proof will ultimately suffice as a warrant for a mathematical proposition.

Many things have happened in the world of mathematics that are changing our opinions about the role of formal proof as Frege envisioned it. Besides the difficulties raised by the proof of the four-color theorem, several of the most important theorems of the second half of this century have proofs in the literature that are very, very long and that required the joint efforts of several people to produce them in their entirety. The net result is that it is getting harder and harder for individuals to envision the essence of their proof in a single mental survey. The *Enormous Theorem,* for example, which gives a complete classification of all finite, simple groups, occupies some 500 discrete fragments in the literature with the result that a person wanting to see the proof in its entirety would have to read 15,000 pages for the entire argument. Imagine how long the proof would be if it were formalized! But think what that means about the role of proof as warrant for knowledge of the result.

Finally, consider what many call the mathematical event of the last decade: In 1993, Andrew Wiles stunned all of us by announcing that he had constructed a proof for Fermat's Last Theorem. It claims that for each integer n greater than 2, there are absolutely no integers x, y, and z (all non-zero) such that $x^n + y^n = z^n$. During the process of reviewing Wiles's proof, specialists found that one important case had been omitted in the argument. Wiles agreed that the objection was correct and continued to work on the problem. Later, in what Wiles described as a moment of revelation, he saw how to complete his argument. Today, the result has been declared proven and has become part of the corpus of mathematics. Yet some mathematicians in the foundations of mathematics are surveying the proof in order to identify just what previous results were used in its construction. (Wiles did what is now common in number theory. He reformulated the problem in the more general setting of cubic curves in algebraic geometry and used some very deep results from a wide variety of areas to get a general result which, when specialized, amounts to a proof of Fermat's Last Theorem.) It is clear to the specialists that Wiles's proof uses theorems that have been proved in set theories with varying axioms, but they would like to know more concretely upon what axioms the proof depends. Again, the warranting power of mathematical proof is of concern here. Furthermore, the episode of Wiles's case illustrates two further points: that the constructions involved in proofs can be as important, if not more, than the targeted result, and that in recent years results have been established by culling together pieces and making major connections amongst previously disparate pieces.

Frege's ideas are decidedly modern. They exude the optimism that one finds in the era of the Enlightenment. Ernest strongly disagrees. As a person interested in the teaching of mathematics, he prefers an epistemology that is in tune with learning as well as knowing. In subsequent chapters we will flesh out more fully what a Christian perspective might bring to bear on various aspects of the discipline of mathematics. We begin in the next chapter with a comparison of how mathematics evolved in three different cultures.

CHAPTER 2

Mathematical Truth: A Cultural Study

Introduction

We saw examples of two very different ways of looking at mathematics in the last chapter. Frege argued for a "Platonic creed" of an independent world of mathematics that is universally applicable at all times and in all cultures. Ernest, on the other hand, views mathematics as existing only in a social context. Others cite recent experimental evidence indicating that basic numerical concepts might be genetically "hard-wired" into our brains at birth.[1] This seems to make mathematics rely upon particular characteristics of its practitioners. Thus, an alien being might not share human mental constructs, and so may have developed an entirely different mathematics, as valid for it as our mathematics is for us.

Indeed, a sidebar to an article by George Johnson in *The New York Times* ("Useful Invention or Absolute Truth: What Is Math?" — February 10, 1998) gave three examples of possible "alien mathematics." The examples are rather unconvincing: with different appendages, they might use a number base different from 10; if all surfaces on the alien planet were spherical then the ratio of the circumference of a circle to its diameter would vary, as the diameter would now be on a curved surface; the numbers 0, 1, 2, 3, . . . might map to unequal intervals on a number line. The first example confuses the admittedly human design of a number system with the relations

1. For example, Stanislas Dehaene, *The Number Sense: How the Mind Creates Mathematics* (Oxford: Oxford University Press, 1997).

and properties that comprise arithmetic, and the third makes a similar error. The second example is more intriguing, since it demonstrates that π (for instance) might play a pivotal role in our mathematics simply because it reflects our physical environment and the choices we make in interpreting our surroundings. Nevertheless, the question of the "Platonic creed" is begged here: regardless of whether or not π is *important* to an alien mathematics, if the question of the ratio of the circumference to the diameter of a circle in a Euclidean plane were posed to the alien, would it not *conclude* that the ratio is 3.14159265 . . . ?

How can we begin to evaluate the competing claims we have heard in a fair and objective manner? We certainly cannot interview aliens from different planets, but at least we can attempt to get a perspective by looking at how mathematics and views of mathematics have evolved. We shall learn that perceptions of the nature of mathematics and its place in systems of knowledge have varied considerably, both between and within cultures. We begin by investigating the origins of axiomatic mathematics in ancient Greece, and argue that the formal reasoning and abstract cosmologies of the Pythagoreans and Platonists conform to current scholarly definitions of religious belief. Second, we explore the interplay between religion and science by looking at the appropriation of Greek philosophy and mathematics in medieval Islam. From this standpoint it may appear that Ernest is correct: culture — including our religious views — does indeed shape our practice of mathematics. On the other hand, our final consideration focuses on an almost independent cultural expression of mathematics, that of pre-modern China. The fact that we find very similar expressions of mathematics in such an isolated culture makes it appear that Frege also has a point when he argues for some kind of universality in mathematics. We will give further attention to this question when we explore the ontological status of mathematical objects in the next chapter.

Ancient Greece

The various enterprises that are considered to be part of mathematics today are diverse, both in content and in method. Operations research, for example, uses a quite different standard for acceptability (defined by publication in research journals) than does abstract algebra. That is, operations research journals normally expect explicit statements of real-world situations to which mathematical results can be applied, allow the use of empirically collected data, and regard techniques like Monte Carlo simulation as appropriate. Abstract algebra journals, on the other hand, expect careful definitions and rigorous, abstract proofs, but without real-world applications, empirical data, or simulations. Nevertheless, it is safe to say that most modern mathematics works toward a shared ideal of formal logical reasoning starting from a small set of axioms. Although subdisciplines vary dramatically in their ability to reach this goal, most mathematicians have the opinion that research quality increases as it approaches this standard. What was the standard in ancient Greece?

The earliest extant appearance of axiomatic-deductive reasoning is Euclid's *Elements* (300 BC). While not flawless from a modern standpoint, it is strikingly familiar despite its age. Beginning with five "postulates" specific to geometry and five "common notions" applicable to all the sciences, Euclid builds a formidable body of 465 propositions, step by step, covering considerable ground in geometry, number theory, and algebra. Much of the contents of the *Elements* is a systematization of subject matter already well known by Euclid's time; the novelty of the work is not in its contents but in its masterly structure. In fact, Euclid did not compose the first *Elements;* that honor (according to Proclus) goes to Hippocrates of Chios (ca. 430 BC), who is most famous for his failed attempts to square the circle, which is unfortunate as he made good progress in finding areas of other curved figures by his method. Regardless, Euclid's work appears to have been the culmination of several such efforts, so successful that his predecessors' works are lost.

Although Euclid's *Elements* had the effect of virtually obliterating from the historical record most previous treatises in mathematics, much of the early period of Greek mathematics can be reconstructed (albeit often with controversy) through its traces in the *Elements* and other sources. It looks like Greek mathematics began with Thales of Miletus, and developed later within the context of the Pythagorean religious sect. The Pythagoreans shared a quest with the Miletian philosophers (Thales, Anaximander, Anaximenes) to grasp a single underlying ultimate reality. To do this, the Miletians focused on conceptions of matter, but the Pythagoreans concentrated on form, or structure.[2] The Pythagoreans believed all of reality could

2. Danie Strauss, "A Historical Analysis of the Role of Beliefs in the Three Foundational Crises in Mathematics," in *Facets of Faith and Science,* vol. 2 (Lanham, Md.: University Press of America, 1996), pp. 218-19.

be reduced to relations between whole numbers, so they asserted that, fundamentally, number governs the universe. From this position, it becomes almost a vacuous statement that mathematics should be the final arbiter of truth. Indeed, this was the Pythagorean position — but then a crisis came.

Believing number to be the essence of reality, the Pythagoreans naturally thought that geometry could be reduced to numerical relations: one could take the sides of a figure such as a triangle, and find a small enough unit so that each side of the triangle is composed of a whole number of these units. The sides of the triangle were said to be *commensurable* (had a "common measure," namely, the small unit length). Geometric relations regarding the sides of the triangle, then, could be discussed in discrete rather than continuous terms. However, it was discovered that there are many pairs of line segments that are *incommensurable;* i.e., they have no common measure. For instance, the ratio between the hypotenuse and one of the sides of a right-angled isosceles triangle is $\sqrt{2}:1$; finding a common measure between these sides would be equivalent to finding a fraction equal to the irrational $\sqrt{2}$. The discovery of line segments that are incommensurable with each other had profound consequences, both for the Pythagoreans and for the course of Greek mathematics: the dictum that the cosmos can be reduced to the whole numbers was dealt a fatal blow. Tradition has it that when the Pythagoreans discovered "irrational" quantities such as $\sqrt{2}$ existed, they demanded their members keep the result a closely guarded secret. Although impossible to verify, an oft-told story is that after one Pythagorean let the secret loose, he was thrown overboard from a boat sailing off the Greek Isles.[3]

Some may argue that this episode with incommensurables explains the focus on geometry in Euclid's *Elements,* which followed Pythagoras by about 200 years. Although the *Elements* contains a great deal of number theory and algebra, the numerical propositions in Books VII-IX and the algebra of Book II are always described in geometrical terms. The Pythagoreans' discovery of the irreducibility of the continuum to the whole numbers makes it tempting to speculate that the Greeks replaced number with geometry at the root of the cosmos. Modern scholarship, however, questions this view.[4] Euclid's recasting of arithmetic and number-theoretical statements in geomet-

3. See, for instance, T. L. Heath, *A History of Greek Mathematics,* vol. 1 (Oxford: Clarendon Press, 1921; reprinted New York: Dover, 1981), pp. 154-57.

4. Wilbur Knorr, *The Evolution of the Euclidean Elements* (Dordrecht: Reidel, 1975), pp. 306-12.

rical garb was necessary to develop a logically consistent theory of incommensurability (Book X), and may not have been motivated by a desire to reverse the ontological relations between number and space. Other Greek texts back this interpretation: Aristotle clearly distinguished between number and magnitude, and Greek number-theoretic texts not dealing with commensurability theory do not reduce number to space. In fact, the existing primary literature on this subject is scanty, and there is no direct proof that outside the Pythagorean sect the Greeks found incommensurables profoundly disturbing.[5]

While the account of the impact of the discovery of incommensurability on the Greek conception of reality might be complicated by modern scholarship, it is nevertheless clear that incommensurability theory changed the course of much of early Greek mathematics (including most of the *Elements*),[6] especially when compared with Chinese and other cultures unaffected by the Greeks, where it seems to have had little or no role whatever. The challenge to the Pythagorean belief in number was serious, eventually fatal; and although it did not bring mathematics to a standstill, it did shape much of the *Elements.* Book X, sometimes described as the pinnacle of the *Elements,* deals exclusively with incommensurables, and the contents of many of the earlier books were completely recast, even rewritten, in order to provide the logical underpinnings for Book X. This seemingly disproportionate Greek interest in commensurability theory is due to the Pythagorean and later Platonic elevation of mathematics as a standard of truth in inquiry. Under pressure from philosophers — and perhaps themselves — to produce unassailable, absolute knowledge, mathematicians restructured their presentation and practice. The effects of this restructuring encompass much of what is considered important about Greek mathematics, and is absent from other cultures: for instance, the banishing of infinite processes led to the method of exhaustion, logical considerations led to Eudoxos' proportion theory, and mathematical precision was elevated over approximation.

Because of this development, we should spend some time looking at the above-mentioned philosophers and their concerns as the driving force

5. David Fowler, *The Mathematics of Plato's Academy: A New Reconstruction* (Oxford: Clarendon Press, 1987), pp. 294-308 (esp. 302-8).

6. For a summary of modern views on the Pythagoreans and incommensurables, see Ken Saito, "Mathematical Reconstructions Out, Textual Studies In: 30 Years in the Historiography of Greek Mathematics," *Revue d'histoire des mathématiques* 4 (1998): 131-42, esp. 132-34.

for formalization in early Greek mathematics. It is here that we begin to consider the influence of religious belief. We note, incidentally, that traditional Greek mythology had little place in the views of the Greek philosophers, even among the early Presocratics. Regardless of how one defines religious belief, the ancient Greek 'gods' were placed so much on the periphery of Greek philosophy that they do not deserve the title, and so we do not consider them here.

What exactly do we mean, then, by a *religious* belief? Paul Tillich argues that religious belief is an "ultimate concern,"[7] by which he means a way of interacting with the absolute, the unconditional, or the infinite. This apparently vague definition has the benefit of encompassing a great diversity of religions and ways of participating in religious experience. Roy Clouser improves on Tillich by attempting to specify what terms like "ultimate" mean:

> A belief is a religious belief provided that:
> (1) it is a belief in something(s) or other as divine, or
> (2) it is a belief concerning how humans come to stand in proper relation to the divine.[8]

Divine is further defined as "having the status of not depending on anything else." We might well criticize this view as an over-intellectualization of religious faith, but Tillich and Clouser at least capture a central notion of the rational aspect of religious belief, even if they miss the ritual and more explicitly spiritual aspects. Nevertheless, Clouser's definition is adequate for our purposes here.

Using our adopted definition, Presocratic thought appears at first glance to be anti-religious. Distinguishing features of their explanations of phenomena are internal, systematic, and economical. They do not appeal to supernatural forces beyond the cosmos. They seek general ordering principles using similar methods, and they invoke a minimum of primitive terms and forces.[9] When we look closely at central concepts, however, we get a different picture. They assumed their universe (κοσμος = *kosmos*) was ordered (it is the meaning of the word), hence accessible to rational analysis. They

7. See, for instance, Paul Tillich, *Dynamics of Faith* (New York: Harper & Row, 1957), pp. 1-40, esp. 8-11.

8. Roy Clouser, *The Myth of Religious Neutrality* (Notre Dame: University of Notre Dame Press, 1991), p. 23.

9. Jonathan Barnes, *Early Greek Philosophy* (London: Penguin, 1987), pp. 17-18.

believed their universe to be inherently simple and reducible to one or a small number of basic elements and first principles *(archai).* Other forces did not further explain these elements; they simply existed, much like matter and energy in a materialist's belief system. Finally, the first principles were assumed to be accessible by the use of reason *(logos).* The purported principles varied considerably from one philosopher to the next, but it was always assumed that logical argument would decide which was the single true system of the *kosmos.*

There is an obvious relation to Greek mathematical method in these beliefs. Later philosophers (Plato prominent among them) asserted that the method of the geometers was the ideal toward which all study should strive, since it is the system of the *kosmos.* As best we can tell, explicitly mathematical *archai* appeared some time in the fourth century BC. Although there are virtually no original documents from this time period, it seems implausible that they could have developed independently of the philosophers' special assignation of primacy to mathematics. Aristotle's systemization of syllogistic logic contributed to the formalization of mathematical thought (although the Greeks rarely if ever used his particular system), which culminated with Euclid. In a sense, then, mathematics — embodied through axiomatic-deductive reasoning — became a cosmological basis of scholarly ancient Greece. We explore this notion in more depth in Chapters 5 through 7.

These occurrences mark the ancient origin of what philosopher Nicholas Wolterstorff calls *foundationalism:* the belief that a theory belongs to genuine science (and thus genuine knowledge) if and only if it is justified by reasoning from a set of foundational, non-inferred truths.[10] Foundationalism might appear to be anti-religious (or at least separate from religion), but we can trace the form it takes in Presocratic philosophy to religious belief, at least according to Clouser's definitions. The structure assumed to exist within the Greek *kosmos* adheres to a particular type of rationality and conforms to a simple set of fundamental principles (the *archai*). Although there was disagreement over which particular candidates for the *archai* were best, there was no concern about the fundamental principle that the *archai* must exist. Turning to the motivations for this belief, we may cite the Pythagoreans' studies of music and perhaps their speculative cosmology as some sort of empirical evidence of a single deductive structure underlying the *kosmos,*

10. See, for example, N. Wolterstorff, *Reason Within the Bounds of Religion,* 2nd ed. (Grand Rapids: Eerdmans, 1984), pp. 28-34.

but it is stretching the point to assert that they based their worldview solely on their sparse explanatory successes. Rather, their commitment in this regard seems closer to the experiential, primarily non-theoretical nature that often characterizes religious belief. It is more like a faith commitment than an empirically based system of thought warranted by its success in dealing with the material world.

Note also that the various proposals for the basic material of the *kosmos,* although they vary considerably (for instance, water or number), are all *internal* explanations: that is, they simply postulate the existence of the material, and do not attempt any explanation that appeals to anything beyond the *kosmos.* These elements, self-existent, incapable of being created or destroyed, are the ultimate stopping points of inquiry, and so satisfy Tillich's and Clouser's criteria of ultimate concern and divinity respectively. More familiar aspects of religious practice occurred among at least some of the adherents of these beliefs; the Pythagoreans, for instance, performed rituals, had moral codes and sacred objects, and so on.

Much more can be said about these matters and others: for instance, the rise of dualism, conceptions of infinity, varying standards of logical rigor, the classification of geometrical methods, and Platonism as a religion of the educated classes. The pursuit of these themes as they developed and were altered within Greek culture and through medieval Islam and Europe would be rich and multifaceted. For now, we leave Greece in the hands of the Presocratics and turn to our second example.

Medieval Islam

Often the scientists of medieval Islam are remembered for preserving the heritage of the ancient Greeks, passing it on more or less intact to Renaissance Europe as it awoke from the slumber of the Middle Ages. In fact, the situation is far more complicated. The religion of Islam burst suddenly onto the world stage in the early seventh century AD, and within decades became the dominant religion and culture of much of the civilized world. The first scholars to emerge from Muslim nations encountered several different intellectual cultures; most prominent were those from Greece (through remnants of ancient texts), India (which bordered the Muslim world on the East), and various indigenous folk traditions. Muslim mathematicians, and scholars in general, faced the daunting task of unifying these widely varying approaches within a common system, and also of reconciling their synthesis with their deep religious theory and practice. The Islamic appropriation and synthesis of Greek, Indian, and folk sciences into a unique science and mathematics is one of the most dramatic occurrences in the history of science, and provides an interesting case study for two crucial questions. First, how can an overtly religiously defined society such as Islam justify the acceptance of a body of knowledge taken from a system of beliefs alien to its own basis? Second, what happens to a foreign worldview implicit in its foreign science when it is embedded in a different religious context? The Islamic experience, while not necessarily a model for science in a Christian context, provides valuable data for the actual occurrence of scientific assimilation, including mathematics, in a deeply and explicitly religious setting.

Medieval Islam is a label that covers a vast expanse of both time and nationalities, and it is possible only to resort to gross generalizations here.[11] Nevertheless, there is sufficient unity to tell a reasonably coherent central story since the academic spectrum was split into two virtually independent spheres. The first, the *religious sciences* (ʿilm), depended on religious precepts, and developed entirely within explicitly Islamic principles. These subjects included theology, but were much broader than that, comprising also the studies of law, literature, and language. The *foreign sciences* (awa'il) included (among others) mathematics, astronomy, the physical and biological sciences, medicine, and optics. In these fields, the acquisition of knowledge from other cultures was not only tolerated, but also often encouraged and funded. The religious sciences were considered to be primary and were studied first in schools. The foreign sciences, however, generally were not relegated to a lower secondary status, at least not to the extent that might be expected. This is because their practitioners were revered and even prized by the rulers in whose service they practiced. Various explanations have been proposed for the benign relation between faith and science in Muslim cultures. One author appeals to the "melting pot effect": Islam, as a religious culture, was formed suddenly by a sweeping merger of a wide variety of cultures with existing scientific and other traditions.[12] The rapid imposi-

11. Sabra refers to this problem as "locality" versus "essence." See "Situating Arabic Science: Locality Versus Essence," *Isis* 87 (1996): 654-70. His article contains a discussion of substantial variations between scientific cultures in various places and times in medieval Islam.

12. Jens Høyrup, "Formation of 'Islamic Mathematics': Sources and Conditions," *Science in Context* 1, no. 2 (1987): 281-329, esp. 296-97.

tion of a new system of beliefs cannot occur without some tacit acceptance, or at least tolerance, of the deeper implicit beliefs held by the assimilated cultures. As a result, the infancy of Islam was in part shaped by a necessity for flexibility. Another significant factor was the fact that "churches" were not central to religious life to the extent that they have been in Christian nations. Religious institutions dominated in education, and legal and administrative office. Thus, the conveying of fundamental religious truths permeated their society and did not come mostly from clerics. As a result, the anti-intellectual attitudes that sometimes occur in religious cultures were discouraged. There seemed little reason to separate the academic world, or any other corner of society, from religious influence; nor was there fear of a power struggle between clerical and other competing influences. Finally, the use of a single scholarly language (Arabic) throughout a region filled with different nationalities and traditions made the fertile exchange of ideas inevitable. Attempts to quash foreign influences in Islamic culture, therefore, would have been virtually impossible.

Explicit defenses of the existence and practice of the foreign sciences within Islam do exist, and they generally follow similar lines. Typically, the foreign sciences were considered to convey truths about the world created by Allah; whether or not a believer discovered them did not affect their ability to reflect Allah's wisdom. Additionally, the foreign sciences were useful tools in the service of Allah and the community at large. Thus, for instance, mathematics was justified through its efficacy in aiding the faithful to observe the Sharīʿah (Islamic law). The study of arithmetic, for example, was justified as a means of settling complex problems dealing with the laws of inheritance; trigonometry and astronomy were used to compute the five daily prayer times, the beginning of the sacred month of Ramadan, and the determination of the *qibla*, the direction of Mecca. These arguments were, however, usually viewed with suspicion: they justified only certain specific aspects of the large intellectual content in mathematics, and even then they seemed to accord little with the actual motives of the practitioners of those areas of mathematics. More often, and less consciously, the sciences (including mathematics) were considered to be helpful servants of the community of the faithful: to aid in practical matters, but also to provide insight into the fundamentally religious cosmos in which they lived.

Within a short time the Greek axiomatic method had become a central feature of Islamic mathematics, especially in geometry. The religious values implicit in the Greek style of mathematics (nature having an autonomous existence apart from a Creator, reason as capable of uncovering ultimate truths about reality with no need of divine revelation) were not lost on foreign cultures exposed to it, and thus it was necessary to reconcile a highly useful, but value-laden, mathematical cosmology and method with religious doctrine of a very different order. This reconciliation was surprisingly smooth. Perhaps as a result of the medieval passion for classification, mathematics (like the other foreign sciences) was assigned its own domain. The Platonic cosmology of Greek geometry still found expression, especially among the purer of the Islamic geometers, but existed more as a convenient presupposition than as anything more deeply reflective of the nature of the world. It was possible to practice mathematics in the Greek style without adopting more than a temporary commitment to its efficacy. The other mathematical traditions, with their more practical concerns, could be used without raising foundational questions.

Inevitably, the mixture of Greek purity and other cultures' utility led to a merger of the theoretical and the practical. This mixture has been called the 'Islamic miracle.' Mathematicians turned pure theory to practical issues, and conversely, scrutinized arithmetic techniques from a logical perspective. Trigonometric techniques once applied almost entirely to the heavens found increasing demand for problems on the earth, including but not limited to requirements of Islamic ritual. Even pure geometry found practical applications, such as in the determination of centers of gravity and the distances to shooting stars.[13] Algebra began to evolve in an arithmetic rather than in a geometric direction, using as its conceptual basis the notion of place-value numeration rather than geometric magnitude. Numerical methods developed to an unprecedented extent, thanks to practical motivation linked with logical analysis and algebraic facility. The interplay between number and geometry that is mathematics today, so natural to the modern student, owes its birth largely to the Islamic miracle.

But did Muslim mathematicians retain this perspective? It depends upon whom you asked. Some Muslim mathematicians appear to have been more Greek than the Greeks, rejecting even some of the ancients' work as insufficient to meet their own standards. It is unclear from the evidence we have, however, whether they generalized axiomatic-deductive principles beyond mathematics. Other scholars were less concerned with formal preci-

13. The leading tenth-century geometer Abū Sahl al-Kūhī composed pure geometric treatises on these topics.

sion. Some religious thinkers, in fact, saw formal mathematics as a deficient science precisely because it was restricted to deductive argument, did not allow fluid terminology, and left basic terms unexplained.[14] Thus, while the values in Greek mathematics may have retained some presence in Islamic mathematics, they carried a considerably smaller intellectual payload: they had no jurisdiction in religious and other matters, and were no longer universal even within mathematics itself.

As we leave our study of these two cultures, we observe that culture strongly affects the practice of mathematics, and that mathematics, in turn, can affect culture. But is there any objective mathematics that transcends culture? To assist in our investigation, we look at the mathematics of an isolated culture.

Pre-modern China

Broadly speaking, three religious expressions have dominated China: Confucianism, Taoism, and Buddhism (especially Pure Land and Zen). Confucius' teachings are moral principles that emphasize ethical commitment, harmonious living in society, and concern and compassion for family and other people. Although Confucius did believe in a "heaven," his teachings focused so much on proper conduct in society that it is debated whether Confucianism can properly be called a religion. Taoism is a mystical atheistic belief system that stresses the natural flow of the cosmos as the ideal toward which humans should strive. Everything in the world has its source in the Way (the Tao), which encompasses all. Intellectual thought is limited, even tantamount to interference with the Tao, especially when it takes on a universal explanatory character. No system of thought can be complete aside from the Tao (since the Tao embraces paradox), and the only path to truth is to follow a route of non-interference with nature. Finally, for Buddhism the only constancy in the universe is change. The path to enlightenment is abandonment of the self through meditation. Especially in Zen Buddhism, this means the emptying of the intellectual mind, accomplished in part by the contemplation of paradox.

Although the three religions in China have separate belief systems,

14. J. Lennart Berggren, "Islamic Acquisition of the Foreign Sciences: A Cultural Perspective," *American Journal of Islamic Social Sciences* 9 (1992): 322.

nevertheless they reflect a shared set of understandings that we may designate as a worldview.[15] The Chinese universe is a natural system in which everything is intimately related to everything else. It is a closed system; there is nothing outside of it. Hence there is a greater emphasis on harmonious life within the world, rather than the attainment of a spiritual plane beyond it. Harmony and balance with family, society, and nature are central goals, and these are not usually achieved by intellectual means. In fact, theoretical thought, especially when conveying an exclusive theory of some aspect of the universe, is viewed with suspicion as limited by human agency and incapable of truly representing the universe.

Given this context, what did Chinese mathematics look like? Getting a clear picture is difficult, as attempts to understand Chinese mathematics encounter various linguistic and other difficulties. Chinese prose, often described as laconic, often translates to long and wordy passages in English. Grammatical differences between Chinese and English often force an English translation to misrepresent the emphases and nuances in the original. The mathematical terminology does not match well with Western languages; for example, the word *tian* refers to fields, but also more generally to a particular classification of geometric figures. Standard terms such as "definition," "axiom," (geometric) "construction," and "theorem" have no Chinese equivalents. In early Chinese editions of Western mathematical texts, "theorem" was translated as "prescription," and "axioms" and "proofs" were translated as "commentaries," and the title of Euclid's *Elements,* in Chinese, suggests a work of practical computation.[16] Even basic terms themselves are not rigidly defined as in our mathematics, but are often dynamically negotiated as the text progresses.

The scarcity of the written record causes further difficulties in the reconstruction of Chinese mathematics. Understanding was conveyed through individual reflection and interaction with the teacher. Justification, less important in China than in Greece, often did not enter the printed text. The lack of large quantities of written texts slowed progress, and later scholars often rediscovered (or, if you will, *re-invented*) earlier achievements. Even

15. This is not to say that conflicts did not exist. In fact, some Chinese reacted negatively to imported Indian Buddhist mathematical textbooks, which to the Chinese contained mystical, vague, and obscure passages. See J. Needham, *Science and Civilization in China,* vol. 3 (Cambridge: Cambridge University Press, 1959), p. 88.

16. Jean-Claude Martzloff, *A History of Chinese Mathematics* (Berlin: Springer Verlag, 1997), p. 115.

so, the list of extant Chinese discoveries is impressive: the binomial theorem, the solution of n-th roots and polynomial equations via Horner's method, the earliest use of negative numbers, combinatorial analysis, Gaussian elimination for the solution of systems of linear equations, solutions of indeterminate integer equations, algebra with infinite series, and finite-difference interpolation methods, provide just a few of the impressive examples. It is interesting, then, that the Chinese developed their mathematics in a starkly different manner than the Greeks and, in some ways, the Muslims, and yet obtained several deep results that were independently also deduced in the West.

The acquisition of mathematical knowledge in Chinese mathematics was multifaceted, reflecting the fact that the goal of "justification" was to persuade in a pedagogical rather than adversarial or skeptical sense. The imparting of understanding could take as many forms as there were people to listen. Hence it is not surprising that, in the words of Ji Kang, "no one has the good method. . . . In this world there are no naturally correct ways, and among methods, no solely good techniques." J. C. Martzloff puts the variety of approaches into the following categories:

- Passage from the particular to the general, based on an example. We find this also in Egyptian and Babylonian mathematics.
- Reasoning by comparison to a similar situation in a different context.
- Use of analogy to extend a current result to a new one: for example, the calculation of an n-th root by appeal to the patterns of calculation in lower-powered roots.
- Use of empirical methods; for instance, the determination of a volume by weighing. The more sophisticated examples of this bear similarities to Archimedes' heuristic techniques in his *On Method.*[17]
- Recourse to heuristic methods, usually involving reduction to known problems.
- Non-linguistic means of communication. In geometry this might include a visual "argument" such as the "Proofs Without Words" in publications of the Mathematical Association of America (such as the *Monthly*), and often involves concrete manipulations of various sorts.

17. Archimedes, *Works, with a supplement "The Method" of Archimedes recently discovered by Heiberg,* ed. Thomas Little Heath (Cambridge: Cambridge University Press, 1897).

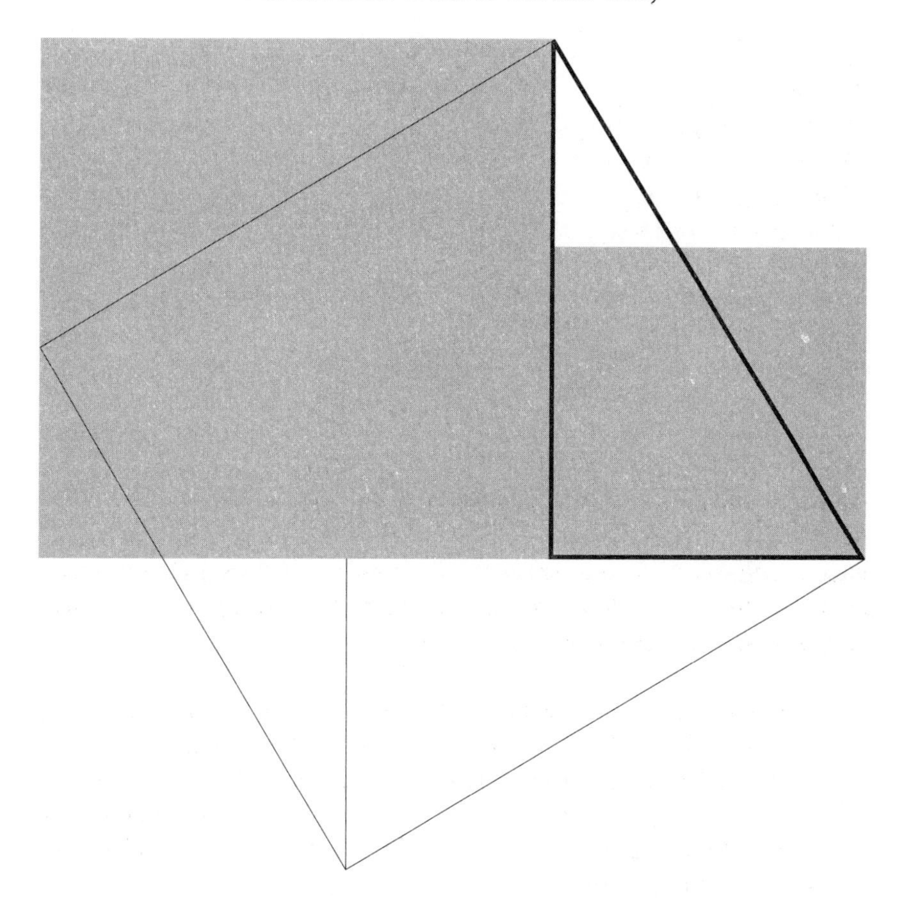

Figure 1: A reconstruction of Liu Hui's demonstration of the Pythagorean Theorem

Liu Hui's "proof" of the Pythagorean Theorem (in his commentary on the *Jiuzhang suanshu*, ca. AD 300) illustrates several of these characteristics. Only a small paragraph accompanies the diagram given as Figure 1 above (which is actually only a possible reconstruction, as it is missing from all known existing texts). Liu Hui simply instructs the reader to apply a technique, which is essentially to cut the pieces of the two smaller squares that fall outside the square on the hypotenuse, and to move them so that they fit within the square on the hypotenuse. He gives no demonstration to determine that the method applies in all possible cases; indeed, the text is so sparse that it is not even possible to determine which of several possible demonstrations he intends.

Curiously, many of the distinctions between Chinese and Greek mathematics tell us more about the ancient Greeks than about China. Features of reasoning that Chinese mathematics shares with other cultures not in the Greek tradition include: (i) the use of, and emphasis on, procedures for solving problems demonstrated by means of a well-chosen example; (ii) connection with concrete examples — a lack of 'total' abstraction; (iii) an acceptance of arguments that are not formally complete or do not necessarily apply in every possible circumstance. The Greek exclusion of these kinds of reasoning in its purest mathematical expressions was necessary to maintain the integrity of mathematics' foundational character. If mathematics is not required to play a foundational role, then valid deductive reasoning becomes less important than insight or pedagogical value, and this shift in emphasis can affect the direction of mathematical growth, though not necessarily the results acquired. A diminished emphasis on logical criteria led, for instance, to some excellent eighteenth-century work by Minggatu and others with infinite series, working with some series expressions that had filtered through from Europe independently of their deductive context.[18] The permissibility of the manipulation of physical objects using a visual scheme led to new discoveries in China, but in Greece its exclusion forced Archimedes to use the method of exhaustion to justify results on areas and volumes that he had already discovered through partly physical arguments. Conversely, the failure to search for axioms or explore the logical foundations of geometry in non-Western cultures left commensurability theory and the parallel postulate undiscovered.

Since mathematics was not the exemplary paradigm in Chinese culture that it was in Greece, one might anticipate the relation between faith and mathematics to be less direct. Certainly the transmission had a different character and was more unidirectional, from religion to mathematics. An obvious connection is the declaration that no single method, even within mathematics, can hope to account comprehensively for any aspect of the world. Openness to variety of approach in mathematics echoes Chinese suspicion of intellectual thought as a narrow exclusive system, and is characteristic of the non-confrontational, harmonious frame of mind sought by Taoism and Confucianism.

18. See, for example, Lĭ Yan, *Chinese Mathematics: A Concise History* (Oxford: Oxford University Press, 1987), pp. 233-40.

Perhaps equally significant is the transmission of core values through societal organization. The firmer structure of Chinese society led to a greater stability and security for scholars there than in Greece. Consequently their methods in all the "sciences," including mathematics, generally aimed for reconciliation, whereas the Greeks were more adversarial and individualistic, and stressed both aggressive and defensive argument.[19] The quest for harmony with family (especially ancestors) and society, reflecting Chinese cosmology, produced a respect for the wisdom of one's teachers, rather than the Greek/modern ultimate reliance on logical argument, separate from personhood. What we might consider to be a failed approach to a problem due to logical errors or limited generality would have been more prominent, placed beside existing work as a worthy, albeit limited, insight: a mark of respect for the author and his idea. Indeed, the Chinese rejection of the universal power of human theory paradoxically led to greater contact between the person and the mathematics.

Universality: Method and Content

At this stage, it is natural to consider whether modern mathematical method, with its obvious successes both in terms of production and global dominance, is inherently superior to all others, and thus independent of cultural values. The sheer bulk of modern mathematics outweighs even the combined efforts of all other approaches, both in content and in depth. In fact, virtually all mathematics developed by other cultures can be mapped "isomorphically," so to speak, into our own framework. Compared to the state of mathematics today, the work produced by the alternatives is primitive. Faced with this perceived superiority, many may conclude that our approach is clearly the most powerful, indeed universal, method.

There can be little doubt that axiomatic-deductive reasoning has produced, and continues to produce, some of the deepest and most comprehensive mathematical insights in human history. Claims of inherent superiority, however, require a presupposed notion of measurability of success in

19. See, for example, G. E. R. Lloyd, "Techniques and Dialectic: Method in Greek and Chinese Mathematics and Medicine," in *Method in Ancient Philosophy,* ed. Jyl Gentzler (Oxford: Clarendon Press, 1998), pp. 354-70; and G. E. R. Lloyd, *Adversaries and Authorities* (Cambridge: Cambridge University Press, 1996).

mathematics. Although we take up the notion of mathematical value in more detail in Chapter 8, some preliminary observations are in order: first, while we do not deny the possibility of relative comparisons, the standard by which one judges performance cannot be completely separated from the aims and central concerns of one's culture. For example, the discovery of non-Euclidean geometry in the nineteenth century arose from a study of geometric axioms. It was valued for its effects on the philosophy of mathematics, and only later for its role in physics. In a context separate from foundational concerns, one might imagine that non-Euclidean geometry would have been considered a relatively unimportant curiosity until its utility in physics was recognized. Likewise, the utilitarian concerns of Chinese mathematics led to a rocky reception of Euclid's *Elements* when this work was introduced into China in the late sixteenth century by Jesuit missionaries: its "dogmatic" structure and topics in pure geometry and commensurability theory were considered strange, of little use, and suspiciously theological.[20] It seems inevitable that criteria to measure mathematical success must inevitably be extra-mathematical; one is forced to judge the merits of another culture's achievements by the values implicit in one's own.

One might assert that an independent criterion is provided by the fact that all other cultures' mathematical achievements are only a tiny subset of Western mathematics. This argument, however, fails to distinguish between the potential of a system and the extent of its actualization. Does historical dominance necessarily imply excellence? The eventual adoption of the Greek style of mathematics by Western nations, and the resulting commitment of vast resources to its pursuit, is inevitably bound up with philosophical, economic, military, and a host of other factors. The Greek philosophical attitude and the Enlightenment view of science necessitate a worldview that requires a certain style of, and role for, mathematics. (We will carefully explore this notion in Chapters 5-7.) The reasons for the widespread success of the entire complexes of Greek or Enlightenment worldviews themselves are complicated by and inextricably interwoven with perspectival issues.

Our cross-cultural reflections, however, may suggest a universality of *content*. Mathematical conclusions on topics that are shared between different cultures do not vary: for instance, the consideration of side lengths of right-angled triangles produced the statement of the Pythagorean Theorem in many cultures, including all those considered in this chapter. Is mathe-

20. Jean-Claude Martzloff, *A History of Chinese Mathematics*, p. 29.

matical content, at least, independent of worldview? By posing this question, we enter a debate, both ancient and modern, on the ontology of the mathematical world, a subject we take up specifically in the next chapter. On one hand, the Platonist or realist view asserts that mathematical content is universal, since mathematical objects have a pure, timeless existence beyond the natural world. On the other hand, the belief that all of knowledge, including mathematics, is based on social norms or linguistic convention[21] is growing in popularity in conjunction with the rise of postmodern thought. The apparent universality of content observed between cultures seems to be a powerful argument against this viewpoint. But what is the reason for this universality? Could it be merely a reflection of our shared experience of physical environment and brain structure, or do mathematical objects such as numbers and sets have some sort of objectively real status that we can somehow discover? This question will be the focus of our next chapter.

Conclusion

We return from the distant past to the distant future: given the results of our tour of three pre-modern mathematical cultures, how alien might an alien's mathematics be? Whether or not mathematical truth is universal, we may find that the alien's mathematics bears little resemblance to our own. The Chinese texts illustrate that fundamental methodological choices can be made differently from our own, and yet produce arguably powerful results such as the binomial theorem, and the solution of polynomial equations via Horner's method. These choices, moreover, are likely to reflect perspectival concerns. The Greeks, for example, placed mathematics at the foundation of their knowledge structure. This led to the acceptance of a certain style of reasoning that was completely different from what we find valued by the Chinese, who had no such esteem for mathematical theories. Finally, the appro-

21. This was essentially Reuben Hersh's thesis in *What Is Mathematics, Really?* (Oxford: Oxford University Press, 1997). Other recent works espousing variations of this general view include William Aspray and Philip Kitcher, eds., *History and Philosophy of Modern Mathematics* (Minneapolis: University of Minnesota Press, 1988), esp. "An Opinionated Introduction" by the editors, pp. 3-60; Thomas Tymoczko, ed., *New Directions in the Philosophy of Mathematics: An Anthology,* revised and expanded edition (Princeton: Princeton University Press, 1998); and Paul Ernest, *Social Constructivism as a Philosophy of Mathematics* (Albany: State University of New York Press, 1998).

priation of the foreign sciences by medieval Islam exemplifies that identical mathematics, practiced in different places and times, can be perceived entirely differently within distinct religious frameworks. Recall, for example, the separation of geometry from algebra in Greek mathematics. This separation also existed in Islamic mathematics, but gradually became blurred. In Chinese mathematics, there never was a separation. So, we may cross paths and agree with the alien at the truth $\pi = 3.14159265\ldots$; however, we may not have arrived at this crossing by the same route, nor may we depart in the same direction, nor even agree on the significance of the stop.

CHAPTER 3

God and Mathematical Objects

Introduction: Models and Modernism

Our historical investigations in the previous chapter uncovered a universality of mathematical content that is independent of culture, although that common content has been expressed in different ways and accompanied by different views of what is important. How do we account for this universality? Part of our answer will depend on how we view mathematical objects such as sets and numbers. A common belief is that mathematical objects have some type of objectively real status that we can access in some way. This would certainly explain why different and isolated cultures have developed similar mathematical theories. Postmodernists, though, typically question the very coherence of claims to knowledge of objective reality, knowledge that is not at least implicitly qualified by reference to culture and context. How might a Christian perspective shape our beliefs about the nature of mathematical objects?

Before we map out a possible position, we must note that there are a number of important insights behind the kind of postmodern skepticism just mentioned, such as the tight connection between claims of objective

The middle sections of this chapter draw heavily from C. Menzel, "Theism, Platonism, and the Metaphysics of Mathematics," *Faith and Philosophy* 4 (1987): 365-82; some of it is lifted verbatim. We would like to express our gratitude to the editor of *Faith and Philosophy* for permission to do so.

knowledge and the exercise of power.[1] At first glance, mathematics may seem less suited to this particular postmodern critique than other fields of knowledge. However, in mathematics, concepts must be precisely defined — things that cannot be precisely defined are excluded from consideration. Furthermore, justification is restricted to proof — things that cannot be proven by a formal argument from axioms cannot be accepted. As a result, people who use mathematical methods to address social (and even some natural science) issues quickly ignore things that cannot be precisely defined and that cannot be quantified. It is a short step from here (and one that has often been taken) to the attitude that anything that cannot be quantified is not real knowledge or is not important knowledge. This attitude became widespread during the past three centuries and its holders came to possess a great deal of power. So this critique does indeed apply to mathematics and the cultural role it has been given. We shall return to this issue in Chapters 5, 6, and 7.

Another insight, the sophisticated thinker might argue, is that mathematics is not obviously *about* anything at all, and is therefore simply its own self-contained discourse with its own set of rules in which a select group of practitioners engage. From this perspective, claims of mathematical objectivity can be dismissed easily. This view, however, does not comport well with the striking applicability of mathematics to our scientific understanding of the world. Nor does it do justice to the common intuition (cf. Frege's thinking) that much of mathematical truth is stable and objective, independent of any particular human mind — on the face of it, it is just objectively and eternally true that $7 + 5 = 12$. Nor does it adequately account for the commonality of mathematical discovery we talked about in our last chapter. Of course, many postmodernists argue that focusing our attention on observations like these leads us to beg all of the big questions at the outset. For belief in the privileged status of scientific knowledge and intuitions about objective and eternal truth simply indicate an unreflective acceptance of the central themes of modernity. They also entail an uncritical acceptance of the modern "grand narrative" that casts humans essentially as autonomous rational knowers, and science as their primary epistemological instrument.

Postmodernists make an excellent point here. Powerful as the point is, however, it does not negate the legitimacy of these observations. The

1. See, e.g., Michel Foucault, *Power/Knowledge: Selected Interviews and Other Writings* (New York: Pantheon, 1980).

postmodernist does not argue that the modernist claim of objective mathematical truth is *false.* Indeed, she cannot so argue, at least not directly, for that would play right into the modernist's hands, as it would itself be a claim to objective truth, one in need of some sort of rational justification. The postmodernist, on the one hand, can point out the inability of any grand narrative — science in particular — to justify its own first principles,[2] and can point out the ease with which grand narratives slip into inflexible dogma that is wielded by those with power to further their own ends.[3] On the other hand, the inability to justify first principles is hardly a revelation. That, after all, is what makes them first principles. It follows from this observation and the tendency of grand narratives to serve dubious interests that whatever credence one accords the grand themes of modernity should be appropriately tempered with philosophical humility: philosophy is hard, and few philosophical theses deserve anything more than tentative, qualified acceptance. Nevertheless, postmodernism gives us no cogent reason to abandon the modernist presuppositions of objective truth and its knowability — in particular, with regard to portions of mathematics. However, it does give us grounds to be extremely wary of the potential for abuse inherent in such a claim.

To this end, then, this chapter develops a "moderate" modernist model of the subject matter of mathematics and argues that it provides a reasonable ground for the possibility of mathematical knowledge. Indeed, we pull out all the modernist stops and place God — perhaps the grandest of modernist grand narratives[4] — in the lead role in the model. We address the

2. This seems to be what Jean-François Lyotard is getting at in the following passage from *The Postmodern Condition* (Minneapolis: University of Minnesota Press, 1984), p. 29:

> Scientific knowledge cannot know and make known that it is the true knowledge without resorting to the other, narrative, kind of knowledge, which from its point of view is no knowledge at all. Without such recourse it would be in the position of presupposing its own validity and would be stooping to what it condemns: begging the question, proceeding on prejudice. But does it not fall into the same trap by using narrative as its authority?

3. See, for example, Lyotard, *The Postmodern Condition,* p. 46.

4. Laplace's haughty dismissal of the "God hypothesis" notwithstanding, the Western philosophical concept of God is in many ways the fullest expression of the Enlightenment conception of humanity — a free, autonomous, conscious, supremely rational, perfectly moral, all-knowing, individual "self." The biblical vision of God includes all this and more — most notably a God whose primary characteristic is love.

more cogent points of the postmodern critique just noted as the development of a *model.* Properly understood, models stand in stark contrast to grand narratives. Unlike grand narratives, models are explicitly tentative, malleable, and limited. They are proposals, suggestions, or "pictures" of how reality might be in certain respects, not dogmatic descriptions of how reality in fact is. Their justification derives from their ability to explain some range of empirical or intuitive phenomena, or to solve philosophical puzzles, not from any sort of extraordinary insight into the nature of reality. But a model's explanations are never complete, and certain questions are inevitably left open. By its nature it is open to modification and extension. It is thus much less prone to become ossified dogma. At the same time, we do not want to lose sight of the fact that models can be *accurate,* that the world can *be* as the model — as far as it goes — *says.* The strength of a claim that a model is true as far as it goes is proportional to the strength of the model's justification. An appropriate measure of sensitivity to postmodern concerns needn't, therefore, push one into a skeptical abyss. That noted, we turn to the construction of our model of mathematical ontology.

The Dilemma of Abstract Objects

In philosophy, ontology is the study of what there is. Typically, except for extraordinary objects like God, ontology concerns itself not with what there is in particular, but what there is in general. That is, ontology concerns itself with what *kinds* of things exist. For Christians, and theists generally, there are several fairly uncontroversial categories of things: physical things, like planets, trees, strands of DNA, and the like; mental things, like pains, thoughts, desires, etc.; and spiritual things, understood broadly to include such things as souls and God. A fourth category is more problematic: abstract things, like universals (e.g., **wisdom, redness** — things in general for which there is more than one instance), relations (e.g., **loving, betweenness**), propositions (e.g., **that 7 and 5 are 12, that God is just**), and, most notably for purposes here, mathematical objects[5] like numbers and sets.

Powerful reasons can be adduced for including abstract objects in one's ontology. Linguists and philosophers of language commonly invoke

5. By the term 'object' here, we mean anything that can be referred to by a singular term; literally any *thing.*

abstract objects of some kind or another to serve as the *meanings* of ordinary language statements and the *objects* of intentional states such as belief, fear, and desire.[6] That is, if one says, "Shoes can be purchased in a department store," the meaning of "department store" is an abstract object, in this case a universal. So also is the meaning of "flying" in the sentence, "Joe is afraid of flying." In particular, linguists, philosophers of language, and philosophers of mathematics often take abstract objects to be the meanings of mathematical statements and, more generally, the things that mathematics is *about.* And, indeed, it is difficult to see how simple mathematical propositions like $7 + 5 = 12$ could be true if there are, in no sense, such things as the numbers 5, 7, and 12 and the relation of addition.

Abstract objects raise a dilemma for the Christian. It is fundamental to Christianity that God is the creator of everything (other than God, of course). Intuitively, however, abstract objects — many of them, anyway — seem to be, like God, eternal. For instance, it is hard to imagine what it could mean for concepts like one, two, and three not to exist.[7] That is, it seems impossible to conceive of a point at which God brought them into being, and hence it is difficult to think of them as created.

Moreover, abstract objects such as 1, 2, and 3 not only exist eternally, but they seem to exist *necessarily;* that is, it seems that even if God had decided not to create pebbles and pomegranates and other countable sorts of things, there still would have been the number 11, as well as the proposition that it is prime.[8] Hence we have the following dilemma for the theist. Either

6. See, for example, J. Barwise and J. Perry, *Situations and Attitudes* (Cambridge, Mass.: MIT Press, 1983); G. Bealer, *Quality and Concept* (Oxford: Oxford University Press, 1982); D. Dowty, R. Wall, and S. Peters, *Introduction to Montague Semantics* (Dordrecht: Reidel, 1981); J. Katz, *Language and Other Abstract Objects* (Rowman & Littlefield, 1981); E. Zalta, *Intensional Logic and the Metaphysics of Intentionality* (Cambridge, Mass.: MIT Press, 1988).

7. This would seem especially hard for believers in the traditional Christian doctrine of the Trinity.

8. Certain types of abstract object are typically thought of as contingent. Sets that contain contingent objects, for example, are usually themselves taken to be contingent. Any set that contains Willard Quine, for example, would not have existed if Quine hadn't. Similarly, on an influential view suggested initially by Bertrand Russell that features prominently in contemporary discussions in the philosophy of language, a proposition that is expressed by a sentence involving a proper name contains the individual denoted by the name as a metaphysical component. Hence, the proposition exists only if the given individual does. Thus, for example, the proposition expressed by 'Quine is a philosopher' would not have ex-

God creates or does not create abstract objects. If the former, there are two problems. First, what sense can it even make to talk about things that have always existed and, moreover, that could not have even failed to exist, as being created? Call this the *coherence* problem. Second, the creation is typically conceived of as *contingent,* the spectacular result of a free and gracious *choice* that God made, something God did not have to do. But if abstract objects are necessary, they *had* to exist. If so, however, then God had no choice in the matter of their creation. Rather than being the result of a free and gracious act, creation seems to be something God was *constrained* to do. Indeed, because he had no choice, God seems to be constrained by something *other* than God, thus raising further problems about God's autonomy and omnipotence. Call this the *freedom problem.*

The other horn of our dilemma is no more attractive. Suppose that abstract objects are uncreated. Then, once again, there seem to be two problems. First, if abstract objects are uncreated, then, far from being the maker of all things visible and invisible, it would appear that God is just one more in a vast array of necessary beings existing independently of God's creative power. Call this the *sovereignty problem.* Second, intuitively, a contingent being (that is, a being that could have failed to exist) needs some sort of explanation for its existence. Since it might have failed to exist, there has to be some reason outside of itself to explain why it exists rather than not. By contrast, a necessary being couldn't but exist. Hence, there is no further need to explain its existence; it exists *a se:* it exists "of and from itself," that is, of its own nature. This property, known as *aseity,* however, has traditionally been associated with God alone; but if abstract objects also exist *a se,* then God's uniqueness in the universe appears now to be compromised as well. God would be just one among many beings that exist of their own nature. Call this the *uniqueness problem.*

Granted, neither horn of the dilemma seems particularly *fatal* to theism. But both force the theist to place severe qualifications on traditional understandings of God's nature, qualifications that most would prefer to avoid. Thus, short of slipping through the horns into nominalism (the view that there are no abstract objects at all — a difficult alternative that we will not

isted had Quine not existed. See Russell's 12/12/04 letter to Frege in G. Frege, *Philosophical and Mathematical Correspondence* (Chicago: University of Chicago Press, 1980), esp. p. 169, and D. Kaplan, "How to Russell a Frege-Church," *Journal of Philosophy* 72 (1975): 716-29, reprinted in M. Loux, ed., *The Possible and the Actual* (Ithaca, N.Y.: Cornell University Press, 1979), pp. 210-24.

pursue here) the theist must either acquiesce in the dilemma and grasp the least desirable horn, or attempt to show that one horn or the other does not actually compromise God's nature in the manner claimed.

We will grasp the creationist horn in this chapter: we argue that we can view abstract objects as created in precisely the same sense in which concrete, contingent things are created. Call this view *AO-creationism* (for "Abstract Object" creationism).

Continuous Creation

The coherence, sovereignty, and uniqueness problems stem in large measure from an inadequate, roughly deistic model of creation. On this model, God brings some initial set of objects into being and sets them going. These objects in turn cause other objects to come to be (as in subatomic particle decay, or biological procreation), though, of course, God could presumably intervene at any point in the process and create any additional objects directly *ex nihilo.* But the key idea on this model is that God initially endows whatever objects he does create with a certain measure of "ontological momentum" at the time of their creation that enables them to remain in existence for some period of time, so to speak, under their own power. On this model, it is difficult indeed to make much sense of the idea of an eternal creature of any sort, abstract or otherwise. For if an object has no beginning in time, there cannot *ipso facto* have been a time at which God first endowed it with its measure of ontological momentum; an object that has always been moving of its own momentum can never have been caused to move.

A more adequate model is the so-called "continuous creation" (CC) model. This model rejects the idea of ontological momentum. Rather, all creatures depend upon God for their existence at every moment, throughout the course of their temporal lifespan. In causal terms, to say that God creates an object x is just to say that, at every moment t at which x exists, God causes x to exist at t (or *sustains* x at t). Creation is thus an ongoing act, consisting in God's continuously sustaining all things.[9] Hence, whether or not there hap-

9. This doctrine is strongly suggested by St. Paul (Colossians 1:16) and the writer of Hebrews (1:3), and is at least implicit in St. Thomas (e.g., *Summa Theologica* I, Q. 45, Art. 3). Its most overt expression in the modern period is found in Descartes (*Principles of Philosophy,* 1, XXI). The doctrine has been defended by J. Kvanvig and H. McCann in "Divine Conservation and the Persistence of the World," in Thomas V. Morris, ed., *Divine and Human*

pens to be a time at which a thing actually comes to be is irrelevant, as the key element of creation is divine sustenance: non-eternal beings can simply mark a point in time at which God's sustenance of them begins. So the eternal character of abstract objects is no barrier to the coherence of the claim that they are creatures.

But precisely the same reasoning applies to necessary beings. The sense in which God can be said to create a necessary being is no different from the sense in which God is said to be creator in general: for God to create some entity *x* is, once again, simply for God to sustain *x* at every moment of its existence. If *x* is a necessary being, then it is simply the case that God *necessarily* sustains *x* at every moment, that he sustains *x*, so to speak, at all times in every possible world. For a non-necessary being *y*, by contrast, this is simply not the case; rather, there are possible worlds and (for non-eternal beings) times in those worlds at which God does not sustain *y* (and hence at which *y* does not exist).

Hence, on the CC model, the coherence problem evaporates. Conceptually, there is nothing incoherent in the idea of necessary but created beings, and hence, in particular, nothing incoherent about the idea that God creates abstract objects.[10] But if the coherence problem is solved, then the sovereignty and uniqueness problems cannot arise, as they are consequences of the thesis that abstract objects are uncreated. It is worth emphasizing that, on the CC model, the inference from necessity to aseity does not follow. Rather, the idea of objects that, although necessary, do not exist of their own

---

Action (Ithaca, N.Y.: Cornell University Press, 1988), pp. 13-49. Note that it is compatible with the view that other beings *collaborate* with God in acts of creation, in the sense that an object's coming to be might also depend causally, in part, at least, upon the actions of some non-divine agent.

10. Indeed, that such a state of affairs exists in the Trinity seems at least hinted at by the Nicene Creed's characterization of the Son's being begotten by the Father, and the Spirit's proceeding from the Father and the Son. If either the begetting or the proceeding relation is in any sense causal, then it would follow that, for the Trinitarian, either the Son or the Spirit is, although necessary, ontologically dependent. Of course, this might appear to be skirting heresy, as we have been taking ontological dependence to be a type of *creation*, and this would make the Son or the Spirit a creature. But one could make qualifications that seem to mitigate the heretical worry. For example, one could take begetting and proceeding to be unique types of sustenance in which the divine nature of the sustaining entity or entities is conveyed to the sustained. One could thus acknowledge a sort of necessary dependence within the Godhead without falling into heresy. See also C. S. Lewis's discussion of begetting in *Mere Christianity* (New York: Collier, 1960).

nature but rather whose existence is explained by something external to them makes perfect sense. Hence, on this model, nothing other than God exists *a se,* of its own nature. Rather, the existence of everything other than God is — at every moment in every possible world — explained by the exercise of God's creative power.

Divine Ideas and the Freedom Problem

The freedom problem still remains for the AO-creationist. As a prelude to addressing it, note that the CC model is programmatic only. That is, it establishes the consistency of AO-creationism by providing a sense in which necessary beings can be considered to be created, and that is certainly an important advance. But the thesis at this point seems empty at best. As things stand, it appears that all the AO-creationist is doing is modifying the traditional, Platonic conception of abstract objects: rather than existing *a se,* abstract objects are now sustained in existence by God. But nothing about the nature of traditional Platonic abstract objects seems to demand this divine sustenance. Thus, a deeper sort of coherence problem remains. What we need is a positive model of the nature of abstract objects that explains how such an object could be both necessary and created.

The model that we adopt in this chapter is simply an updated and refined version of Augustine's doctrine of divine ideas, a view we call *theistic activism,* or just *activism,* for short.[11] Very briefly, activism views abstract objects as the contents of a certain kind of divine intellective activity in which God is essentially engaged. Roughly, they are God's thoughts, concepts, and perhaps certain other products of God's "mental life." This divine activity is thus causally efficacious: the abstract objects that exist at any given moment, as products of God's mental life, exist *because* God is thinking them; which is just to say that God creates them. Moreover, in the case of non-contingent abstract objects, it is simply the case that God *necessarily* thinks them, that is, that God creates them necessarily.

An answer to the freedom problem stems from the claim that God

11. For a detailed exposition and defense of the view, see T. Morris and C. Menzel, "Absolute Creation," *American Philosophical Quarterly* 23 (1986): 353-62, reprinted in Thomas V. Morris, *Anselmian Explorations* (Notre Dame: University of Notre Dame Press, 1987).

conceives the abstract objects essentially, that is, by virtue of God's nature. It is God's power that sustains abstract objects. It is, moreover, not possible that God withhold that power. However, this is only to note that God is rational; God necessarily thinks and conceives, and moreover, necessarily thinks and conceives the same things. Note that this is not to say that God necessarily *believes* (and hence knows) the same things; what God believes will depend in part on contingent facts, e.g., how many people there are at any given moment. However, the *thought* "There are *n* people," for any given *n*, is conceived by God regardless of whether or not it is true. Consider, by analogy, the fact that a grammar of English (ideally) generates all and only the grammatical sentences of English regardless of their truth or falsity. Similarly, God's intellect can be thought of (in part) as a metaphysical "grammar" that "generates" all logically well-formed thoughts and concepts, regardless of their truth or falsity. That God cannot refrain from generating them is no more an infringement on God's nature than is God's inability to believe a contradiction or commit a sinful act. Being perfectly good, God cannot sin. Likewise, being perfectly rational, God cannot but conceive all logically well-formed thoughts and concepts. They are the natural issue of God's maximally perfect intellect.[12]

On the activist model, then, seen as natural and essential byproducts of God's own intellect, the necessary existence of abstract objects does not seem to imply any kind of infringement on God's autonomy and omnipotence. Hence the problem of freedom appears to be dissolved.

Activism, Numbers, and Sets

The activist model provides a coherent, substantive account of the sort of activity in which God is engaged that gives rise to abstract objects. On the face of it, though, it is not exactly obvious where mathematical objects fit into the picture. Before we can address this issue we need to clarify some terminology. A *proposition* is what is expressed by a declarative sentence — the sentence's meaning. Thus, "It is raining" and "Es regnet" are different sentences but they

12. Admittedly, it does seem to be within God's power, in some abstract sense, to sin, that is, to bring about some intrinsically evil, morally unwarranted state of affairs. But God will necessarily not exercise that power. Similarly, perhaps it is within God's power, in some abstract sense, to prevent God from conceiving all possible thoughts and concepts. It's just that, necessarily, God won't.

express the same proposition. A *predicate* is what remains when we remove one or more names from a declarative sentence. That is, "Henri eats soup with a fork" is a proposition, but "eats soup with a fork" is a predicate. A *property* is a characteristic or feature of something. It's the meaning of a predicate; in our example, it is the characteristic of being a person who eats soup with a fork. A *relation* is like a property except that it holds for two or more things; in the sentence "Ivan loves Maria," "loves" expresses a relation. We can think of propositions as zero-place relations and properties as one-place relations. *Thoughts* and *concepts* are ideas in someone's head. Finally, *particulars* are individual things — the opposite of universals.

Since the early twentieth century, mathematicians have sought to base their technical vocabulary in set theory, so terms like *predicate* and *relation* have a special mathematical meaning somewhat different from that given above. However, for philosophers, this vocabulary is expressed in terms of linguistic entities. Because philosophers primarily developed the ideas we want to address here, we will stick to the philosophical usage.

The activist model proposes that abstract objects are the products of God's intellective activity — God's thoughts and concepts. On this model, thoughts that are expressible by declarative sentences (e.g., "Kim is a biology major") correspond naturally to propositions. Concepts are expressed by terms like 'being happy,' 'being a good mathematician,' and 'loving' and so correspond naturally to properties and relations. But note a key point: in saying this, we are implicitly assuming a correspondence between our language and God's thoughts. We will address this assumption in some detail near the end of this chapter, after completing the development of our model.

What about numbers and sets? Traditionally, philosophers have not thought of these entities as properties, relations, or propositions (PRPs) of any kind, and hence they find no clear place in the story thus far. We want to argue that a place can be found, though the search is going to lead us into some fairly deep waters. Let's begin with numbers.

A natural view of numbers, as we will argue below, is that they are properties. For the past century, however, the dominant view of numbers has been that they are abstract *particulars* of one kind or another. (By *abstract,* we mean something that is not concrete, that is, something that does not occupy space and time. So, examples of abstract particulars would be 'the equator,' and 'the North Pole.') The philosophical roots of this view go back to Frege, whom we discussed in Chapter 1. Frege held that no property could be denoted by a singular term. However, in mathematics the numbers *are* in fact denoted by singular terms — for example, there is only one 'two,' one 'four,' one 'thirteen hundred and seven.' Thus, it follows for Frege that numbers are not properties but objects, or, loosely once again, particulars. In other words, Frege viewed any number, such as 3, as an object, because it is the denotation of a singular term. Thus it follows that for Frege numbers are not properties.

However, our ordinary usage itself doesn't easily square with Frege's doctrine. For there exist a prodigious number of singular terms in natural language that, to all appearances, refer straightforwardly to properties: abstract singular terms ('wisdom,' 'redness'), infinitives ('to dance,' 'to raise chickens'), gerunds ('being faster than Lance Armstrong,' 'running for president'), even mathematical terms like 'circularity,' etc. There are thus good reasons to question Frege's doctrine, and hence to suggest that there are no cogent *a priori* reasons for rejecting the idea that numbers are properties.

There are two plausible accounts that identify numbers with properties, both of which trace their origins back to the beginnings of contemporary mathematical logic. The first extends back to Frege himself. Frege clearly saw that statements of number typically involve forming a predicate from a numerical property of some kind. For example, the statement 'There are four moons of Jupiter' associates the property **having four instances** (which is expressed by the quantifier 'There are four') with the concept **moons of Jupiter.** However, Frege was unwilling to take this property itself to be the number four. As we've just seen, though, one needn't follow Frege here, and hence, in general, can take the number *n* to be the property of having *n* instances.[13]

Cantor suggested a different though related view in his (often entirely opaque) discussions of the nature of number. Cantor's insight was that the notion of number is understood most clearly in terms of a special relation between *sets:* we associate the same number with two sets just in case they are "equivalent," that is, just in case a one-to-one correspondence can be established between the members of the sets. In assigning the same number to two sets then, we are isolating a common property the two sets share; such properties (roughly) Cantor identifies as the cardinal numbers. In his words, the cardinal number of a given set **M** is "the general concept under which fall all and only those sets that are equivalent to the given set."[14]

13. This can be understood noncircularly in the usual Fregean/Russellian way. This is essentially the analysis of number developed in Bealer, *Quality and Concept,* ch. 6.

14. Quoted in M. Hallett, *Cantorian Set Theory and the Limitation of Size* (Oxford:

There are thus at least two ways of understanding numbers to be properties, both of which are natural and appealing. The first, quasi-Fregean account has a semantical bent, focusing in particular on the fact that statements of number involve predicates. The Cantorian account emphasizes the intuitive connection between number and relative size in the more general, abstract form of one-to-one correspondence between sets. Both, however, provide us with good reasons for thinking that numbers are properties of some kind. If so, we have found room for numbers within our theistic framework as it stands.

Of course, if we adopt the Cantorian view of number, then it obviously remains to explain how *sets* fit into our picture. We could avoid this question by choosing instead the quasi-Fregean picture, as it makes no appeal to sets. We will not so choose, however, for we want to be able to give our picture the broadest possible scope, and hence we want it to encompass all manner of abstract flora and fauna whose existence Platonism might endorse.

So what then are we to say about sets? Are they also assimilable into our framework as it stands? Formally, yes. Both George Bealer and Michael Jubien, for instance, have developed theories in which sets are identified with certain "set-like" properties; roughly, a set {*a*, *b*, . . .} is taken to be the property **being identical with *a* or being identical with *b* or**. . . .[15] Given sufficiently powerful property theoretic axioms, one can then show that analogs of the usual axioms of Zermelo-Fraenkel *(ZF)* set theory hold for these set-like properties, and hence that one loses none of *ZF*'s mathematical power.

But this is not an altogether happy move. For instance, as Cocchiarella has noted, intuitively, sets are just not the same sort of thing as properties. Sets are generally thought of as being wholly constituted by their members. That is, a set "has its being in the objects which belong to it."[16] This conception is deeply at odds with the view of PRPs that underlies metaphysical realism, on which PRPs exist independently of their instances. Because it is this very idea of sets as having their being in their members that motivates the

---

Oxford University Press, 1984), p. 122. Hallett, we should note, disputes the idea that Cantor held that cardinal numbers are concepts or properties.

15. See Bealer, *Quality and Concept*, ch. 5, and M. Jubien, "On Properties and Property Theory," in G. Chierchia, B. Partee, and R. Turner, eds., *Properties, Types and Meanings*, Vol. 1: *Foundational Issues* (Dordrecht: Kluwer, 1989), pp. 159-75.

16. N. Cocchiarella, "Review of Bealer, *Quality and Concept*," *Journal of Symbolic Logic* 51 (1983).

axioms of *ZF*, and since this is an inappropriate conception of properties, there seems to be no adequate motivation for a *ZF*-style property theory.

So there is at least ground for suspicion of the thesis that sets are just a species of property. It would be desirable, then, if our account could respect the intuitive distinction between the two sorts of entity. In particular, we should like to trace the origin of sets to a different, and more appropriate, sort of divine activity than that to which we've traced the origin of PRPs. But what sort? Here we have a fairly rich (though often unduly obscure) line of thought to draw upon from the philosophy of set theory. A common idea one often encounters in expositions of the notion of set is that sets are "built up" or "constructed" in some way out of previously given objects.[17] Though generally taken to be no more than a helpful metaphor in explaining the contemporary iterative conception of set, a number of thinkers seem to have endorsed the idea at a somewhat more literal level. Specifically, their writings suggest that sets are the upshots of a certain sort of constructive *mental* activity.

Hints of this idea can be seen in the very origins of set theory. Cantor himself held that the existence of a set was a matter of *thinking* of a plurality as a unity.[18] His distinguished mentor Dedekind may be embracing a similar line when he writes that

> [i]t very frequently happens that different things . . . can be considered from a common point of view, can be associated in the mind, and we say they form a *system S*. . . . Such a system (an aggregate, a manifold, a totality) as an object of our thought is likewise a thing.[19]

Hausdorff, Fraenkel, and more recently Schoenfield, Rucker, and Wang express comparable thoughts.[20] Of these it is Wang who seems to develop the idea most extensively. He writes:

17. Cf., e.g., G. Boolos, "The Iterative Conception of Set," *Journal of Philosophy* 68 (1971): 215-31.

18. G. Cantor, *Gesammelte Abhandlungen* (Berlin: Springer, 1932), p. 204.

19. R. Dedekind, *Essays in the Theory of Numbers,* trans. W. W. Beman (New York: Dover, 1963), p. 45.

20. F. Hausdorff, *Grundzuge der Mengenlehre* (Leipzig: von Veit, 1914); A. Fraenkel, *Abstract Set Theory* (Amsterdam: North Holland, 1961); J. Schoenfield, *Mathematical Logic* (Reading, Mass.: Addison-Wesley, 1967); R. Rucker, *Infinity and the Mind* (Boston: Birkhäuser, 1982); H. Wang, *From Mathematics to Philosophy* (London: Routledge & Kegan Paul, 1974).

> It is a basic feature of reality that there are many things. When a multitude of given objects can be collected together, we arrive at a set. For example, there are two tables in this room. We are ready to view them as given both separately and as a unity, and justify this by pointing to them or looking at them or thinking about them either one after the other or simultaneously. Somehow the viewing of certain objects together suggests a loose link which ties the objects together in our intuition.[21]

We may interpret this passage in the following way. Wang here is keying on a basic feature of our cognitive capacities: the ability to direct our attention selectively to certain objects and collect or gather them together mentally, to view them in such a way as to "tie them together in our intuition"; in Cantorian terms, to think of them as a unity. Wang stresses the particular manifestation of this capacity in *perception,* one of a number of related human perceptual capacities emphasized especially by the early Gestalt psychologists.[22] It is best illustrated for our purposes by a simple example. Consider the following array:

Think of the dots as being numbered left to right from 1 to 9, beginning at the upper left-hand corner. While focusing on the middle dot 5, it is possible to vary at will which dots in the array stand out in one's visual field (with perhaps the exception of 5 itself), e.g., [1, 5, 9], [2, 4, 5, 6, 8], or even [1, 5, 8, 9]. The dots thus picked out, we take Wang to be saying, are to be understood as the elements of a small "set" existing in the mind of the perceiver.

The account of the nature of sets obviously won't do as it stands. The

21. H. Wang, *From Mathematics to Philosophy,* p. 182.

22. Cf., e.g., W. Köhler, *Gestalt Psychology* (New York: The New American Library, 1947).

axiom of extensionality (that sets are identical if they have the same elements), for instance, seems not to hold on this picture. In other words, if different people direct their attention to the same dots, then each person has his or her own separate "set," despite the fact that each set has the same members. Moreover, these entities do not have the requisite ontological stability — they drift in and out of existence with the vagaries of individual collecting activities. Finally, human cognitive limitations put a severe restriction on the number and size of sets there can be. Wang is well aware of this, and hence builds an account of the nature of sets based on an *idealized* notion of collecting. Irrespective of the success of Wang's efforts, such an idealization can be of use to us here in developing a model of divine activity that works sets into our activist framework.

The idea is simple enough: we take sets to be the products of a collecting activity on God's part, which we model on our own perceptual collecting capacities. Consider first all the things that are not sets in this sense. While the number of these "first order" objects that we can apprehend at any given time is extremely limited, presumably God suffers from no such limitations; all of them fall under his purview. Furthermore, we can suppose that his awareness is not composed of more or less discrete experiential episodes the way ours is, and hence that he is capable of generating, not just one collection of first order objects at a time, but all *possible* collections of them simultaneously.[23] We suppose next that, once generated, the "second order" products of this collecting activity on first order objects are themselves candidates for "membership" in further collectings, and hence that God can produce also all possible "third order" collections that can be generated out of all the objects of the first two orders. The same of course ought to hold for these latter collections and for the collections generated from them, and so on, for all finite orders. Finally, in a speculative application of the doctrine of divine infinitude, we postulate that there are no determinate bounds on God's collecting activity, and hence that it extends unbounded through the Cantorian infinite.[24]

Identifying sets with the products of God's collecting activity, then,

23. That is, possible in the sense of all the collections he can form from all the nonsets there happen to be; we don't want to suggest that there could be sets containing "merely possible objects."

24. The collecting described occurs in a logical rather than a temporal sense. These ideas are by no means trivial, and deserve much further discussion for which there is simply no room here.

and supposing that God in fact does all the collecting it is possible for him to do, what we have is a full set theoretic cumulative hierarchy as rich as in any Platonic vision. In this way we locate not just the Platonist's PRPs, but the entire ontology of sets as well firmly within the mind of God.

A Volatile Ontology

Let's stop for a moment and take stock of where we are. We set out to address the problem of the nature of mathematical objects. We have developed a model that gives God the lead role in creating and sustaining all abstract objects. That is, in our model, God is rational and (among other things) conceives of all possible well-formed thoughts and concepts; we assume these correspond naturally to properties, relations, and propositions. Thus PRPs are components of human languages that correspond to entities in the mind of God. We placed numbers into this scheme by showing that they can be regarded as properties. We included sets also by arguing that they can be regarded as the product of a collecting activity on God's part.

Alas, but as is so often wont to be the case in matters such as this, things are more complicated than they appear. The lessons of the last hundred years are clear that caution is to be enjoined in constructing an ontology that includes sets and PRPs. Too easily the abstract scientist, eager to exploit the philosophical power of a Platonic ontology, finds herself engaged unwittingly in a metaphysical alchemy in which the rich ore of Platonism is transmuted into the worthless dross of inconsistency. Our account thus far is a laboratory ripe for such a transmutation. There are several paradoxes that, on natural assumptions, are generated in the account as it stands.

The first is essentially identical to a paradox originally discovered by Russell and reported in section 500 of the *Principles.* In the cumulative or "iterative" picture of sets we've developed here, all the things that are not sets form the basic stuff on top of which the cumulative hierarchy is constructed. Now, the chief intuition behind the iterative picture, one implicit in our theistic model above, is that any available objects can be collected into a set; and the "available" objects at any stage form the basis of new sets in the next stage. Thus, since propositions are not sets, all the propositions that exist are among the atoms of the cumulative hierarchy; and since all the atoms are available for collecting (God, after all, apprehends them all "prior" to his collecting), there is a set *S* of all propositions. We adopt a "fine-grainedness" principle for propositions: that for any entities x and y and any property **P**,

$$(*) \text{ if } x \neq y, \text{ then } [\lambda\ \mathbf{P}(x)] \neq [\lambda\ \mathbf{P}(y)],$$

where the notation means that if x is not identical with y, then the proposition **that x is P** (denoted by $[\lambda\ \mathbf{P}(x)]$), is not identical with the proposition **that y is P** (denoted by $[\lambda\ \mathbf{P}(y)]$).[25] [26]

Consider, then, any property you please, for example, the property SET of being a set. In this context, the notation $[\lambda\ \mathbf{SET}(s)]$ denotes the proposition **that s is a set.** By (*), there is a one-to-one correspondence between Pow(S) (the power set of S) and the set $T = \{[\lambda\ \mathbf{SET}(s)]: s \in \text{Pow}(S)\}$. But T is a subset of S, because T is a set of propositions. Therefore, Pow(S), being in a one-to-one correspondence with T, is no larger than S, contradicting Cantor's theorem.[27]

A strictly analogous paradox arises for properties (and, in general, relations as well). This is true in particular if one holds that every object a has an *essence*, that is, the property **being *a***, or perhaps **being identical with *a*.**[28] For the same reasons we gave in the case of propositions, properties (and relations) are also among the atoms of the cumulative hierarchy, and hence there is a set M of all properties. Essences being what they are, we have, for any x and y, that

25. In general, $[\lambda x_1 \ldots x_n\ \varphi]$ is an n-place relation that holds among entities $a_1, \ldots, a_n$ just in case the proposition φ is true when x_i takes the value a_i for $i = 1, \ldots, n$. When $n = 0$, this is just the proposition that φ, so that in statement (*), φ is the symbol P(x) in the first case, and **P**(y) in the second case. Roughly, binding no variables, the symbol λ translates to *that.* When binding variables $x_1, \ldots, x_n$ it can be read as "being objects $x_1, \ldots, x_n$ such that."

26. This principle is natural to those conceptions of PRPs that tend to individuate them on psychological grounds, e.g., that properties are identical if it is not possible to conceive of the one without conceiving of the other. Cf., e.g., R. Chisholm, *The First Person* (Minneapolis: University of Minnesota Press, 1981), ch. 1; A. Plantinga, *The Nature of Necessity* (Oxford: Oxford University Press, 1974). For more formal developments cf. Bealer, *Quality and Concept,* and C. Menzel, "A Complete Type-free 'Second-order' Logic and Its Philosophical Foundations," Report No. CSLI-86-40, Center for the Study of Language and Information, Stanford University, 1986. Please note that we will be abusing metalanguage/object language and use/mention distinctions mercilessly throughout this chapter.

27. Recall Cantor's theorem asserts that the power set of a given set cannot be put into a one-to-one correspondence with a subset of that given set.

28. Cf., e.g., Plantinga, *The Nature of Necessity,* ch. 5.

$$(**)\ \text{if } x \neq y, \text{ then } E_x \neq E_y,$$

where E_z is the essence of z. Consider then Pow(M). By (**) there is a one-to-one correspondence between Pow(M) and the set $E = \{E_z: z \in \text{Pow}(M)\}$. But E is a set of properties, and Pow(M) is in a one-to-one correspondence with E, which is a subset of M, contradicting Cantor once again.

We consider three quick replies to these related paradoxes. The first is to question the fine-grainedness principles (*) and (**). Certainly there are views of PRPs on which this would be appropriate. To understand these views, though, we need some terminology: by the *extension* of a PRP, we mean the class of things that satisfy it; two PRPs are *coextensional* if they are satisfied by the same things. Possible worlds theorists in the tradition of Montague, for example, define PRPs such that they are identical if necessarily coextensional. This perspective represents a "coarse-grained" view incompatible with the fine-grained view we are advocating here. Similarly, views that might be broadly classified as "Aristotelian" hold that properties and relations exist first and foremost "in" the objects that have them, not separate from them, and are "abstracted" somehow by the mind. Such views rarely find any need for PRPs any more fine-grained than are needed to distinguish one state of an object, or one connection between several objects, from another. Whatever the appeal of these alternatives, the problem is that they are out of keeping with our activist model. If we are pushing the idea that PRPs are literally the products of God's conceiving activity, then it would seem that properties which intuitively differ in content, that is, which are such that grasping one does not entail simultaneously grasping the other, could not be the products of exactly the same intellective activity and hence must be distinct. This is especially pronounced in the cases of singular propositions and essences that "involve" distinct individuals, such as those with which we are concerned in (*) and (**); it is just not plausible, for example, that singular propositions "about" distinct individuals could nonetheless be the products of the same activity. To abandon these principles in the context of our present framework, then, would be unpalatable.

The second reply is simply to deny the power set axiom. After all, one might argue, many set theorists find the axiom dubious; so why suppose it is true in general, and in the arguments at hand in particular? Over the years, mathematical logicians and philosophers of mathematics have indeed called the power set axiom into question. The root cause of this disaffection, however, has always been the radically nonconstructive character of the axiom.

Mathematicians are not in general able to specify any sort of general property or procedure that will enable them to pick out every arbitrary subset of a given set. In this sense, it is the Platonic axiom *par excellence*, declaring sets to exist in utter spite of any human capacity to grasp or "construct" them.

It should be clear that any sort of objection on these grounds, as with the previous objection, is just out of place here. For obviously we are far from supposing that set existence has anything whatever to do with human cognitive capacities. Quite the contrary; on our model, the puzzle would rather be how the power set axiom could *not* be true. For supposing that God has collected some set *S*, since each of its members falls under his purview just as the elements of some small finite collection of our own construction fall under ours, how could he not be capable of generating all possible collections that can be formed from members of *S* as well? So this response to the paradoxes is ineffective.

The third reply is that, since there are at least as many propositions (and properties) as there are sets, it is evident that there is no set *S* of all propositions any more than there is a set of all sets; there are just "too many" of them. Hence the argument above breaks down. The same goes for properties, so the second paradox fares no better.

Briefly, the problem with this reply is that how many of a given sort of thing there are in and of itself has nothing whatever to do with whether or not there is a set of those things.[29] The reason there is no set of all sets is not because there are "too many" sets, but rather because there is no "top" to the cumulative hierarchy, no definite point at which no further sets can be constructed. In our model as it stands, however, all propositions and properties (as nonsets) exist "prior" (in a conceptual sense) to the construction of all the sets. Hence, they are all equally available for membership. But if so, there seems no reason for denying the existence of the sets *S* and *M* in the above paradoxes.[30] So an appeal to how many PRPs there are won't turn back the arguments.

29. See C. Menzel, "On the Iterative Explanation of the Paradoxes," *Philosophical Studies* 49 (1986): 37-61.

30. Of course, the paradoxes above aside, in such a picture the axiom of replacement would have to be restricted in some way, else replacing on, e.g., the set $\{[\lambda\ \mathbf{SET}(x)] : x \text{ is a set}\}$ would yield the set of all sets. See "On the Iterative Explanation."

A Russell-type Solution

Though always discomfiting, the discovery of paradox needn't necessarily spell disaster. As in the case of set theory, it may rather be an occasion for insight and clarification. Russell's original paradox of naive set theory was grounded in a mistaken conception of the structure of sets that was uncovered with the development of the iterative picture. Perhaps, in the same way, the paradoxes here have taken root in a similar misconception about PRPs. There are two avenues to explore.

The final paragraph in the last section uncovers a crucial assumption at work in the paradoxes: that all PRPs are conceptually prior to the construction of the sets; or again, that all the PRPs that exist are among the atoms of the hierarchy. The Russellian will challenge this assumption, arguing that one cannot so cavalierly divide the world into an ordered hierarchy of sets on the one hand and a logically unstructured domain of nonsets on the other. For although they are not sets, the nonsets too fall into a natural hierarchy of logical *types*. More specifically, in the simple theory of types, concrete and abstract particulars, or "individuals," are the entities of the lowest type, usually designated 'i'. Then, recursively, where $t_1, \ldots, t_n$ are types, let $(t_1, \ldots, t_n)$ be the type of n-place relation that takes entities of these n types as its arguments. So, for example, a property of individuals would be of type (i); a 2-place relation between individuals and properties of individuals would be of type $(i,(i))$; and so on. The type of any entity is thus, in an easily definable sense, higher than the type of any of its possible arguments.[31] By dividing entities into disjoint levels based on the height of their types, we arrive at a hierarchy of properties and relations analogous to the (finite) levels of the cumulative hierarchy.

To wed this conception with our current model we could propose that *both* sets and PRPs are built up *together* in the divine intellect, so that we have God both constructing new sets *and* conceiving new PRPs in every level of the resulting hierarchy. Thus, at the most basic level are individuals; at the next level God constructs all sets of individuals and conceives all properties and relations that take individuals as arguments; at the next level God constructs all sets of entities of the first two levels and conceives all properties

31. Specifically, let the order ord(i) of the basic type i be 0; and if t is the type $(t_1, \ldots, t_n)$, let $\mathrm{ord}(t) = \max(\mathrm{ord}(t_1), \ldots, \mathrm{ord}(t_n))+1$; then we can say that one type t is higher than another t' just in case $\mathrm{ord}(t) > \mathrm{ord}(t')$.

and relations that take entities of the previous (and perhaps both previous) level(s) as arguments; and so on. Thus, as there are new PRPs at every level, it is evident that, for example, there will be no level at which there occurs the set of *all* properties, and hence it seems we can explain the paradoxes above in much the same way as Russell's original paradox.[32]

Easier said than done. Serious impediments stand in the way of implementing these ideas. First of all, there are several well-known objections to type theory that are no less cogent here than in other contexts. For example, in a typed conception of PRPs, there can be no universal properties, such as the property of being self-identical, since no properties have all entities in their "range of significance."[33] The closest approximation to them are properties true of everything of a given type. But, thinking in terms of our model, even if many PRPs *are* typed, there seems no reason why God shouldn't also be able to conceive properties whose extensions, and hence whose ranges of significance, include all entities whatsoever.

Along these same lines, type theory also prevents any property from falling within its own range of significance, and in particular it rules out the possibility of self-exemplification. Thus, for example, there can be no such thing as the property of being a property, or of being abstract, but only anemic, typed images of these more robust properties at each level, true only of the properties or abstract entities of the previous level.

Standard problems aside, much more serious problems remain. In many simple type theories, including our brief account above, propositions are omitted altogether. Those that make room for them[34] lump them all together in a single type (quite rightly, in the context of simple type theory). This clearly won't do on the current proposal since the entities of any given type are all at the same level and hence form a set at the next level, thus allowing in sufficient air to revive our first paradox.

A related difficulty is that this proposal is still vulnerable to a modified version of the second paradox as well. Consider any relation that holds be-

32. Note that to pull this off in any sort of formal detail one would have to assign types to sets as well. Since sets on the cumulative picture would be able to contain entities of all finite types, we would also have to move to a transfinite type theory. However, as we will see, the issue is moot.

33. To use Russell's term; see his "Mathematical Logic as Based on the Theory of Types," in J. van Heijenoort, ed., *From Frege to Gödel* (Cambridge, Mass.: Harvard University Press, 1967), p. 161.

34. E.g., the theory in E. Zalta, *Abstract Objects* (Dordrecht: D. Reidel, 1983), ch. 5.

tween individuals u and sets s of properties of individuals, for example, the relation **I** that holds between u and s just in case u exemplifies some member of s. Let A be the set of all properties of individuals. For each $s \in \text{Pow}(A)$, we have the property $[\lambda x\, \mathbf{I}xs]$, which we can read as **being an entity *x* such that *x* bears the relation I to *s*.** For example, if s is the set {being tall, being male}, then a person exemplifies $[\lambda x\, \mathbf{I}xs]$ just in case he or she exemplifies one of the members of {being tall, being male}, that is, just in case he or she is either tall or male. By a simple generalization of the fine-grainedness schemas (*) and (**), for all $s, s' \in \text{Pow}(A)$ we have that

$$(***)\ \text{If } s \neq s', \text{ then } [\lambda x\, \mathbf{I}xs] \neq [\lambda x\, \mathbf{I}xs']$$

Consider now the set $I = \{[\lambda x\, \mathbf{I}xs] : s \in \text{Pow}(A)\}$. By (***) there is a one-to-one correspondence between Pow(A) and I. But I is a subset of A, because I is a set of properties of individuals. Hence, Pow(A) is no larger than A, contradicting Cantor's theorem.

A little reflection reveals a feature common to both paradoxes that seems to lie at the heart of the difficulty. First, we need an intuitive fix on the idea of (the existence of) one entity *presupposing the availability of* another. The idea we're after is simple: for God to create (that is, construct or conceive) certain entities, he must have "already" created certain others; the former, that is to say, presuppose the availability of the latter. For sets this is clear. Say that an entity e is a *constituent* of a set S just in case it is a member of the transitive closure of S.[35] Then we can say that a set S presupposes the availability of some entity e just in case e is a constituent of S. For PRPs we need to say a little more. As suggested above, there seems a clear sense in which PRPs, like sets, can be said to have constituents. Thus, a set-like "singleton" property such as $[\lambda x\ x = \text{Kripke}]$ contains Kripke as a constituent. But not just Kripke; for the identity relation too is a part of the property's make-up, or "internal structure"; it is, one might say, a structured composite of those two entities. (We will develop this idea in somewhat more detail shortly.) Combining the two notions of constituency (one for sets, one for PRPs), we can generalize the concept of presupposition to both sets and PRPs: one entity e presupposes the availability of another e^* just in case e^* is a constituent of e.

35. That is, intuitively, just in case it is a member of S, or a member of a member of S, or a member of a member of a member of S, or . . .

Now, even though the properties [λx Ixs] are properties of individuals, if we look at their internal structure, we see that many of these properties presuppose the availability of entities which themselves presuppose the availability of those very properties, to wit, those properties **P** = [λx Ixs] such that **P** ∈ s. (Analogously for those propositions **p** = [λ **SET**(s)] such that **p** ∈ s.) Call such properties *self-presupposing;* this notion, independent of the power set axiom, is sufficient for generating Russell-type paradoxes. Conjoined with the power set, the possibility of self-presupposing properties can be held responsible for the sort of unrestrained proliferation of PRPs of (in general) any type that fuels the Cantor-style paradoxes as well.

The source of all our paradoxes, then, in broader terms, lies in a failure so far adequately to capture the dependence of complex PRPs on their internal constituents. What we want, then, is a model that is appropriately sensitive to internal structure, but which at the same time does not run afoul of any of the standard problems of type theory.

A Constructive Solution

Let's review. Our excursion into type theory was prompted by doubts over the idea that PRPs are conceptually prior to the construction of sets. Type theory suggested an alternative: PRPs themselves form a hierarchy analogous to the cumulative hierarchy of sets such that a PRP's place in the hierarchy depends on the kind of arguments it can sensibly take. The idea then was to join the two sorts of hierarchy into one. However, even overlooking the standard problems of type theory, we found that the resulting activist model was still subject to paradox. Our analysis of these paradoxes led us to see that our problems stemmed from the fact that our models were insensitive to the dependence of PRPs on their internal constituents.

How, then, do we capture this dependence? Here we can draw on some recent ideas in logic and metaphysics. Logically complex PRPs are naturally thought of as being "built up" from simpler entities by the application of a variety of logical operations. For example, any two PRPs can be seen as the primary constituents of a further PRP, their conjunction, which is the result of a *conjoining* operation. Thus, in particular, the *conjunction* of two properties **P** and **Q** can be thought of as the relation [λxy **P**(x) & **Q**(y)] that a given object a bears to another b just in case **P**(a) and **Q**(b). A further operation, *reflection,* can be understood to act so as to transform this relation into the property $[\lambda x\ \mathbf{P}(x)\ \&\ \mathbf{Q}(x)]$ of having **P** and **Q**. We can take related operations to yield complements (e.g., $[\lambda x\ {\sim}\mathbf{P}(x)]$), generalizations (e.g., $[\lambda\ \exists x\mathbf{P}(x)]$), and PRPs that are directly "about" other objects such as our set-like property $[\lambda x\ x = \text{Kripke}]$, or the "singular" proposition [λ **MATH**(Erdös)] **that Erdös is a mathematician.**[36]

On this view, then, the constituents of a complex PRP are those entities that are needed to construct the PRP by means of logical operations, just as the constituents of a set are those entities that are needed to construct the set. It is in this sense that a complex PRP is dependent on its constituents. This then suggests that, analogous to sets on the iterative conception, PRPs are best viewed as internally "well-founded," or at least, noncircular, in the sense that a PRP cannot be one of its own constituents.[37]

This picture of PRPs is especially amenable to activism. For as with set construction, the activist can take the logical operations that yield complex PRPs to be quite literally activities of the divine intellect. This leads us to a further, more adequate model of the creation of abstract entities. At the logically most basic level of creation we find concrete objects and logically simple properties and relations (whatever those may be). The next level consists of (i) all the objects of the previous level (this will make the levels cumulative), (ii) all sets that can be formed from those objects, and (iii) all new PRPs that can be formed by applying the logical operations to those objects. The third level is formed in the same manner from the second, and similarly for all succeeding finite levels. As in our initial models, there seems no reason to think this activity cannot continue into the transfinite. Accordingly, we postulate a "limit" level that contains all the objects created in the finite levels, which itself forms the basis of new, infinite levels. And so it continues on through the Cantorian transfinite.

36. These are ideas with syntactic roots in W. V. Quine, "Variables Explained Away," reprinted in his *Selected Logic Papers* (New York: Random House, 1966), and P. Bernays, "Über eine natürliche Erweiterung des Relationenskalkuls," in A. Heyting, ed., *Constructivity in Mathematics* (Amsterdam: North Holland, 1959), pp. 1-14, and have close algebraic ties to L. Henkin, D. Monk, and A. Tarski, *Cylindrical Algebras* (Amsterdam: North Holland, 1971). For fuller development within the context of metaphysical realism, cf. Bealer, *Quality and Concept,* Zalta, *Abstract Objects,* and Menzel, "A Complete Type-free 'Second-order' Logic." Several categories of logical operations have been omitted here to simplify exposition.

37. This claim is severely called into question in J. Barwise and J. Etchemendy, *The Liar: An Essay on Truth and Circularity* (Oxford: Oxford University Press, 1987). A full defense of the claim would have to deal at length with their challenge.

Now, how do things stand with respect to our paradoxes? As we should hope, they cannot arise on the current model. Consider the first paradox. Since there are new propositions formed at every level of hierarchy, there cannot be a set of all propositions any more than there can be a set of all non-self-membered sets. Similarly for the second paradox: since essences (as depicted above) contain the objects that exemplify them in their internal structure and hence do not appear to be simple, they too occur arbitrarily high up in the hierarchy and hence also are never collected into a set. What about the two new paradoxes above? The first of these is just a type-theoretic variant on the original paradoxes, and so poses no additional difficulty. And although the concept of self-presupposition can be reconstructed in our type-free framework, the corresponding paradox still cannot arise since there can be no set of all non-self-presupposing properties, as the paradox requires.[38] We seem at last to have found our way out of this dense thicket of Cantorian and Russellian paradoxes.

But our task is still not quite complete. Recall that one of our first orders of business was to work (cardinal) numbers into the activist framework. We opted for the Cantorian-inspired view that numbers are properties shared by equinumerous sets. But just where do they fit into our somewhat more developed picture? Intuitively, numbers seem to be logically simple; for example, they do not appear to be conjunctions or generalizations of other PRPs. Hence, they seem to belong down at the bottom of our hierarchy. It follows that there is a set C of all numbers at the next level, according to our model. But this supposition, of course, assuming the truth of the axioms of *ZF*, leads to paradox in a number of ways. For example, one can use the axiom of replacement on C to prove that there is a set of all cardinal numbers. It has been argued that, on certain conceptions of the abstract universe that might allow "overly large" sets, it is appropriate to restrict this axiom to sufficiently "small" sets, and such a restriction would not permit its use here.[39] This

38. In type-free terms, for any n+1-place relation **R**, we define the condition $\mathbf{TSP_R}$ such that $\mathbf{TSP_R}(y)$ if for some z, $y = [\lambda x_1 \ldots x_{i-1}x_{i+1} \ldots x_{n+1}\, \mathbf{R}x_1 \ldots x_{i-1}zx_{i+1} \ldots x_{n+1}]$ and $y \in z$. It is easy to see that, on the present model, nothing — in particular, no n-place relation — satisfies this condition. For any n-place relation of the form $[\lambda x_1 \ldots x_{i-1}x_{i+1} \ldots x_{n+1}\, \mathbf{R}\, x_1 \ldots x_{i-1}sx_{i+1} \ldots x_{n+1}]$, where s is a set, contains s as an internal constituent, and hence on our current model can only have been constructed after the construction of s. Thus, there is no set $s^* = \{y : \sim\mathbf{TSP_R}(y)\}$ (since this would be the universal set) and so no property $y^* = [\lambda x_1 \ldots x_{i-1}x_{i+1} \ldots x_{n+1}\, \mathbf{R}x_1 \ldots x_{i-1}s^*x_{i+1} \ldots x_{n+1}]$.

39. See C. Menzel, "On the Iterative Explanation of the Paradoxes."

would still not redeem the situation, however, as our model seems to require that for every set there exists a definite property, which is its cardinal number. For it appears quite impossible that God should construct a set without also conceiving its cardinality, the property it shares with any other set that can be put into one-to-one correspondence with it. Hence, the set C of all numbers must itself have a cardinality k, and so $k \in C$. But it is easy to show (with only unexceptionable uses of replacement) that k is strictly greater than every member of C, and hence that $k > k$.[40] Once more we have to confront paradox.

Happily, there is a simple and intuitive solution to this paradox. How many numbers must we say there are? Given our understanding of numbers as properties of sets, and our reasoning in the previous paragraph, if we divide up the universe of sets according to size, then there must be as many numbers as there are divisions. Numbers are thus in a certain sense dependent on sets in a way that other sorts of properties are not. This suggests a natural way of fitting numbers into our hierarchy in such a way as to avoid paradox: a given number is not introduced into the hierarchy until a set is constructed whose cardinality is that number. The number is then introduced at the next level; God, we might say, doesn't conceive the number until he "has to." Since there are larger and larger sets at every new stage in our hierarchy, there will be no point after which new numbers are no longer introduced, and thus there can be no set of them. Hence, our numerical paradox above cannot get started.

As our model is informal, the only rigorous way of demonstrating that it is indeed paradox-free is to formalize the picture of the abstract universe it yields and then to prove the consistency of the resulting theory. This can be done. The universe of the activist model can be formalized in a first-order theory that includes all of *ZFC* and a rich logic of PRPs that embodies all the fine-grainedness principles above; and this theory is provably consistent relative to *ZF*.

40. One can, for example, use the theorem (which requires only replacement on ω) that there are arbitrarily large fixed points in the mapping $\aleph$ of (von Neumann) ordinals onto (von Neumann) cardinals. This entails that for any $n \in C$ there is an $m > n$ such that $|\{j \in C : j < m\}| = m$. But $\{j \in C : j < m\}$ is a subset of C, hence $|C| = k \geq m > n$.

Divine Ideas, Platonic Objects, and Mathematical Knowledge

The possibility of mathematical knowledge has always been a problem for Platonism. Knowledge of objects of any kind seems to require some sort of causal connection between the objects known and the knower. How, then, is mathematical knowledge possible, if it consists in knowledge of nonphysical, causally inert entities like numbers and pure sets? How do we know that our number theoretic and set theoretic axioms describe *those* things? It is not clear, at first sight anyway, that the theistic activist is any better off. For, on the activist view, it would seem that mathematical knowledge consists of knowledge of objects in the mind of God. Such objects seem no more epistemically accessible than classical Platonic objects. In response, the activist could postulate that every mind has access to the mind of God. But it seems implausible in the extreme that, say, in first learning of the number 2, a child is accessing the contents of the divine mind. Fortunately, it isn't clear that direct access is needed.

The activist model builds on the intuition that sets and numbers (being properties of sets) are grounded in our own capacities for perceptual aggregation. Following Wang, we took this capacity to give rise to set-like entities in the mind. There were several reasons why we could not identify these entities with sets directly: notably, failure of extensionality (as different minds aggregating the same objects give rise to different mental entities), lack of ontological stability, and limitations on the size and number of sets there can be. But these objections are really almost beside the point. For it is not so much the individual constructions themselves that ground set theory, but their *possibility.* Suppose we are given some urelements (elements that are not sets themselves). Different agents might all generate set-like constructions by mentally aggregating those urelements. Or they might not. Either way, it doesn't matter. For each such construction is an instance of a more general, overarching idealization — the possibility of such a construction — and *that* is the true object of any set theory based upon those urelements.

Constructivism, on the face of it, seems not to suffer from Platonism's epistemological problems. (We will discuss educational implications of a form of constructivism in some detail in Chapter 12.) For mathematical objects, if perhaps not *identical* to the products of our own mental activity, are at least intimately tied to those products. Hence, mathematical knowledge can be grounded in the objects of our own construction. But a problem with most all constructivist accounts — both "strict," intuitionist accounts as well as "looser," classically based accounts such as those of Kitcher and Chihara[41] — is: What, exactly, is an idealization, the *mere possibility* of a construction? And for whom is the construction possible? For us? For possible humans? For an ideal agent of some ilk? And how do such idealizations serve as the subject matter of mathematics? How, in particular, can mathematical statements be *true* if these idealizations do not exist in any sense? How can an existentially quantified statement be true in virtue of the mere possibility of an idealized construction that has not in fact ever been carried out?

The activist model, of course, answers this question by proposing that the idealized constructions that mathematics is about are in fact actual in the divine intellect, and hence that the objects of mathematics can be identified with divine constructions — God's collectings and God's concomitant concepts. In one sense, of course, collectings and concepts in the divine mind are just one more bunch of specific mental constructions. But the identification in question is far from arbitrary. God's is not just one more mind among many, but the fullest possible realization of everything that mind can be and do, and indeed, the very source of mind and consciousness. Moreover, as God is a necessary being, divine constructions have all the stability of existence of traditional Platonic objects, and indeed, as God's is the only necessary intellect, they are the only possible constructions with this sort of stability. The identification of the idealized constructions of mathematics with actual divine constructions is thus natural and justifiable. Given both the stability of these objects and their constructive character, this identification satisfies both Platonic intuitions about mathematical existence and constructivist intuitions about the connections between mathematical existence and mentality.[42] In addition, however, the activist model seems not

41. P. Kitcher, *The Nature of Mathematical Knowledge* (Oxford: Oxford University Press, 1983); C. Chihara, *Constructibility and Mathematical Existence* (Oxford: Oxford University Press, 1990).

42. That said, it must be noted that the idea that all possible constructions are actual in the divine intellect is, in fact, somewhat problematic. Given the constructive nature of collecting, it always seems possible to collect *more* than what one has already collected. This seems true for the divine intellect no less than for ours. Thus, suppose God has in fact done all the collecting that he can at the current time, so that all the collections there can be (at the current time) are now actual and fixed in the divine intellect. Does it not still seem possible, given the "dynamic" nature of collecting, that God could now *continue* his collecting ac-

to be prone to the epistemological liabilities of Platonism mentioned earlier. Baldly, the idea is this. What we do when we construct a set or form a concept is *like* what God does. Hence, our set-like constructions and concepts are like his. We thereby gain basic knowledge of mathematical objects in virtue of knowledge of our own perceptions and concepts, and of their similarity to those in the divine mind.

In a bit more detail, although it is a fertile ground for skepticism, there seem to be good scientific and philosophical grounds for the common sense belief that human beings have similar capacities for organizing and aggregating perceptions and for abstracting concepts. In our last chapter we saw this ability manifested in that different civilizations independently have obtained the same mathematical results. We seem also to have observed this in particular in our common ability to aggregate different subgroups of the nine dots in the figure given earlier. Indeed, in a certain clear sense, we can talk about seeing the *same* thing — by which we mean that our numerically distinct perceptions have identical, or at least very similar, structural features. Of course, this capacity cannot be explicitly verified — we cannot literally compare the features of someone else's perceptions with our own. But it is reasonable for us to infer such structural similarity from our respective behavior. We describe our perceptions in similar ways, appear to understand one another when we talk about shifts in figure and ground in a perceptually ambiguous image, and so on. In the same way, we seem to be able to say of one another that we possess the same concept of one thing or another. For example, we all seem to have a concept of a *pair,* or, more abstractly, of *two-ness.* Once again, our concept of two-ness — the thing, or information, or whatever it is in our head that encodes what we understand when we have this concept — is not literally identical with someone else's.

---

tivity and generate a collection that contains all of the collections that he has in fact generated? If so, then it seems that it makes no sense to say that God, at any given time *t*, has done all the collecting that it is possible for him to do at *t*. We will not address this issue in detail here, but perhaps the most promising activist line to pursue draws upon so-called "reflection" principles. (See, e.g., A. Levy, "Principles of Reflection in Axiomatic Set Theory," *Fundamenta Mathematicae* 49, 1-10.) Very roughly, these principles can be thought of as saying that, for certain large ordinal numbers κ, the initial segment V_κ of the iterative hierarchy already contains all of the interesting structure of the entire hierarchy. Thought of in terms of our model, such ordinals can be considered natural "stopping points" in the constructing process — in any possible world, God always constructs as far as some natural stopping point.

Nonetheless, we can reasonably say that we both have the *same* concept. As before, the grounds for such an assertion will be facts of verbal behavior (we both use the word "two" when we wish to identify a pair) as well as such higher-order facts as that we are both conscious, rational beings with the ability for perceptual discrimination and aggregation as well as conceptual abstraction. And since we can share the "same" perceptions and concepts, we can reasonably be said to have *knowledge* of another's perceptions and concepts.

Admittedly, God presumably does not come to have perceptions and concepts in the same way that we do. Nonetheless, it is a presupposition of most robust brands of theism that we are "created in God's image," that is, that we are *like* God in certain important respects such as consciousness and rationality that are especially important here. Given this, it seems reasonable to infer, in particular, that our perceptions and concepts are akin to God's. (Granted, we may not want strictly to say that God has *perceptions,* insofar as these depend on some sort of physical perceptual apparatus. But what is important in the notion for purposes here is the idea of *focused awareness* of some kind, and that certainly seems like a mode of consciousness, however it is realized in fact, that God would possess.) Thus, when we aggregate some objects together in perception or in thought, it is reasonable to say that our resulting collection is the "same" (that is, has the same members) as one in God's mind. When we abstract a concept of *pair* or *two-ness,* it seems reasonable to say that God possesses the "same" concept, a concept with similar content that applies to the same things as our concept, namely, aggregates of two things.

Now, our basic set theoretic axioms are, arguably, based upon facts about, and operations that we can perform mentally on, our set-like perceptions. Insofar as God has the "same" perceptions, then, these same axioms apply to collections in the divine mind. Thus, insofar as the axioms correctly describe our own mental collections they also describe God's. Moreover, intuitively, these axioms describe any possible collections whatever, regardless of size. Consequently, we can be said to have knowledge of the collections in the divine mind for which there are no corresponding collections in any human mind, that is, according to activism, we have knowledge of *sets per se.* By the same token, our basic number theoretic axioms correctly describe our own concepts. Insofar as God has the "same" concepts, they describe his as well, and hence we can be said to have knowledge of God's number concepts, that is, according to activism, knowledge of the *numbers per se.* Unlike classical Platonism, then, activism seems to provide us with an epistemology that explains our knowledge of mathematics, that is, of mathematical objects.[43]

Conclusion

The postmodernist's skeptical critique of objective knowledge is not so much an argument but a general rejection of a certain philosophical stance, of a certain way of looking at the world. As we noted in our introductory section, there are some important insights behind this rejection. The appropriate response, however, seemed to be the adoption of both a certain tentativeness toward the acceptance of any philosophical doctrine and an openness toward modes of knowing, thinking, and expressing knowledge that do not depend on precise language and formal reasoning. We therefore suggested that the development of philosophical positions, especially in metaphysics, be thought of as the development of *models.* For models are by their nature limited, fallible, and revisable.

Like any model, then, the one developed in this chapter is incomplete. There are a number of outstanding issues we have not addressed. Moreover, the account gives rise to new questions of its own. However, these caveats noted, insofar as the model does provide an appealing solution to the problem of abstract objects for the believer there seems no reason to deny that we cannot reasonably, albeit tentatively, deem the model to be accurate as far as it goes.

Granted, the idea of a model being accurate as far it goes could use some fleshing out that we will not provide here. But the rough idea would be that a model is true as far as it goes if its salient structural features map in a natural way onto the thing being modeled. Thus, a model of an airplane is accurate as far as it goes insofar as whatever represents wings in the model

43. The line here is analogous to one developed by Penelope Maddy in "Perception and Mathematical Intuition," *Philosophical Review* 89 (1980): 163-96. Maddy, being a naturalist, does not invoke God in her account. The similarity between the two accounts is in the idea of grounding general set theoretic knowledge in the perception of (or, in our case, the construction in perception of) small, finite collections. Specifically, Maddy argues that sets of concrete objects are themselves concrete, and hence that we have perceptual knowledge of small, finite concrete sets. From this knowledge we extract our basic set theoretic axioms, which then ground our knowledge of sets in general.

can be mapped in some structure-preserving way to the actual wings on the aircraft, and so on for certain other major elements of the actual airplane. Moreover, there ought not to be any additional features of the model that do not correspond to any structural features of the airplane. Lots of things might be missing from the model — that's the "as far as it goes" part. But a model that meets the stated conditions seems reasonably thought of as accurate as far as it goes.

Some believers may not be happy to leave things in such a tentative and uncertain state. But that our beliefs about the world *are* in fact tentative and uncertain is perhaps one of the themes of postmodern philosophy with which thoughtful Christians might well agree.

CHAPTER 4

The Pragmatic Nature of Mathematical Inquiry

Our last chapter gave plausible arguments for a realistic view of mathematical objects, locating them in the mind of God. The rationale we suggested for our ability to know something about those objects can explain, at least in part, the universality of mathematical content across cultures that we discussed in Chapter 2. Not everyone, of course, would affirm the model we put forth, but even if there were universal agreement, the way we can apply God's thoughts to mathematical ideas is not immediately obvious. For example, while most people would probably agree that God's thoughts are consistent, many may question whether the mathematical community has a sound basis for the conviction that certain well-established portions of mathematics, like Euclidean geometry and number theory, are consistent. In this chapter we will argue that this belief is guided by largely pragmatic concerns. More generally, we will look at mathematical inquiry, generally, and argue that it has a pragmatic component. We examine, for example, the role that *effort,* as opposed to logic, plays in the guiding of how we direct our mathematical inquiries. We begin with a look at our confidence in mathematical consistency.

Is Our Confidence in Mathematical Consistency Legitimate?

In 1926 Hermann Weyl's *Philosophy of Mathematics and Natural Science* appeared in Oldenbourg's *Handbuch der Philosophies.* At the time Hilbert's formalist program to "eradicate via proof theory all the foundational questions of mathematics" was in full swing. As a pupil of Hilbert, Weyl was looking to the complete and ultimate success of Hilbert's program, a confidence evident in Weyl's treatment of the foundations of mathematics in the original version of *Philosophy of Mathematics and Natural Science.* But in an appendix to that same text appearing twenty years later, Weyl admitted that this confidence was misplaced:

> Gödel showed that in Hilbert's formalism, in fact in any formal system M that is not too narrow, two strange things happen: (1) One can point out arithmetic propositions F of comparatively elementary nature that are evidently true yet cannot be deduced within the formalism [Gödel's first theorem — the incompleteness theorem]. (2) The formula Ω that expresses the consistency of M is itself not deducible within M [Gödel's second theorem]. More precisely, a deduction of Φ or Ω within the formalism M would lead straight to a contradiction in M.[1]

The aim of Hilbert's "Beweistheorie" was, as he declared, "die Grundlagenfragen einfürallemal aus der Welt zu schaffen" [i.e., the aim of Hilbert's "proof theory" was to "eradicate all the foundational questions" of mathematics]. In 1926 there was reason for the optimistic expectation that by a few years' sustained effort he and his collaborators would succeed in establishing consistency for the formal equivalent of classical mathematics. The first steps had been inspiring and promising indeed. But such bright hopes were dashed by a discovery in 1931 due to Kurt Gödel that questioned the whole program. Since then the prevailing attitude has been one of resignation. The ultimate foundations and the ultimate meaning of mathematics remain an open problem; we do not know in what direction it will find its solution, nor even whether a final objective answer can be expected at all.

Weyl's assessment of mathematical foundations after Gödel is perhaps too pessimistic. In particular, just how decisive Gödel's theorems are in overthrowing Hilbert's program remains open to question. Gerhard Gentzen's proof of the consistency of arithmetic using transfinite methods,[2] though

1. Hermann Weyl, *Philosophy of Mathematics and Natural Science,* revised and augmented English edition, based on a translation by Olaf Helmer (Princeton: Princeton University Press, 1949).

2. Gerhard Gentzen, "Die Widerspruchsfreiheit der reinen Zahlentheorie," *Mathematische Annalen* 112 (1936): 493-565.

overstepping the finitary requirements of Hilbert's program, nevertheless shows that consistency can be proved if we are willing to extend our methods of proof.[3] More recently Michael Detlefsen has argued that a finitistic interpretation of the universal quantifier can lead to cases where consistency becomes provable — this time as Hilbert would have it by finitary means (however, the resulting finitistic proof theory is not a subsystem of the classical proof theory).[4]

Although the epistemological significance of Gödel's theorems is still a matter of debate among philosophers, the practical effect of Gödel's theorems on the mathematical community is more easy to discern. On the question of completeness, given a conjecture C and axioms **B**, mathematicians admit the following possibilities:

(1) C is provable from **B**.
(2) The negation of C is provable from **B**.
(3) It can be proven that neither C nor its negation is provable from **B** (C is provably undecidable, or if you will, decidably undecidable).
(4) It can't be proven that neither C nor its negation is provable from **B** (C is unprovably or undecidably undecidable).

Statement (4) involves the greatest admission of ignorance. Statements (3) and (4) together are a far cry from Hilbert's confident rejoinder to Du Bois-Reymond that "in mathematics there is no *ignorabimus*."[5] Individual mathematicians have always recognized that open mathematical problems might well lie beyond their mathematical competence. In some cases the requisite mathematical machinery for solving an open problem has had to wait millennia (e.g., the role of Galois theory in resolving such problems as

3. Extensions of Gentzen's work on consistency can be found in Wilhelm Ackermann, "Zur Widerspruchsfreiheit der Zahlentheorie," *Mathematische Annalen* 117 (1940): 162-94, and Gaisi Takeuti, "On the Fundamental Conjecture of GLC I," *Journal of the Mathematical Society of Japan* 7 (1955): 249-75.

4. Michael Detlefsen, "On Interpreting Gödel's Second Theorem," *Journal of Philosophical Logic* 8, no. 3 (1979): 297-313.

5. "We hear within us the perpetual call. There is the problem. Seek its solution. You can find it by pure reason, for in mathematics there is no *ignorabimus*." See Constance Reid, *Hilbert-Courant* (New York: Springer, 1986), p. 72. Hilbert was responding to Emile Du Bois-Reymond, who in the nineteenth century had vented his epistemological pessimism with the watchword *ignoramus et ignorabimus* — we are ignorant and shall remain ignorant. Hilbert vehemently opposed this attitude.

squaring the circle and trisecting an angle). Hilbert's confidence, however, did not rest with the individual mathematician, but with the nature of mathematics and with the scope and power of mathematical proof. Hilbert had believed in the capacity of proof to access any nook of mathematical ignorance. Gödel showed that nooks exist from which proof is forever barred. Mathematicians nowadays recognize that their research problems may not only be beyond the scope of their ingenuity, but also beyond the scope of their mathematical methods. This awareness can be credited to Gödel's incompleteness theorem.

Although incompleteness limits what mathematicians can prove, it in no way destroys the mathematics they have to date proven. The same cannot be said for inconsistency. Consider Weyl's comments about consistency from the 1926 version of *Philosophy of Mathematics and Natural Science:*

> An axiom system must under all circumstances be free from contradictions, in which case it is called *consistent;* that is to say, it must be certain that logical inference will never lead from the axioms to a proposition *a* while some other proof will yield the opposite proposition ~*a*. If the axioms reflect the truth regarding some field of objects, then, indeed, there can be no doubt as to their consistency. But the facts do not always answer our questions as unmistakably as might be desirable; a scientific theory rarely provides a faithful rendition of the data but is almost invariably a bold construction. Therefore the testing for consistency is an important check; this task is laid into the mathematician's hands.[6]

At the time Weyl was waiting for a demonstration of the consistency of classical mathematics, a demonstration that was to depend on nothing more than basic arithmetic. Basic arithmetic, the mathematics of the successor operation, presumably the simplest of all mathematical theories, was to ground the consistency of all of mathematics, including basic arithmetic itself. Now whatever else we might want to say about Gödel's second theorem, it did show that basic arithmetic is inadequate for demonstrating this consistency.

The need to go beyond this minimalist basis to demonstrate consistency has therefore left mathematicians with less than deductive certainty regarding the consistency of their mathematical theories. Mathematicians

6. Weyl, *Philosophy of Mathematics and Natural Science,* p. 20.

are deductively certain that $2 + 2 = 4$ inasmuch as they can produce a deductive proof for this result (e.g., from the Peano axioms). On the other hand, mathematicians have no deductive certainty that their theories are consistent. Indeed, the typical mathematician will be hard pressed to direct the earnest inquirer to a convincing proof of consistency for his or her favorite mathematical theory.

The theorems of a mathematical theory concern questions internal to the theory. Consistency, on the other hand, poses a question external to the theory. To decide the consistency of a given mathematical theory *T* in a way that is mathematically rigorous (and therefore leads to deductive certainty), it is first necessary to embed *T* in an encompassing mathematical framework *U* within which the consistency of *T* can be coherently formulated. For any nontrivial theory *T*, however, mathematicians lack a canonical method for first determining *U* and then embedding *T* in *U*. Gödel's second theorem provides one such embedding (the one in which Hilbert had hoped to prove consistency, namely *U* = basic arithmetic), but then demonstrates that this embedding is inadequate for determining consistency.

Mathematicians are confident when they affirm or deny claims internal to their theories since such claims either are axiomatic, or follow by some logically acceptable consequence relation from the axioms. Their confidence is the confidence people place in a properly working machine. If the machine is at each step doing what it is supposed to do, its overall functioning will presumably be satisfactory. So too in mathematics, if both background assumptions (axioms) and consequence relation (inference rules) are uncontroverted, then the theorems and proofs that issue from this machine will be uncontroverted as well. This is the beauty of the formalist picture. To accommodate consistency within this picture it is necessary to embed the machine we hope is consistent (i.e., our original theory) in a bigger machine whose consistency we don't question. Gödel's bigger machine was basic arithmetic. This machine was inadequate for the task. Since then other machines have been proposed, but none has gained universal acceptance.

Thus while mathematicians have mathematically compelling reasons for accepting the theorems that make up their theories, they lack mathematically compelling reasons for accepting the consistency of these theories. How then do they justify attributing consistency to their theories? Whence the confidence that mathematics is consistent, if this confidence cannot be justified through mathematical demonstration?

Weyl's view of consistency still prevails, even if this fact is advertised less now than in times past. Whether openly or tacitly, mathematicians agree that a mathematical theory "must under all circumstances be free from contradictions." Indispensable to the success of mathematics is the method of indirect proof — reductio ad absurdum. Given axioms **B**, a conjecture C, and a contradiction that issues via a logically acceptable consequence relation from **B** and ~C taken jointly, the method of indirect proof allows us to conclude C. This method is so powerful that the mathematical community is loath to give it up. In fact, whenever constructivists try to limit the method of indirect proof, they are in practice ignored. This is not to say that constructivists have nothing interesting to say about the foundations of mathematics. But the working mathematician whose living depends on proving "good theorems" simply can't afford to lose a prize tool for proving them.

Because reductio ad absurdum is a basic tool in the working mathematician's arsenal, a single contradiction is enough to ruin a mathematical theory. The problem here is that the contradiction follows from the axioms **B** alone — without the aid of conjectures like C which lie outside **B**. In the previous example the contradiction arose by looking at the consequences of **B** and ~C taken together. But this time the contradiction arises from **B** itself (i.e., the very axioms that are supposed to constitute the secure base for all our subsequent reasonings). Since a contradiction springs from **B** itself, any C together with **B** entails a contradiction. Hence by the method of indirect proof, an inconsistent system proves everything and rules out nothing.

In the history of mathematics a notable example of such ruin occurred when Frege learned of Russell's paradox. As Frege put it,

> Hardly anything more unfortunate can befall a scientific writer than to have one of the foundations of his edifice shaken after the work is finished. This was the position I was placed in by a letter of Mr. Bertrand Russell, just when the printing of this volume was nearing its completion. It is a matter of my Axiom (V).[7]

This remark appears in the appendix to volume II of Frege's *Grundgesetze der Arithmetik*. Russell's paradox had demonstrated that inherent in Frege's system was a contradiction. The history of logicism subsequent to

7. Gottlob Frege, *Translations from the Philosophical Writings of Gottlob Frege*, 3rd edition, ed. Peter Geach and Max Black (Oxford: Basil Blackwell, 1985), p. 214.

Frege's *Grundgesetze* can be viewed as an attempt to salvage the offending Axiom (V).

Logicism sought to ground mathematics in self-evident logical principles, thereby making mathematics a branch of logic. Axiom (V) was supposed to be one such principle in the logical grounding of mathematics. Nevertheless, Axiom (V) was responsible for a contradiction. For logicism therefore to succeed, the logical legitimacy of Axiom (V) had to be discredited. To mitigate the force of Russell's paradox, Frege questioned the self-evidence of Axiom (V): "I have never disguised from myself its lack of the self-evidence that belongs to the other axioms and that must properly be demanded of a logical law."[8] Having finished the *Grundgesetze* only to discover a contradiction, Frege confined himself to identifying and then discrediting the offender responsible for the inconsistency. In the *Principia Mathematica* Russell and Whitehead then took positive steps to salvage Frege's program. This they did by introducing their theory of types and postulating an infinite number of individuals of lowest type. Using types and actual infinities, Russell and Whitehead were able to accomplish the work of Axiom (V) while at the same time avoiding a provable inconsistency. Nevertheless, there was a cost. Indeed, they had to sacrifice the principal claim of logicism — that mathematics is a branch of logic. Indeed, as we partially discussed in Chapter 3, it has never been clear that types and actual infinities are primitives of logic.[9]

The point to recognize in this historical digression is not so much that mathematicians strive at all costs to save consistency, but rather that they have a strategy for saving consistency, a strategy that hinges on the method of indirect proof. An inconsistent mathematical system with axioms **B** is as it stands worthless because by reductio ad absurdum it entails everything. Nevertheless, since mathematicians have typically devoted time and effort to the system, the usual strategy is to save as much of the system as possible. The strategy is therefore to find as small and insignificant part of **B** as possi-

8. Frege, *Translations from the Philosophical Writings of Gottlob Frege*, p. 214.

9. As William and Martha Kneale observe in their exhaustive history of logic, "In *Principia Mathematica* the axioms are all supposed to be necessary truths. . . . Admittedly Russell has misgivings about his axiom of reducibility and his axiom of infinity, but he still thinks that if they are to be accepted at all they are to be accepted as [necessary] truths, and he therefore puts forward such considerations as he can produce to convince the reader or at least make him sympathetic." See William and Martha Kneale, *The Development of Logic* (Oxford: Oxford University Press, 1988), p. 683.

ble which, if removed, removes the known inconsistency: prune **B** down to **B′** and call the leftovers C. The hope is that **B′** does not lead to a contradiction. Still to be preferred is that a supplement C′ be found to **B′**, which plays the same role as C, but without introducing the inconsistency for which C is held responsible. Since **B′** and C together (= **B**) lead to a contradiction, C becomes the offender guilty of producing the contradiction inherent in **B**. Note that this strategy for saving consistency underdetermines the choice of C and C′: **B** can typically be pruned and supplemented in various ways to save consistency. In line with our previous example, Frege identified the offending C with Axiom (V) whereas Russell and Whitehead offered their theory of types as the preferred supplement C′.

The readiness of mathematicians to employ the foregoing strategy to save consistency supports Weyl's claim that "an axiom system must under all circumstances be free from contradictions." Nevertheless, inherent in this strategy is the disturbing possibility that pruning and salvaging might continue interminably because of an unending chain of contradictions. The worst case scenario has **B** leading to a contradiction, requiring that **B** be reduced to a proper subset **B′**, which after some time in turn leads to a contradiction, requiring that it be reduced to a proper subset **B″**. . . . This process might continue until nothing is left of the original **B**. One way to avert this possibility is to produce a consistency proof of the type Hilbert was seeking. Yet even if Gödel's second theorem doesn't demonstrate that the search for such a consistency proof is vain, the lack of a universally recognized consistency proof leaves open the possibility that mathematics is a hydra which, however many contradictions we lop off, will never cease to sprout further contradictions.

How then can we account for the conviction in the mathematical community that certain well-established portions of mathematics, like Euclidean geometry and number theory, are consistent? Mathematicians may, contrary to Gödel, leave open the possibility that Euclidean geometry and number theory are inconsistent. But their confidence that these theories won't sprout contradictions is analogous to the lay person's confidence that the sun will rise tomorrow. The lay person's confidence rests on an induction from past experience (supplemented perhaps by theoretical support from the lay person's physical understanding of the world). Similarly, the mathematician's confidence in consistency rests on an induction from mathematical experience.

Weyl himself was aware that this type of induction goes on within mathematics. In describing the axiom of parallels from Euclidean geometry, Weyl noted:

> From the beginning, even in antiquity, it was felt that [the axiom of parallels] was not as intuitively evident as the remaining axioms of geometry. Attempts were made through the centuries to secure its standing by deducing it from the others. Thus doubt of its actual validity and the desire to overcome that doubt were the driving motives. The fact that all these efforts were in vain could be looked upon as a kind of inductive argument [*N.B.*] in favor of the independence of the axiom of parallels, just as the failure to construct a perpetuum mobile is an inductive argument for the validity of the energy principle.[10]

The continued efforts of mathematicians to derive the axiom of parallels from the remaining axioms of Euclidean geometry supported the claim that this axiom is in fact underivable from the remaining axioms. Of course, interest in such inductive support evaporated with the discovery of non-Euclidean geometries — here then finally was a proof that the axiom of parallels is underivable from the remaining axioms.

Generally, of course, mathematical arguments are not arguments from ignorance. The inability of one or even several mathematicians to establish a result does not mean that the result is impossible to establish. Nevertheless, the inability of the mathematical community as a whole even to make progress on, much less establish, a given result over an extended period of time can lead to a conviction within the mathematical community that the result is impossible to establish. It is worth noting how often this conviction has in the end been justified deductively. The problems of trisecting an angle and squaring a circle date back to antiquity. Their "solution" in the nineteenth century (through the work of Galois and his theory of groups) simply confirmed the vain efforts of previous generations, namely that with ruler and compass these problems are insoluble. In this light, confidence in the consistency of Euclidean geometry, number theory, and other well-established mathematical theories can be viewed as the failure of the mathematical community to discover a contradiction from these theories — despite sustained and arduous efforts to discover such a contradiction. In fact, what makes these theories well-established is precisely this

10. Weyl, *Philosophy of Mathematics and Natural Science*, p. 21.

failure despite sustained effort (what Weyl calls "the fact that all these efforts were in vain").

Arend Heyting picks up this train of thought in his book *Intuitionism.* There he presents a delightful dialogue in which proponents of various philosophical positions on the nature of mathematics argue their views. In this dialogue Heyting places the pragmatic view of consistency we are describing in the mouth of an interlocutor named Letter. Letter advocates a philosophy of mathematics nowadays referred to derisively as "if-thenism": "Mathematics is quite a simple thing. I define some signs and I give some rules for combining them; that is all."[11] Among current philosophers of mathematics if-thenism is rightly rejected as too incomplete and simplistic an account of mathematics. If-thenism simply leaves too many questions unanswered, in particular the initial choice of axioms and the indispensability of mathematics for the natural sciences.[12] Nevertheless, when the interlocutor known as Form (= the Hilbertian formalist) demands "some modes of reasoning to prove the consistency of your formal system," Letter's response, particularly in light of Gödel's second theorem, seems entirely appropriate.

> Why should I want to prove [consistency]? You must not forget that our formal systems are constructed with the aim towards applications and that in general they prove useful; this fact would be difficult to explain if every formula were deducible in them. Thereby we get a practical conviction of consistency which suffices for our work. (Heyting, p. 7)

Whence this practical conviction of consistency? In our mathematical exertions we continually try to deduce contradictions in the following way: Reductio ad absurdum is a mathematician's stock in trade. To prove C from axioms **B**, it is enough to derive a contradiction from ~C and **B**. In trying to derive a contradiction from both **B** and some auxiliary hypothesis ~C, however, mathematicians are a fortiori trying to derive a contradiction from **B** itself. Hence mathematicians are ever checking for contradictions inherent in **B**. Perhaps we have overstated things a bit here, but it seems clear, nevertheless, that in claiming consistency for the mathematical theory entailed by

11. Arend Heyting, *Intuitionism: An Introduction,* 3rd rev. ed. (Amsterdam: North Holland, 1971), p. 7.

12. Penelope Maddy, *Realism in Mathematics* (Oxford: Oxford University Press, 1990), p. 25.

B, mathematicians are making an induction similar to that which is practiced by natural scientists.

What sort of induction is this? Corresponding to any inductive generalization is what Steve Meyer calls a *proscriptive generalization.* By this, he means a generalization that is dangerous, or that can get us into trouble. Moreover, corresponding to the inductive support for an inductive generalization is what can be called *proscriptive support* for the proscriptive generalization. A celebrated example of an inductive generalization concludes from the observational claim "all observed ravens have been black" to the general claim "all ravens are indeed black." These two claims, however, are respectively equivalent to "no observed ravens have been non-black" and "no ravens are non-black." Now the move from "no observed ravens have been non-black" to "no ravens are non-black" can be viewed as proscriptive support for a proscriptive generalization, the proscriptive support being that no observed ravens have been non-black and the proscriptive generalization being that no ravens whatsoever are non-black.

Within mathematics this sort of move from proscriptive support to proscriptive generalization occurs all the time when the consistency of a mathematical theory is in question: from "no contradiction has to date been derived from **B**" (the proscriptive support) mathematicians conclude that "no contradiction is in fact derivable from **B**" (the proscriptive generalization). Just as the grounds for concluding that no ravens are non-black is the failure in practice to discover a non-black raven, so the grounds for concluding that no contradiction can be derived from **B** is the failure in practice to discover a contradiction from **B**.

The failure in practice to discover a thing may or may not provide a good reason for doubting the thing's existence. Consider the familiar "God of the gaps" objection to miracles. Some strange phenomenon M is observed ("M" for miracle). A search is conducted to discover a scientifically acceptable explanation for M. The search fails. Conclusion: no scientifically acceptable explanation exists, and what's more God did it. There is a problem here. As physicist and philosopher of religion Ian Barbour aptly notes,

> We would submit that it is *scientifically stultifying* to say of any puzzling phenomenon that it is "incapable of scientific explanation," for such an attitude would undercut the motivation for inquiry. And such an approach is also *theologically dubious,* for it leads to another form of the

> "God of the gaps," the *deus ex machina* introduced to cover ignorance of what may later be shown to have natural causes.[13]

Or as C. A. Coulson puts it, "When we come to the scientifically unknown, our correct policy is not to rejoice because we have found God; it is to become better scientists."[14]

Barbour and Coulson are right to block lazy appeals to God within scientific explanation. The question remains, however, how long are we to continue a search before we have a right to give up the search and declare not only that continuing the search is vain, but also that the very object of the search is nonexistent? The case of AIDS suggests that certain searches must never be given up. The discovery of the cause of AIDS in HIV has proved far easier than finding a cure. Yet even if the cure continues to elude us for as long as the human race endures, we trust the search will not be given up. There is of course an ethical dimension here as well — certain searches must be continued even if the chances of success seem dismal.

There are times that searches must be continued against extreme odds. There are other times when searches are best given up. To illustrate, consider some examples from Greek mythology. Despite Poseidon's wrath, Odysseus was right to continue seeking Ithaca. Sisyphus, on the other hand, should long ago have given up trying to roll the rock up the hill. We no longer look kindly on angle trisectors and circle squarers. Purported perpetuum mobile devices amuse us. We deny the existence of unicorns, gnomes, and fairy godmothers. In these cases we don't just say that the search for these objects is vain; we positively deny that the objects exist.

We do not have a precise line of demarcation for deciding when a search is to be given up and when the object of a search is to be denied existence. Nevertheless, we can offer a necessary condition. The failure in practice to discover a thing is good reason to doubt the thing's existence *only if* a diligent search for the thing has been performed. If we are to be convinced on the basis of observational evidence that no ravens are non-black, we must first be convinced that a diligent search for a non-black raven has been conducted. If ravens can conceivably be found in a trillion different places and if only a small fraction of those places can, given our resources, be examined,

13. Ian Barbour, *Issues in Science and Religion* (London: SCM Press, 1966), p. 390.

14. C. A. Coulson, *Science and Religion: A Changing Relationship* (Cambridge: Cambridge University Press, 1955), p. 2.

we should still want to see full use made of those limited resources. What's more, we should want to see those resources used to obtain as representative a sample of ravens as possible (e.g., our search for non-black ravens should not be confined to just one locale). A full and efficient use of our resources for discovery should be made before we accept a proscriptive generalization.

If all our efforts to discover a thing have to date been in vain, then our practical conviction that the thing doesn't exist is proportional to how much (seemingly wasted) effort has been expended to discover the thing. This is one way of characterizing proscriptive generalizations, though in the natural sciences this type of induction is usually described in the language of confirmation. Unfortunately, in mathematics claims about practical conviction tend to get short shrift. The mathematical community is so used to operating by analytic standards of rigor and proof that inductive justifications of mathematical claims are typically regarded as no more than precursors to precise analytic demonstrations.[15] Thus even though the collective experience of mathematicians for two thousand years supported the independence of the axiom of parallels from the remaining axioms of Euclid, only when Bolyai and Lobachevsky produced their non-Euclidean geometries (and Gauss and Riemann accepted those works as legitimate) were mathematicians finally satisfied.

This attitude of mathematicians to prefer analytic demonstration over inductive justification is generally healthy. For a mathematical claim, analytic demonstration is always a firmer support than inductive justification. In this light Hilbert's program can be seen as the grand endeavor to assimilate all of mathematics to analytic demonstration — a worthy goal if feasible. In this way analytic demonstration would always have supplanted inductive justification. Gödel's theorems, however, rendered Hilbert's program doubtful and in the process left open the need for inductive justification within mathematics.

What happens when our analytic methods continually fail to produce a given result? When a mathematical research program is just beginning, mathematicians often share practical convictions about claims they hope will eventually be decided analytically through their program. Thus as Weyl might put it, mathematicians come into the research program looking at past efforts as supplying "a kind of inductive argument" for claims they want

15. This attitude is now changing because of the computer and the proliferation of problems in the physical sciences that admit no exact mathematical solution.

later to prove rigorously. Inductive arguments, however, are second-class citizens in the mathematical hierarchy of justification. Weyl's reference to inductive argument in mathematics was made at a time when Hilbert's program still seemed promising. Inductive arguments for consistency and independence of certain axioms were therefore pointers to the rigorous demonstrations that Hilbert's program was to produce. As Hilbert's program ran out of steam, however, it became apparent that rigorous demonstrations for claims previously supported only by inductive justifications would not be forthcoming, at least not from the program. What was left was only the original, inductive justification.

The precise relation between analytic demonstration and inductive justification is therefore an open problem. The history of mathematics confirms that inductive justifications (Weyl's "kinds of inductive argument") have always played an important role in mathematics. Moreover, when mathematical programs have sought to eliminate inductive justifications by superseding them with analytic proofs, they have not always been successful. In fact, we submit that the history of mathematics supplies ample evidence for the ineliminability of inductive considerations from the actual content of mathematics.

Earlier we suggested that our practical conviction that a thing doesn't exist is in part proportional to how much (seemingly wasted) effort has been expended to discover the thing. Let us now tailor that suggestion to mathematics: Apart from a precise analytic demonstration, our practical conviction toward a mathematical claim is in part proportional to how much effort mathematicians have expended trying to decide the claim. Mathematicians expend effort whenever they deduce consequences from axioms and theorems. Thus, the if-then picture of the mathematician cranking away at an inference engine is at least part of what we are calling effort. Note that this is an empirical picture. Observations and experiments that make up the picture are deductive arguments — chains of reasonings issuing from background assumptions and proceeding according to a logically acceptable consequence relation. The data comprise everything from student problem sets to the articles in mathematical journals to computer simulations. Within this picture a mathematical theory can be empirically adequate only if no expenditure of effort has to date discovered an inconsistency.

Lacking as they do analytic demonstrations for the consistency of their mathematical theories, mathematicians accept the consistency of their theories out of a practical conviction that springs from their persistent failed efforts (even though indirect) to discover a contradiction. The type of induction responsible for this pragmatic conviction is nothing new to contemporary mathematicians. After repeated failures at trying to solve a problem, mathematicians come to believe that the failure is in the nature of the problem and not in their competence. Then the search is on to provide an analytic demonstration that the problem has no solution. Yet this search can fail as well. Repeated failure here then yields the practical conviction that the problem has no solution — despite the absence of strict analytic proofs. The point to realize is that in circumstances where no analytic resolution is in fact possible, practical convictions of this sort are all that remain to the mathematician. The history of mathematics simply does not support the hope that practical conviction can always be turned into mathematical certainty by means of analytic proof.

Commenting on failed attempts to prove the axiom of parallels from the other axioms of Euclid's geometry, Weyl writes, "The fact that all these efforts were in vain could be looked upon as a kind of inductive argument in favor of the independence of the axiom of parallels."[16] The independence of the axiom of parallels was in the end provable, so that all the failed efforts over thousands of years to disprove independence could at length be disregarded. However, in instances where not only "all these efforts were in vain," but also no strict demonstration is forthcoming, mathematicians can frequently do better than simply admit the continued failure of their efforts to establish a claim. Having made this admission, they can advance a proscriptive generalization whose support is precisely this "vain" expenditure of effort.

Conjecture Conditionals

We next consider a class of conditionals that has only recently gained the attention of the mathematics and computer science communities, a class we will refer to as *conjecture conditionals.*[17] These are conditionals (see the example below) whose antecedents are conjectures and whose consequents are computational results. The problem with conjectures is, of course, that they might be false. The beauty of computational results, on the other hand, is

16. Weyl, *Philosophy of Mathematics and Natural Science,* p. 21.
17. We owe this phrase to Mark Wilson.

that they have immediate, straightforward applications. Such conditionals introduce an intriguing tension between uncertain antecedents and readily applicable consequents. Mathematicians exploit this tension by adopting an attitude toward these conditionals for which the usual logical modes of analysis, viz., truth and proof, frankly fail to give an account.

The conjecture conditionals that will interest us most have a famous conjecture in the antecedent, and therefore assume the following form:

FAMOUS CONJECTURE $\Rightarrow$ COMPUTATIONAL RESULT

For our purposes it is useful that the conjecture be famous, since this guarantees that considerable effort (for now taken intuitively) has already been expended trying to decide its truth. Moreover, because it still is a conjecture, all this effort has till now been expended in vain. For concreteness we state one such conditional as it appears in the mathematical literature:

> If the Extended Riemann Hypothesis is true, then there is a positive constant K such that for any odd integer $n > 1$, n is prime just in case for all $a \in Z_n$ satisfying $a < K(\log n)^2$, then $a^{(n-1)/2} \equiv a \bmod n$.[18]

This conditional is a theorem of computational number theory. Let us represent it more compactly as

$$RH \Rightarrow C,$$

where RH denotes the Extended Riemann Hypothesis, and C the computational result stated in the consequent. Let us stress that for our purposes the precise statement of RH is unimportant. What is important is that RH $\Rightarrow$ C is a conditional whose antecedent is a conjecture (a claim whose truth or falsehood has yet to be established and may in fact never be established) and whose consequent is a computational result having straightforward applications.

At the level of truth and proof it is difficult to make sense out of conditionals like RH $\Rightarrow$ C in a way that satisfies philosopher and mathematician

18. This is a slightly modified version of Theorem 2.18 in Evangelos Kranakis, *Primality and Cryptography* (Stuttgart: Wiley-Teubner, 1986), p. 57. This theorem is significant to computational number theorists for its relation to the Solovay-Strassen deterministic test for primality, a result useful among other things in cryptography.

alike. The ordinary logic of truth and proof results in an analysis of conditionals which, for convenience, we'll call the orthodox analysis. According to the orthodox, analysis conditionals are material conditionals and therefore logically equivalent to disjunctions. Now, because RH $\Rightarrow$ C is a theorem, according to the orthodox analysis we know that at least one of ~RH and C is true. Yet because RH is a conjecture, we have no idea which is true. Thus, the orthodox analysis asks us to rest content with a proven disjunction (~RH$\vee$C) whose disjuncts both remain unproven.

As far as it goes, the orthodox analysis is unobjectionable. Unfortunately, for RH $\Rightarrow$ C the analysis doesn't get us very far. In particular, the orthodox analysis fails to account for how computational number theorists actually use conditionals like RH $\Rightarrow$ C in practice. Computational number theorists are not content to analyze conditionals like RH $\Rightarrow$ C by replacing them with their logically equivalent disjunctions (in this case ~RH$\vee$C), looking up the truth table that applies to the disjunction, and thereafter resting easy with the knowledge that at least one of the disjuncts is true (which one is true we don't know since RH is a conjecture). Instead, computational number theorists take the bold step of accepting C as *provisionally* true — even though the actual truth of C remains strictly speaking a matter of ignorance.

To justify this move computational number theorists offer the following line of reasoning: "I don't know whether the famous conjecture is true or false. But that doesn't matter. If it's true, I can use the computational result to my heart's content and never get in trouble. If it's false, the worst that can happen is that I apply the computational result and obtain an error. But what a precious error! As a counterexample to the computational result, this error will demonstrate that the famous conjecture is false. I'll be world-famous, having resolved a celebrated open problem." Put another way: either our computation goes through without a snag, or our computation goes awry and we become world-famous, having unintentionally resolved an outstanding open problem. How sweet it is — fame by *modus tollens*!

RH is a famous conjecture in part because the best mathematicians have racked their brains trying to solve it — to date without success. A great deal of effort has been expended trying to prove or disprove RH. On the other hand, to show that the computational result C follows from RH is easy, requiring little effort (the proof is about five lines). Mathematicians therefore feel justified in freely applying the computational result since any single computation will require little effort and therefore seems unlikely to resolve a famous conjecture on which so much effort has already been expended. It is a question of effort: much effort in trying to decide the conjecture without success, little effort in establishing the computational result from the conjecture, and little effort in applying the computational result in practice.

Computational number theorists understand RH $\Rightarrow$ C not ultimately in terms of truth and proof, but in terms of effort relations that give a pragmatic justification for freely using the consequent C. Indeed, as soon as a conjecture conditional like RH $\Rightarrow$ C becomes a demonstrated mathematical theorem, C gains independence from the conjecture RH that entails it, and becomes a computationally useful stand-alone result. On the orthodox analysis, the logical status of C remains as uncertain as ever. Yet from the point of view of effort, C has gained substantial pragmatic support. This perhaps should alert mathematicians to look for a proof of C that does not depend on assuming RH.

Our use of effort here has been a bit loose, but perhaps the general point is clear enough.[19] What is perhaps not so clear, however, is whether we are fairly representing the ideal mathematician — the sincere seeker after mathematical truth. Maybe we are merely representing opportunistic mathematicians, vain seekers after self who think the worst that can happen if we accept the consequences of a famous unproven conjecture is that we refute the conjecture and become world-famous. Perhaps the worst case scenario is really this: we accept the consequences of a famous conjecture that is itself false, but that we cannot prove to be false, and so we wind up with a lot of false beliefs.[20]

Even if we have painted an accurate picture of how the mathematical community handles conjecture conditionals (and some may wish to dispute this claim), the philosopher has every right to wonder whether the mathematical community is correct in its handling of conjecture conditionals. It seems that what concerns the philosopher most about mathematicians' cavalier attitude toward conjecture conditionals centers on the risk that mathematicians assume when they accept the consequences of a famous conjec-

19. A precise account of effort can be developed in terms of computational complexity. See Jan Krajicek and Pavel Pudlak, "Propositional Proof Systems: The Consistency of First Order Theories and the Complexity of Computations," *Journal of Symbolic Logic* 54, no. 3 (1989): 1063-79.

20. Note that this objection presupposes precisely what's at issue in this discussion, namely, whether mathematical knowledge is limited to what is true and provable. It is precisely this point that we are questioning.

ture. The risk is real, since accepting the consequences of a famous conjecture does indeed make us vulnerable to winding up with a lot of false beliefs. Because the picture of mathematics as a haven for deductive certainty is so entrenched, it is hard to imagine mathematics harboring uncertainties, not just about its future progress, but also about its present state. The fact is, however, that mathematicians assume such risks all the time.

Indeed, the mathematical community as a whole risks the consistency of mathematics on conjectures known as axioms. The working mathematician accepts the consistency of a mathematical theory as a provisional truth. As we saw earlier, no mathematical system can bear the strain of a contradiction — hence the backpedaling and reshuffling of axioms whenever an inconsistency is found. It's possible that a well-established mathematical theory is inconsistent. So too, it's possible that C is false. But to trouble oneself over accepting potentially false mathematical beliefs that serve us well, that require more effort than we are able now or perhaps ever to expend on deciding their truth, that are consequences of conjectures whose solution is nowhere in sight; and then to pretend that the entire edifice into which these individual beliefs are embedded is secure — an edifice that is always threatened by the possibility of contradiction — strikes us as problematic.

If RH should at some point be refuted, our acceptance of C would change, at least if assuming RH were the only means we had for establishing C. Similarly, if a mathematical theory should at some point lead to a contradiction, our acceptance of the relevant axioms would change. The latter change is certainly more far-reaching than the former, but both are changes of the same kind. History bears this out: when the axioms of mathematics lead to a contradiction, they are either adjusted or discarded to avoid the contradiction. In section 1 we considered Frege's response to Russell's paradox as a case in point. Frege's Axiom (V) led to a contradiction and therefore had to be trashed. Riemann's celebrated conjecture, on the other hand, has yet to issue in a contradiction.

At the level of truth and proof we have no warrant for accepting the consequences of a famous conjecture or the consistency of a mathematical theory. At this level the best we can do is wait for a contradiction. Thus at the level of truth and proof we are in the uncomfortable position of being unable to reject C or consistency until it is too late, i.e., until the conjecture RH or the axioms of the relevant mathematical theory are known to have issued in a contradiction. At the level of effort, on the other hand, there can be plenty of warrant for accepting both C and the consistency of a mathematical theory. Both beliefs are confirmed by an expenditure of effort; moreover, the degree of confirmation depends on the amount of effort expended. C is entailed by the conjecture RH on which much effort has been expended trying to decide it — as yet to no avail. A mathematical theory comprises what consequences have to date been deduced from its axioms, axioms on which even more effort has been invested to deduce a contradiction — again, to no avail.

Of course mathematicians don't view themselves as consciously trying to find contradictions in their mathematical theories. But since reductio ad absurdum is basic to the working mathematician's arsenal, plenty of occasions arise for proving contradictions. Mathematicians are therefore ever on the alert for contradictions that might arise from the axioms of their theories. For this reason we have no qualms saying that mathematicians have invested even more effort trying to decide the consistency of their mathematical theories than trying to decide RH. The computational number theorist's confidence in C and the mathematician's confidence in consistency are parallel beliefs whose degree of confirmation in both instances is proportional to the effort expended trying to decide those beliefs. Expended effort is capable of confirming mathematical beliefs that cannot be confirmed via strict proof.

A final objection to accepting the consequences of a famous conjecture needs now to be addressed. The problem of deciding RH is the problem of either proving or disproving RH, that is to say, either proving RH or proving ~RH.[21] It therefore follows that deciding RH and deciding ~RH are one and the same problem. Hence the effort expended trying to decide a conjecture like RH is identical with the effort expended trying to decide its negation ~RH. The question therefore arises: Why merely accept the computational results that are deductive consequences of RH? Why not accept the computational results that are deductive consequences of ~RH as well? We have urged accepting C because RH is a conjecture with much effort expended on it, and because RH $\Rightarrow$ C is an easily proved theorem. But ~RH is just as much a conjecture, with just as much effort expended on it as RH. Why not accept a computational result D as provisionally true whenever ~RH $\Rightarrow$ D is a theorem?

A thoroughgoing pragmatist might well say, "Go right ahead. If D runs

21. Although in practice the effort expended on one may produce more methods and specialized results than the other.

afoul, you'll get a Field's Medal[22] for having demonstrated that RH is true; if C runs afoul, you'll get a Field's Medal for having demonstrated that RH is false. In either case you'll be world-famous, having resolved the Riemann Hypothesis." Yet the more likely scenario is that neither C nor D will run afoul when we run the computations, and that the Field's Medal will continue to elude us.

Since this pragmatic line is likely to offend more traditional sensibilities, let us offer an alternative line. When confronted with opposite conjectures like RH and ~RH, mathematicians invariably make a choice, though a choice that depends neither on truth, nor proof, nor effort. The sort of choice we have in mind comes up frequently in set theory. It often happens that set theorists want to add some additional axiom to their theory of sets. Such axioms typically serve either to proscribe certain pathological sets (cf. the axiom of foundation) or to guarantee the existence of certain desired sets (cf. the axioms having to do with large cardinals). Before adding a new axiom A to the old axioms for set theory, however, it is desirable to know two things: (1) that A is consistent with the old axioms; (2) that ~A is consistent with the old axioms. The former guarantees that adding A won't ruin our theory of sets, the latter that adding A won't be redundant. If cases (1) and (2) hold, we say that A is independent of our original axioms. Of course, independence is a symmetric notion, and hence ~A will be independent of our original axioms as well. Any choice that favors A over ~A, or vice versa, is therefore dictated by considerations other than consistency. In practice mathematicians make such choices by looking to such things as simplicity, beauty, fruitfulness, interest, and purposes at hand.[23]

Of course, it may happen that neither A nor ~A can be proved from the original axioms, and that the independence of A from the original axioms cannot be proved either. Thus, despite a vast expenditure of effort, the logical status of A might remain completely indeterminate. In this case, considerations of simplicity, beauty, fruitfulness, interest, and purposes at hand must again be invoked to elicit a choice. Often mathematicians have strong preferences. Often they would like things to be a certain way. And barring any compelling reasons to the contrary, they are willing, at least provisionally, to accept that things are that way. Now, RH is a much nicer hypothesis

22. The Field's Medal is the highest honor the international mathematics community bestows on its members. This is the Nobel Prize of mathematics.

23. See Maddy, *Realism in Mathematics.*

than ~RH. RH says that the zeros of a certain class of analytic functions fall in a certain neat region of the complex plane. ~RH says that they also fall outside that neat region. Presumably it is this nice property of RH that is responsible for RH having interesting computational consequences like C. ~RH, on the other hand, appears to have no interesting computational consequences.

For reasons then that ultimately have nothing to do with either truth or proof or effort, given a choice mathematicians prefer RH over ~RH. Nevertheless, effort is a precondition for this choice being possible: without all that "wasted" effort expended in trying to decide RH, we might suspect that RH or its negation has a simple proof. Having "wasted" this effort, mathematicians feel justified investing the computational consequences of RH with a certain confidence. If they were inclined, mathematicians could invoke effort and invest the computational consequences of ~RH with the same confidence. But for reasons extrinsic to both logic and complexity (truth and proof being logical notions, effort being a complexity-theoretic notion), mathematicians prefer RH. Hence even though effort by itself confers equal weight to the computational consequences of both RH and ~RH, factors outside logic and complexity favor RH, inducing mathematicians to accept its computational consequences, while neglecting those of ~RH.

Effort and the Possibility of Mathematical Knowledge

Anyone who is not a complete skeptic about mathematical knowledge can ask the following question: What are the conditions for the possibility that mathematical knowledge exists? More simply, how is mathematical knowledge possible? While this question may be very Kantian in form, our aim in asking it is very far from Kantian in spirit. We are not, for instance, interested in exploring those properties of the intuition and understanding that make mathematical knowledge possible. This is not to suggest that the properties of the intellect aren't important to the question we are asking. In raising this question, however, we are motivated by pragmatic rather than theoretical concerns.

In exploring the possibility of mathematical knowledge, it is striking how much knowledge mathematicians are actually capable of attaining. This sort of practically attainable mathematical knowledge needs to be distinguished from the strictly in-principle mathematical knowledge that may exist in the mind of God (as we discussed in Chapter 3) but as a practical matter is beyond our ken. We shall concern ourselves with the former, practical sense of possibility rather than the latter, theoretical sense of possibility. For this reason we must first unpack practical possibility in terms of the effort mathematicians expend in trying to prove things.

We start by noting that the connection between effort and practical possibility is neither new nor artificial. Richard von Mises, for instance, used effort to distinguish degrees of possibility:

> Ordinary language recognizes different degrees of possibility or realizability. An event may be called possible or impossible, but it can also be called quite possible or barely possible according to the *amount of effort* that must be expended to bring it about. It is only "barely possible" to write longhand at 40 words per minute; impossible at 120. Nevertheless it is "quite possible" to do this using a typewriter. . . . In this sense we call two events equally possible if the *same effort* is required to produce each of them.[24]

Two features stand out in the way von Mises relates effort and possibility. The first is their inverse proportion: the more effort is needed or must be expended to bring about a state of affairs, the less possible is that state of affairs. The second is that the effort required to obtain a state of affairs varies with the resources at hand: typewriters make for faster longhand than pens alone.

Bradley and Swartz make much the same point, only this time with reference specifically to mathematical knowledge and its limitation:

> There are . . . some propositions the knowledge of whose truth, if it is humanly possible at all, can be acquired only by an enormous investment in inferential reasoning [cf. expenditure of effort]. The proofs of many theorems in formal logic and pure mathematics certainly call for a great deal more than simple analytical understanding of the concepts involved. And in some cases the amount of investment in analysis and inference that seems to be called for, in order that we should know

24. As quoted in Ian Hacking, *The Emergence of Probability* (Cambridge: Cambridge University Press, 1975), p. 123. The italics are ours.

> whether a proposition is true or false, may turn out to be entirely beyond the intellectual resources of mere human beings.
>
> As a case in point consider the famous, but as yet unproved, proposition of arithmetic known as Goldbach's Conjecture, viz., Every even number greater than two is the sum of two primes. . . . Goldbach's Conjecture is easily understood. In fact we understand it well enough to be able to test it on the first few of an infinite number of cases. . . . [But] for all we know, it may turn out to be unprovable by any being having the capacities for knowledge-acquisition which we human beings have. Of course, we do not *now* know whether or not it will eventually succumb to our attempts to prove it. Maybe it will. In this case it will be known ratiocinatively. But then, again, maybe it will not. In that case it may well be one of those propositions whose truth is not known because its truth in *unknowable.* At present we simply do not know which.[25]

The "enormous investment in inferential reasoning," the "intellectual resources of mere human beings," and "the capacities for knowledge-acquisition which we human beings have" can all be unpacked in terms of the effort mathematicians expend in proving things. An infinitely powerful problem solver is able to settle the Goldbach Conjecture, either by providing a counterexample (i.e., an even integer greater than 2 that is not the sum of two primes), or by running through all the even integers greater than 2 and in each case finding a pair of primes which sums to it (this is of course a brute force approach, unlikely to win prizes for elegance; but then again this is the virtue of an infinitely powerful problem solver — the ability to solve everything by albeit inelegant means). Once the problem solver is limited, however, the question about resources and their optimal use cannot be avoided. Mathematical propositions are widely held to be noncontingent, being either necessarily true or false. Nevertheless, the capacity of rational agents to establish propositions is contingent, depending on their resources for establishing propositions. This capacity, or alternatively this practical possibility of attaining mathematical knowledge, seems therefore inextricably tied to the complexity of the problem under consideration and the effort that is available and can be expended to try to solve it.

Is this a deep insight or a mere tautology? If to accomplish task B we

25. Raymond Bradley and Norman Swartz, *Possible Worlds: An Introduction to Logic and Its Philosophy* (Indianapolis: Hackett, 1979), pp. 147-49.

need resources A, then for B to be a practical possibility A must be available. Unpacking practical possibility in terms of effort appears therefore something of a tautology. Yet if it is a tautology, it is one that nevertheless does some philosophical work, pointing up certain limits to mathematical knowledge. Take for instance Wittgenstein's ideas about the perspicuity and surveyability of mathematical proof. In his remarks on the foundations of mathematics Wittgenstein considers the problem of truth-preserving correspondences between alternate ways of representing numbers:

> Now let us imagine the cardinal numbers explained as 1, 1+1, (1+1)+1, ((1+1)+1)+1, and so on. You say that the definitions introducing the figures of the decimal system are a mere matter of convenience; the calculation 703000 × 40000101 could be done in that wearisome notation too. But is that true? — "Of course it's true! I can surely write down, construct, a calculation in that notation corresponding to the calculation in the decimal notation." — But how do I know that it corresponds to it?[26]

Wittgenstein is pessimistic about our capacity to demonstrate the correspondence between 703000 × 40000101 and its unary form. The problem for him is perspicuity: "One cannot command a clear view of it."[27] Recall, also, our imaginary dialogue with Professor Thomas in Chapter 1.

But it seems this correspondence is perspicuous just in case the effort that must be expended to demonstrate the correspondence (i.e., to translate faithfully from one notation to the other) is available. Wittgenstein's intuitions are therefore on the mark: he uses big numbers to probe the perspicuity question, for it is calculations with big numbers, rather than small numbers, that test the limits of our computational resources and thus determine the effort needed to carry out these calculations. Hence, right after the preceding passage Wittgenstein invokes big numbers again:

> Now I ask: could we also find out the truth of the proposition 7034174 + 6594321 = 13628495 by means of a proof carried out in the first nota-

26. Ludwig Wittgenstein, *Remarks on the Foundations of Mathematics*, revised edition, ed. G. H. von Wright, R. Rhees, and G. E. M. Anscombe, trans. G. E. M. Anscombe (Cambridge, Mass.: MIT Press, 1983), p. 144.

27. Wittgenstein, *Remarks on the Foundations of Mathematics*, p. 145.

> tion [i.e., the unary notation]? — Is there such a proof of this proposition? — The answer is: no.[28]

But what if we take a different tack and invoke small numbers instead of big numbers? "Could we also find out the truth of the proposition" 2 + 2 = 4 "by means of the proof carried out in the first notation"? The answer is of course yes: (1+1) + (1+1) = ((1+1)+1)+1. A problem of demarcation therefore confronts us: Where do we draw the line between 7034174 + 6594321 = 13628495, which according to Wittgenstein is intractable under the first notation, and 2 + 2 = 4, which plainly is tractable?

Let us consider more closely whether Wittgenstein is right in claiming that there is no proof of 7034174 + 6594321 = 13628495 in the first notation (i.e., the unary notation). Wittgenstein is probably right in denying that there is such a proof for humans limited to paper and pencil. But for humans with access to computers, a proof in the first notation is a very modest computation by present standards. Richard von Mises' example contrasting writing speed for pencil and paper as opposed to typewriters springs to mind. The perspicuity of equivalent claims in alternate arithmetic notations depends on the resources we have for demonstrating the equivalence. These resources in turn determine whether we can expend enough effort to establish the equivalence.

In this vein consider the more general claim of Wittgenstein that mathematical proofs must be perspicuous and surveyable (terms he seems to use interchangeably). At times Wittgenstein seems to be saying no more than that proofs must be exactly reproducible: "'A mathematical proof must be perspicuous.' Only a structure whose reproduction is an easy task is called a 'proof.'"[29] But when he deals with the question of representing and calculating with numbers, perspicuity and surveyability look more like a capacity of rational agents to take in calculations at a single glance.

What is it for a proof to be surveyable? Is the 10,000-page proof of the classification theorem for finite simple groups, scattered as it is throughout hundreds of journal articles, surveyable?[30] Group theorists certainly con-

28. Wittgenstein, *Remarks on the Foundations of Mathematics,* p. 145.

29. Wittgenstein, *Remarks on the Foundations of Mathematics,* pp. 95, 143ff.

30. What makes this theorem the more remarkable is that it can be stated on a single page. As a classification theorem, its statement is just a list — "Here are all the finite simple groups . . ." — where the ellipsis signifies the complete list of groups that are both finite and simple. This list includes the cyclic groups of prime order, the alternating groups on n ele-

sider the classification theorem as proven even though it is certain that no one mathematician has a complete grasp of its proof. Is the classification theorem therefore provable without its proof being surveyable? We would say the proof is surveyable to the mathematical community of group theorists, but not to any individual group theorist. Moreover, it seems that what makes the proof surveyable to the community but not to any individual is that the community is able to expend enough effort to prove it and then check the proof, whereas no one individual has the resources even to begin checking it, much less prove it.

Wittgenstein relativized perspicuity and surveyability to the individual. Thus for Wittgenstein what is perspicuous is perspicuous to the individual and what is surveyable is surveyable to the individual. Once, however, perspicuity and surveyability are understood in terms of the availability of effort and the technologies by which effort can be expended, the individual is no longer paramount. What becomes important is whether enough effort can be expended to establish the mathematical result in question. The actual agent that expends the required effort is now left open. Certainly the agent can be a single individual. But the agent can also be a computer, or an individual working with a computer, or a community of mathematicians. Effort thus provides a way of unpacking Wittgenstein's notion of perspicuity and surveyability.

As another example of how effort elucidates questions about the limits to mathematical knowledge, consider how effort dissolves a distinction of Norman Malcolm's between mathematical knowledge in the strong and weak sense. Malcolm lays out the distinction as follows:

> I have just now rapidly calculated that 92 times 16 is 1472. If I had done this in the commerce of daily life where a practical problem was at stake, and if someone had asked "Are you sure that $92 \times 16 = 1472$?" I might have answered "I *know* that it is; I have just now calculated it." But also I might have answered "I know that it is; but I will calculate it again to *make sure*." And here my language points to a distinction. I say that I

ments for $n \geq 5$, various sporadic groups, etc. If one considers how little room it takes to state the theorem and how much room it takes to prove it (not to mention the fifty years mathematicians have actively worked on it), the ratio of statement length to proof length is the smallest we know. Spanning fifty years, the project of classifying all finite simple groups was finally completed in 1982. For more on this achievement consult Daniel Gorenstein, *The Classification of Finite Simple Groups*, vol. 1 (New York: Plenum, 1983).

know that $92 \times 16 = 1472$. Yet I am willing to *confirm* it — that is, there is something that I should *call* "making sure"; and, likewise, there is something that I should *call* "finding out that it is false." If I were to do this calculation again and obtain the result that $92 \times 16 = 1372$, and if I were to carefully check this latter calculation without finding any error, I should be disposed to say that I was previously mistaken when I declared that $92 \times 16 = 1472$. Thus when I say that I know that $92 \times 16 = 1472$, I allow for the possibility of a *refutation*, and so I am using "know" in its weak sense.

Now consider propositions like $2 + 2 = 4$ and $7 + 5 = 12$. It is hard to think of circumstances in which it would be natural for me to say that I know that $2 + 2 = 4$, because no one ever questions it. Let us try to suppose, however, that someone whose intelligence I respect argues that certain developments in arithmetic have shown that $2 + 2$ does not equal 4. He writes out a proof of this in which I can find no flaw. Suppose that his demeanor showed me that he was in earnest. Suppose that several persons of normal intelligence became persuaded that his proof was correct and that $2 + 2$ does not equal 4. What would be my reaction? I should say "I can't see what is wrong with your proof; but it *is* wrong, because I *know* that $2 + 2 = 4$." Here I should be using "know" in its strong sense. I should not admit that any argument or any future development in mathematics could show that it is false that $2 + 2 = 4$.

The propositions $2 + 2 = 4$ and $92 \times 16 = 1472$ do not have the same status. There *can* be a demonstration that $2 + 2 = 4$. But a demonstration would be for me (and for any average person) only a curious exercise, a sort of *game*. We have no serious interest in proving that proposition. It does not *need* a proof. It stands without one, and would not fall if a proof went against it. The case is different with the proposition that $92 \times 16 = 1472$. We take an interest in the demonstration (calculation) because that proposition *depends* upon its demonstration. A calculation may lead me to reject it as false. But $2 + 2 = 4$ does *not* depend on its demonstration. It does not depend on anything! And in the calculation that proves that $92 \times 16 = 1472$, there are steps that do not depend on any calculation (e.g., $2 \times 6 = 12$; $5 + 2 = 7$; $5 + 9 = 14$).[31]

31. Norman Malcolm, "Knowledge and Belief" (1952), in *Knowledge and Belief*, ed. A. Phillips Griffiths (London: Oxford University Press, 1967), pp. 73-74.

We sympathize with Malcolm's distinction, but not for reasons Malcolm would accept, for it seems he confuses a matter of degree with a difference in kind. Malcolm's distinction appears plausible, at least, when he contrasts $2 + 2 = 4$ with $92 \times 16 = 1472$, because a matter of degree looks like a difference in kind if we attend to the extremes. Is $2 + 2 = 4$ really beyond the reach of all our efforts to refute it? What about the other examples Malcolm cites: $2 \times 6 = 12$, or $5 + 2 = 7$, or $5 + 9 = 14$? If we represent these equations in a unary notation, it is by no means clear that calculation and proof become superfluous:

$$2 \times 6 = 12: \qquad || \times |||||| = ||||||||||||$$
$$5 + 2 = 7: \qquad ||||| + || = |||||||$$
$$5 + 9 = 14: \qquad ||||| + ||||||||| = ||||||||||||||$$

Perhaps we err here on the side of too cumbersome a notation. For the moment let's therefore grant that $2 + 2 = 4$, $2 \times 6 = 12$, $5 + 2 = 7$, and $5 + 9 = 14$ represent instances of knowledge in Malcolm's strong sense. From here it is a small step to claim that the addition and multiplication tables we learned in grammar school constitute knowledge in the strong sense as well. A further extrapolation is possible if we question the sanctity of base ten numerical representations, and ask whether as grammar school students we might not equally well have learned our addition and multiplication tables for bases larger than ten. If so, at how large a base should we stop? If we stop at base 93, then $92 \times 16 = 1472$ would represent an instance of knowledge in the strong sense. Again, we run into a problem of demarcation.

To the question that started Malcolm's discussion, "Are you sure that $92 \times 16 = 1472$?" we would respond as follows: "We know $92 \times 16 = 1472$ because we've expended enough effort in calculation to check it to our satisfaction. We could, however, expend still more effort to check it, and thereby render it still more secure." The degree to which an arithmetic equation is securely established (in Malcolm's words, "made sure") depends both on how complicated the equation is and on how much effort was expended to check it. Without having done the calculation ourselves, we would feel more confident about the correctness of $92 \times 16 = 1472$ if it were checked five times by an accountant as opposed to only one time by a second grader. Against $92 \times 16 = 1472$, $2 + 2 = 4$ has the advantage of being less complicated, and therefore requiring less effort to check. As Bradley and Swartz argue,

> How does one go about checking for mistakes . . . ? First and foremost, we recheck the process carefully. Then if we wish still further corroboration, we might repeat the process, i.e., do it over again from the beginning. Also we might enlist the aid of other persons, asking them to go through the process themselves, and then comparing our results with theirs. And finally, we might make our results public, holding them up for scrutiny to a wider audience, and hoping that if there is a mistake, the joint effort of many persons will reveal it.[32]

Given his distinction of mathematical knowledge into strong and weak senses, Malcolm sees $92 \times 16 = 1472$ as needing confirmation, but $2 + 2 = 4$ as being immune to disconfirmation. We argue that both $92 \times 16 = 1472$ and $2 + 2 = 4$ need to be confirmed by an expenditure of effort — the former requiring a greater expenditure than the latter.

Although further examples could be given of how effort helps elucidate questions about the practical possibility of mathematical knowledge, we hope enough has been said to draw the following conclusion: The degree to which a mathematical claim can be securely established is proportional to the amount of effort that can be expended by the relevant community of mathematicians to check or refute the claim. Moreover, whether enough effort can be expended to establish the claim depends on the amount and nature of the resources available for expending effort (e.g., computers allow for greater expenditures of effort in less time than pencil and paper).

Epilogue

The social scientist Thomas Hern has indicated that often it is more enlightening to examine people's garbage than their public pronouncements. For example, consider a consumer economics researcher trying to discover the amount of alcohol consumed within a given community. From a door to door survey it appeared that the community was less under the influence of liquor than it was under the influence of the local temperance league. Naturally the researcher questioned whether the community was consuming as little liquor as it claimed. To check this suspicion the researcher decided to

32. Bradley and Swartz, *Possible Worlds: An Introduction to Logic and Its Philosophy,* p. 156.

rummage through people's garbage by night. The garbage revealed that liquor was flowing far more freely than had been claimed publicly.

Perhaps there is a parallel in mathematics. What mathematicians show the world differs significantly from their desultory scribblings and reflections. What ends up in mathematics journals is hardly ever historical reconstruction. Try to publish a mathematics paper that describes motives for approaching a problem in a particular way or recounts several dead ends attempted before success, and the journal's editor will immediately command excisions. Mathematics journals want to save space, and historical reconstruction is the place to start. The emphasis is ever on positive results and concise verifications of those results. If a result is in question, either its proof, its disproof, or a proof that there is no proof is about all that will make it into the journals.

It would be interesting to map the reaction of the mathematical community if things such as motive appeared in journal articles. Whether or not this ever happens, our point remains that there is a lot more to mathematics than appears in print, and one of the things that tends not to appear in print is effort. Consider this example. A few years ago William Thurston, the premier mathematician in low-dimensional topology, wouldn't check a supposed proof of the Poincaré conjecture in dimension 3, not merely because the methods used in the proof were in his opinion passé, but more importantly for this discussion because it would have taken him several months to work through the details of the proof. The problem of checking the proof was therefore left to his students. As it turned out, his intuition was correct. By refusing to devote his effort to checking an incorrect proof, he was able to expend his effort more profitably elsewhere.

Granted, this is a purely sociological point about the practice of mathematics. But it underscores why the role of effort in mathematics is difficult to grasp unless one has actually worked within the mathematical community. The mathematical community's emphasis is on finished products. The world sees the finished products and rightly stands in awe. Unfortunately, the effort involved in attaining these finished products tends to get short shrift. The formalist picture of mathematics is as guilty on this point as any. Mach's positivist ideas about economy of thought fare no better, since for him economy in mathematics consists in "its evasion of all unnecessary thought and on its wonderful saving of mental operations."[33] Mach's econ-

33. Ernst Mach, *Popular Scientific Lectures* (LaSalle, Ill.: Open Court, 1986), p. 195. For

omy is the economy of a perfected mathematics, not the economy of a mathematics struggling to develop.

A view of mathematics that takes effort seriously is a view not wedded to traditional logical theory. Traditional logical theory has concerned itself with the conditions under which a mathematical proposition is true or provable, but hasn't concerned itself much with the quite different problem of determining the conditions under which enough effort is available even to start addressing the question whether the proposition is true or provable. Thus, for mathematical propositions, and especially for computational problems, where the only reason for ignorance may be our inability or unwillingness to expend sufficient effort, effort and not traditional logical theory seems to provide the right mode of analysis.

By being sensitive to the role of effort in mathematics, we obtain a picture of mathematics quite different from the Platonic picture of timeless unchanging mathematical forms, the Machian picture of a perfect mental economy,[34] or the formalist picture of an inference engine chugging along. Rather we come to view mathematics as a dynamic entity, struggling to create new techniques and technologies in order to increase the resources it has for expending effort and thereby to facilitate its continued growth and flourishing, which in turn involves the creation of still newer and better techniques and technologies to continue the cycle of growth. This is certainly a pragmatic view of mathematics. But it is also a developmental and organismic view of mathematics.

This view of mathematics is perfectly compatible with Christian theism. It does nothing to undercut the existence of mathematical truth or God's knowledge of mathematical truth. It does point up, however, that God's knowledge of mathematical truth is very different from ours. God's knowledge of mathematical truth may be a direct intuition, in which he grasps the totality of relations among mathematical claims in one direct act of intuition. As finite rational agents we don't. We must build our mathematical edifices piecemeal. What's more, we must build our mathematical edifices without the guarantee that they won't come tumbling down because

---

Mach's theory of economy as it relates to both mathematics and science see John T. Blackmore, *Ernst Mach: His Work, Life, and Influence* (Berkeley: University of California Press, 1972).

34. Cf. Mach's claim that "the greatest perfection of mental economy is attained in that science which has reached the highest formal development, and which is widely employed in physical inquiry, namely mathematics" (*Popular Scientific Lectures*, p. 195).

of some hidden inconsistency. The effort we expend in building and testing our mathematical edifices gives us confidence that they are secure and lay hold of mathematical truth. This confidence, however, must always fall short of Cartesian certainty. It is an inductive confidence, one that hinges on our own efforts as well as on our faith that God is guiding those efforts.

SECTION II

THE INFLUENCE OF MATHEMATICS

CHAPTER 5

Mathematization in the Pre-modern Period

Introduction

Mathematics pervades all aspects of our modern world. We take the presence of quantities for granted; they are almost as much a part of life as the air we breathe. Practical applications and theoretical connections of mathematics can be found everywhere. Some are quite mundane, such as the simple measurements and calculations used in baking cakes or building houses. Others, such as the mathematics that underlies modern telecommunication systems, are highly specialized and may be known only to the people who use them. By means of abstract notions, systematic procedures, visual representation schemes, symbolic language, and logical derivations, mathematics brings a measure of order, precision, coherence, and objectivity to many areas of our experience. Mathematics enables us to abstract, model, express, predict, and control certain dimensions of behavior in fields that at first glance appear quite unrelated to it. It seems to furnish us with a secure basis for performing everyday tasks and making informed decisions. We may not be aware of them, but mathematical theories are now intimately involved in all parts of our life and provide enormous benefits.

However, some applications of mathematics are not so beneficial. Mathematics can be used to impersonalize relations and obscure essential aspects of situations. Through advances in modern technology, mathematics has enabled humans to assert control over issues of life and death, matters that we are ill prepared to regulate, if indeed we should. These developments have led some in our postmodern age to react against domination by science, and have reminded us that extra-scientific perspectives direct the pursuit of science and mathematics.

As mathematicians, we humbly and thankfully acknowledge that mathematics can help us better understand and control the world around us. However, at the same time, we reject mathematical imperialism — viewing everything through a quantitative filter. While we affirm the importance of mathematics and may even strive to expand its scope and power in our work as mathematicians, we also recognize that mathematics has inherent limitations and that it must be complemented and contextualized by other approaches. Though thoughtful mathematicians admit this,[1] it has been far from universally upheld. In fact, quite the opposite is true. Modern Western civilization has long advocated a strong program of mathematization, something no other culture has done to the same extent or with the same fervor and pride. Our goal is to put this part of our cultural heritage in clearer perspective and provide a solid basis for evaluating it from a Christian perspective. Thus, in the next few chapters we will show how mathematization was promoted by Western culture and became firmly embedded in it.[2] We begin our story by sketching some of the historical highlights we will examine in detail.

The origin of mathematization is found in ancient Greek culture. Various philosophers put versions of programs for mathematization forward, and the beginnings of several mathematical sciences were worked out toward the end of the Greek era, though they were not grounded in extensive experimentation. Some Christian thinkers kept this ideal alive during medieval times, but few advances were made in mathematizing science until the fourteenth century, when an abstract theory of motion was proposed. Again, though, it failed to connect strongly with experiential reality.

In the late medieval and Renaissance periods, the process of mathematization became more practical and less theoretical. Thinkers and artisans turned from rational speculation to down-to-earth quantifying and measuring. Mathematization put down deep roots in the affairs of civic and everyday life and made important connections with the arts. Advances were made in astronomy, mechanics, and algebra as Renaissance humanists at-

1. For a critical evaluation of mathematization in Western culture, see Philip J. Davis and Reuben Hersh, *Descartes' Dream: The World According to Mathematics* (San Diego: Harcourt Brace Jovanovich, 1986).

2. For a pioneering (though somewhat dated) treatment of this same theme, see Edwin Arthur Burtt, *The Metaphysical Foundations of Physical Science*, second edition (New York: Doubleday Anchor Books, 1954).

tempted to restore classical mathematics. Around the same time, a revival of speculative and mystical philosophies promoted mathematics as the key for unlocking the secrets of the universe. The story of these philosophies will conclude Chapter 5.

Chapter 6 will examine the early modern period. Mathematical scientists in the seventeenth century (Kepler, Galileo, Descartes, Newton, and Leibniz) drew upon pre-modern mathematization trends and developed them in various ways. Their work and thought firmly established the role of mathematics in science and convinced many people of the central importance of mathematics in developing a modern scientific outlook on the world. Thus, the methods and outlook of mathematics began to play a larger role in other areas of human thought and activity, such as economics. Many eighteenth-century Enlightenment thinkers valued mathematical science so highly that they consciously sought to extend its rational approach to cover all aspects of life. Mathematization was in this way securely implanted in the modern worldview. Though some rebelled against such an outlook, mathematical science became the dominant model for all human knowledge. A prominent late nineteenth-century physicist put it this way: "when you can measure what you are speaking about and express it in numbers you know something about it; but when you cannot measure it, when you cannot express it in numbers, your knowledge is of a meager and unsatisfactory kind."[3] Mathematization is as much alive today as it was in the early modern period. Chapter 7 focuses on this expansion of mathematization from the domain of natural science into other areas of life.

In the introduction, we expressed our conviction that the capacity to do mathematics is a good gift of God intended to reveal his glory and to serve his benevolent purposes for humanity. Nevertheless, while such capacities have been used in ways that serve these purposes, they have also been widely used in ways counter to these purposes. We hope to demonstrate the following theses through our analysis of the process of mathematization:

1. that as a philosophical outlook, the mathematical science ideal had its source in a pagan Greek conception of the universe and of the role that humans should play within the world,

3. This quote is due to William Thompson, Lord Kelvin (1891). It is cited on page 225 in Alfred W. Crosby, *The Measure of Reality: Quantification and Western Society, 1250-1600* (Cambridge: Cambridge University Press, 1997).

2. that this ideal was taken over and shaped by medieval Christians who were insufficiently critical of its pagan moorings,
3. that it was secularized by Renaissance and early modern thinkers, who made quantitative features of reality the only sure source of true knowledge about the world, and
4. that this ideal eventually became the core of an outlook that asserted human independence of God and sought to extend predictive mastery over nature and society by means of knowledge whose certainty was guaranteed by mathematics.

Mathematization and Natural Science in Ancient Greek Culture

Although the character and use of mathematics have varied from place to place and from time to time, thinkers in many cultures have affirmed its importance. The *Rhind Mathematical Papyrus,* an Egyptian problem-solving manual written about 1800 BC, pays homage to mathematics in its opening words: "Accurate Reckoning: the entrance into the knowledge of all existing things and all obscure secrets." Egyptian scribes who learned basic computational procedures and problem-solving techniques from this source were thereby assured that nothing would remain outside their mental grasp as they engaged in various administrative and engineering projects in service of the Pharaoh. Babylonian culture also held mathematics in high esteem, both for its practical workaday benefits and for its ability to generate astronomical knowledge of celestial events that influenced human life. Ancient Hindu culture likewise exhibited a deep respect for mathematics. Hindu authors considered calculation eminently useful in all worldly and religious matters. "Like the crest of a peacock, like the gem on the head of a snake," asserts the oldest Hindu astronomical text from around 500 BC, "so is mathematics at the head of all knowledge."[4]

It was ancient Greek culture, however, that developed this idea in a more systematic and thoroughgoing way. Greek veneration of mathematics is well known. Greek thinkers came to revere mathematics for the certainty of its deductive, axiomatic method. The most renowned example of this

4. Quoted in George Gheverghese Joseph, *The Crest of the Peacock: Non-European Roots of Mathematics* (Harmondsworth: Penguin Books, 1991), v.

method is Euclid's *Elements* (300 BC). However, even before Euclid, many had touted the importance of mathematics' content and point of view for understanding the natural world.

The Pythagorean sect of the sixth century BC is the source of many later ideas about mathematics. They speculated that *all is number.* Number is the source and meaning of things; everything in the cosmos exists and is known only through number. This outlook was largely mystical in character, though it also contained some scientific elements. The number ten, for instance, is the most holy number and the number that represents the whole universe, for ten is composed of the first four numbers, which represent points, lines, surfaces, and solids respectively. The regular pentagon with its diagonals is a sacred Pythagorean symbol because it contains ten segments. Ratios of numbers were also important to the Pythagoreans. They observed that when strings having the same tension are plucked, those with lengths in the ratio of two whole numbers produce harmonious sounds. Their success in analyzing harmony using mathematics may have confirmed the Pythagoreans in their view that all phenomena can be accounted for in terms of whole numbers and their ratios. This was the case even when the numbers involved were not concrete quantities or measurements.

The atomistic philosopher Democritus a century after Pythagoras drew upon mathematics in a slightly different way. He proposed that all matter was composed of homogeneous atoms, and that qualitative differences between things were due to the mathematical features (quantity, size, and shape) of their constitutive atoms.

In the early fourth century BC, Plato analyzed and expanded these ideas concerning the relation between mathematics and the real world. He saw mathematical truths as necessary, eternal verities whose recollection and contemplation prepared men for even higher intellectual activity as well as for virtuous living. Mathematical ideas, according to the later development of this outlook by Neoplatonist philosophers (AD third century), were in the mind of the Deity when he created the world. They provided the patterns by which material things were formed and structured, and they were implanted as seeds of knowledge in human minds. Hence, understanding mathematics gave one access to the very core of nature. Proclus (AD mid-fifth century) put it this way:

> Mathematics alone can revive and awaken the soul to the vision of being, can turn her from images to realities and from darkness to the light

> of intellect, can release her from the cave, where she is held prisoner by matter. . . . For the beauty and order of mathematical discourse, and the abiding and steadfast character of this science, bring us into contact with the intelligible world itself and establish us firmly in the company of things that are always fixed. . . . Mathematics also . . . reveals the orderliness of the ratios according to which the universe is constructed and the proportion that binds things together in the cosmos.[5]

In Neoplatonic/Neopythagorean philosophy, this knowledge was often associated with mystical or occult thought. Much later, in the early modern period of European history, it helped to stimulate a more mathematical treatment of natural science.

Mathematization was more an ideal than a reality at first, since few Greeks actually used mathematics in their science. In fact, Aristotle largely rejected a mathematical approach to nature in the mid-fourth century BC. This stance is important for our topic, because Aristotle's views on natural philosophy were to become the dominant position in Western Europe during the late Middle Ages. Nevertheless, Archimedes, Ptolemy, and others accepted a quantitative outlook to some extent and successfully applied mathematics to several fields of study. Their scientific practice, once it was recovered, inspired many in the Renaissance and early modern period to take a similar approach. We will thus look briefly at both of these traditions in Greek natural philosophy.

For Aristotle, "science" meant demonstrative knowledge of the natural world. Scientific knowledge proceeds from universally valid first principles intuited from experience (in ways that Aristotle never fully clarified) to results deduced from these principles via a chain of logical consequences. In this respect, axiomatic mathematics provides the prototype for all true science. For later followers of Aristotle, Euclid's *Elements* (written a generation after Aristotle's death) together with an Aristotelian philosophy of science modeled the proper approach to natural science. Knowledge generated in this fashion consists of absolutely certain truths about the world. Strange as it may seem to us today, the Greeks considered mathematics truly scientific because of its deductive methodology, while physics was less so, since it was based on experiential data and, hence, was less certain.

5. Proclus, *A Commentary on the First Book of Euclid's Elements. Book I, Chapter VIII,* trans. Glenn R. Morrow (Princeton: Princeton University Press, 1970), pp. 17-19.

The subject matter of mathematics, according to Aristotle, is abstract quantity, while natural science focuses on the inner essence or form of things. Thus, physics and mathematics are and should remain two distinct areas of thought. From this perspective, mathematics contributes only general method, not specific content, to natural science. Knowledge of physical phenomena must take into account the nature of the material objects involved, their essential properties. Natural philosophy is mainly concerned with how things develop in order to fulfill their purpose or achieve their potential. Mathematics abstracts from the particularity of objects and considers only their quantitative features. For Aristotle, mathematics does indeed remain tied to human experience of material bodies, but its objects (circles, for instance) are separated in thought from the corporeality of what is experienced (from circular wheels, say). Physics investigates underlying causes of natural phenomena, and these are usually due to their non-quantitative features. A rock falls to the earth, for example, because of its intrinsic heaviness; it seeks its natural place at the center of the world. Physics is only marginally concerned with how rocks fall. Each field of science, then, Aristotle said, has its own peculiar subject matter and basic principles and should proceed by logical demonstration from these alone, without confusing matters by importing rational content from elsewhere.

However, there are certain passages in Aristotle's philosophical writings that discuss "subordinate" sciences, such as optics (regarded by Aristotle as subordinate to geometry). Optics is concerned with the path of light rays, and to that extent depends on what geometry says about lines and angles. Thus in spite of Aristotle's admonition that mathematics and physics ought to be separated, his remarks on subordinate sciences encouraged some to mix mathematics with physical theories, both at the time and two thousand years later, near the end of the Renaissance.[6] Several areas in Greek thought exemplify this tradition of mathematical science: we will look at music, optics, mechanics, astronomy, and geography.

We do not tend to think of music any more as a mathematical science. However, the Greeks did, and so did most scholars in Western culture from the time of the Romans into the seventeenth and eighteenth centuries, when it separated from the study of what we now call acoustics. As we have seen,

6. For a discussion of subordinate and mixed sciences, see Richard D. McKirahan, "Aristotle's Subordinated Sciences," *British Journal for the History of Science* 11, no. 39 (1978): 197-220.

the Pythagoreans first subjected music to numerical treatment. Later treatments of music remain indebted to this tradition.

Whereas music used number theory for analyzing physical phenomena, optics combined geometry and physics. Mathematicians of the time were familiar with the law of reflection, but they were unable to determine the law of refraction, which governed the path of light passing from one medium into another. Both Euclid and Ptolemy, among others, wrote elementary texts on optics. In addition to mathematical results about light rays and perspective, books on optics put forward theories of how sight occurred, usually deciding in favor of a cone of rays proceeding from the eye to the subject instead of the reverse.

Mechanics was also a synthesis of geometry and physical science. The law of the lever was known by Aristotle's time: bodies are in balance when their distances to the fulcrum are inversely proportional to their weights. About a century later, Archimedes gave this principle its first genuine proof. He then used it as the basis of a highly sophisticated investigation of centers of gravity for planar figures. He provided a masterful treatment of this topic, following the standard mathematical mode of deductive presentation and arriving at some important results, all without having integral calculus to discover or prove them. A mathematical theory of floating bodies (hydrostatics) also originated with Archimedes.

The most highly developed and successful mathematical science in ancient times was astronomy, which applied both spatial and numerical ideas to celestial phenomena. Early theories of stellar and planetary motion by Greek thinkers were strictly geometrical, in contrast to Babylonian astronomy, which was primarily numerical. Aristotle himself put forward a theory that was a refinement of an earlier one by the Greek mathematician Eudoxus. Using 56 spheres, he could qualitatively model the rotation of all the planets and stars about a stationary Earth located at the center of the universe. By AD 150, due to the work of Ptolemy and his predecessors, astronomy had become quantitative as well, combining the Babylonian approach with that of the earlier Greeks. In developing his system of astronomy, Ptolemy used cycles and epicycles (circles on circles), as well as other calculating devices and various results that we now consider part of trigonometry, to predict the motion of the planets. The final system designed by Ptolemy was more mathematical than physical, however. It did not function well as a model of how the heavens actually work, though it provided an impressive theoretical tool for calculating positions of heavenly bodies.

Ptolemy was also responsible for developing an Earth-based geometry (geography). Though the knowledge of land and sea available at that time was inadequate for mapping out the world with any degree of accuracy, his idea of representing the surface of the globe on a flat surface proved very fruitful. Maps were spatially coordinatized by means of a rectangular grid, using meridians and parallels to represent longitude and latitude. In the fifteenth-century age of discovery, the recovery of Ptolemy's *Geography* with its explicit mapping instructions and examples supplied navigators and their patrons with an important resource. Cartographers, sailors, and surveyors joined forces at that time to make maps quantitative rather than symbolic.

In Greek times, then, we can see the strong beginnings of a mathematical method for both justifying and systematizing scientific results. The method included (1) deductions from an axiomatic basis of first principles and (2) a quantitative and spatial approach to describing and modeling physical behavior. On the first point, people were convinced that it sufficed to pattern their work after Euclid, in accordance with Aristotle's prescriptions on scientific method. On the second point, however, questions persisted about how relevant mathematics was to physical science. Pythagoras, Plato, and other philosophers had speculated about the necessity of mathematics for probing the inner workings of the world, and various mathematicians had achieved significant results in quantitative physical science. Though the mathematization of natural science was still more talk than action, it produced a viable viewpoint and a small number of successes, and these would induce later European natural philosophers to push the program of mathematization further. While it faced some stiff opposition from traditional Aristotelian natural philosophy, a high regard for mathematics was nurtured by Greek sources and moved out to become an established feature of Western intellectual culture. This program of mathematization shared many characteristics with religious belief — it involved faith and commitment and assumed the existence of something whose existence and nature did not depend on anything else but which determined the nature of other things.

Mathematization and Natural Science in the Early Middle Ages

In spite of the phenomenal achievements of Euclid, Aristotle, Archimedes, and other Greek thinkers, Greek mathematics seriously declined after about

200 BC. After Ptolemy's work (AD mid-second century), no major Western contributions to the process of mathematization occurred for about twelve hundred years. The early Middle Ages (AD 400-1000) were an important transitional period, however. Thus, in this section, we will briefly sketch some of its main features that are relevant to our topic.

Greek ideas on mathematics and natural philosophy were often inaccessible to later thinkers. Already in Greek times, intimate knowledge of certain philosophical systems, such as Aristotle's, was restricted to its adherents and was not readily available to the outside public or even to those who belonged to a competing school of thought. Relatively few copies of any given manuscript were made, and these were often located only in a very limited number of places. Social and economic upheaval, such as that caused by the fall of Rome, also militated against the continuous development of scientific scholarship.

The church fathers and Christian ecclesiastics in the early Middle Ages did little to promote science and mathematics.[7] They were primarily concerned with the spread and defense of the gospel and not with the pursuit of worldly wisdom. Yet, it was largely through their transcription and use of various ancient manuscripts in monastery schools that Greek intellectual traditions were preserved in the West at all. Administering the affairs of the church and teaching biblical doctrine required a learned clergy. The trivium, composed of grammar, rhetoric, and logic, was obviously important for furthering the cause of Christian religion, but uses were also found for rudimentary natural philosophy and even mathematics. A moderate amount of natural knowledge was necessary in order to understand the meaning and truth of the Scriptures. In addition, astronomical computations were important for various reasons, notably calculating movable feast days, such as Easter. The texts used for teaching the seven liberal arts of the curriculum, which besides the trivium included the quadrivium of arithmetic, geometry, music, and astronomy, were translations or paraphrases of Greek and Roman works. Edited portions of Euclid, Aristotle, and Plato were made available for study in this way.

The level of mathematics and natural science during the early Middle Ages, however, was extremely low. Those parts that were important for prac-

7. See David C. Lindberg, *The Beginnings of Western Science: The European Scientific Tradition in Philosophical, Religious, and Institutional Contexts, 600 BC to AD 1450* (Chicago: University of Chicago Press, 1992).

tical affairs or ecclesiastical matters were studied by a few, but there was no interest in pursuing knowledge of these subjects for their own sake. The depressed state of geometric knowledge is indicated, for example, by the fact that the fifth proposition in Euclid was given the colorful name *Pons Asinorum,* the Bridge of Asses. Euclid's proof of this theorem, which asserts that the base angles of an isosceles triangle are equal, was considered too difficult for all but the brightest. Geometry was often presented as a series of bare propositions, without any proofs. Given the dearth of scientific and mathematical ability at this time, there was no possibility that the more complex knowledge in these fields would even be understood, much less furthered.

Logic and philosophy fared somewhat better because of their value in formulating theological dogma, but these, too, remained in a subservient status. The Bible and church doctrine were the final authority on all matters on which they spoke. Platonic and Neoplatonic views about the world were deemed less dangerous to Christian thought than other pagan viewpoints, so these found acceptance by many Christian thinkers in the early medieval period. Many viewed certain Platonic doctrines about God and the soul as providentially preparing the way for Christianity. Aristotle, on the other hand, was known primarily as a logician at this time; his systematic philosophy remained largely unknown in the West until the thirteenth century.

St. Augustine, who wrote around AD 400, was one of the most influential thinkers of the early Middle Ages. After his conversion to Christianity, he sought to make faith foundational to all of his thought. Pagan philosophy and science, he exhorted, should be used only insofar as they assist and deepen Christian belief. He even went so far as to say: "The good Christian should beware of mathematicians, and all those who make empty prophecies. The danger already exists that mathematicians have made a covenant with the devil to darken the spirit and to confine man in the bonds of Hell."[8] Augustine's warning here, however, was not against mathematicians, in our sense of the term, but against those who make astrological predictions or engage in the practice of magic. His view of mathematics was actually quite exalted. Mathematics can provide helpful ideas for understanding passages of Scripture in which patterns of numbers are involved. It is also important for understanding nature. Augustine followed the Neoplatonists in ascribing the

8. Taken from Augustine's work *The Literal Meaning of Genesis* II.xviii.37. Similar sentiments are expressed by Augustine in *On Christian Doctrine* II.21. Before Augustine's conversion, he felt quite differently about astrology, however; see his *Confessions* IV.III.

highest place to mathematics for giving order to the creation. Mathematical truths are necessary, unchangeable, and universal. God in his wisdom made use of mathematical ideas in structuring the world at the beginning of time. Humans can study the great book of nature to appreciate the order and beauty of God's handiwork, but care must be taken to worship the Creator and not the creation. Augustine's ideas on mathematics and natural reality thus provided a Christianized version of mathematical rationalism that was to resonate in many later thinkers. Medieval Christian scholars, following St. Augustine, were fond of quoting from the apocryphal *Wisdom of Solomon*, which they interpreted as stipulating God's reliance on mathematics in structuring the creation: "You have ordered everything in measure and number and weight."[9] This passage was taken as biblical support for a Neoplatonist attitude toward mathematics.

Later medieval educators such as Boethius (AD 500) and Isidore of Seville (AD 600) bolstered Augustine's view of the importance of mathematics. While they compiled only snippets of mathematical texts, they continued to promote the Pythagorean ideal of mathematics as the highest form of knowledge available to humans. "Take away number from everything," Isidore claims, "and everything perishes. Deprive the world of computation, and it will be seized by total blind ignorance."[10] Without a tradition of ongoing research in mathematical science, these statements read like hollow epigrams, but they did keep alive the notion that mathematics provides the chief avenue to understanding nature.

Mathematization and Natural Science in the Later Middle Ages

In the eleventh and twelfth centuries, Europe underwent a period of great economic, social, and intellectual development. The Crusades reestablished the power and influence of the Church and opened the way for European peoples to interact with other cultures. It was a time of increased urbanization and an expansion of money-based commerce. Along with all these developments came a somewhat revitalized interest in mathematics, both

9. Wisdom 11:20b.

10. Quoted in Jens Høyrup, *In Measure, Number, and Weight: Studies in Mathematics and Culture* (Albany: State University of New York Press, 1994), p. 130.

practical and theoretical. Several Latin translations of Euclid's *Elements* from Arabic sources became available by the middle of the twelfth century. In his arithmetic of 1202, Fibonacci proposed using Hindu-Arabic numeration and reckoning for commercial computations in place of Roman numerals, something he had become acquainted with through his travels.

Education also improved. Monastery schools gave way to cathedral schools, and early in the thirteenth century, universities were officially established in Italy, France, and England. The curriculum was broadened and deepened due to the recovery of numerous Greek manuscripts in Arabic translation and Arabic writings based on Indian scholarship. These included works in philosophy, mathematics, and mathematical science, in addition to literary and other works.

Plato's cosmological speculations received a boost in the twelfth century with the translation of his dialogue *Timaeus*. However, Plato's influence on natural philosophy was curtailed as the full range of Aristotle's thought became known. The recovery of Aristotle's philosophy posed a monumental challenge to Christian thinkers. The enormity, brilliance, and encyclopedic character of his system demanded that Christians respond to it, yet the pagan outlook that tainted many of his ideas meant that they could not be accepted at face value. Aristotle believed, for instance, that the material world was eternal and uncreated, although immortality was denied to humans as persons. He also viewed God as an impersonal "unmoved mover" who had no providential concern for created reality. For a short time in the thirteenth century, his works were banned from the university curriculum in Paris, but this was soon modified. The church then suggested that Aristotle's philosophy be purged of all offending ideas that impinged on Christian theology, but this proved impractical. In 1277, a long list of controversial Aristotelian theses was condemned, but this ban was not strictly enforced and failed to stem the tide. In the meantime, various medieval philosophers had attempted to sanctify Aristotle by synthesizing his thought with Christian ideas. Thomas Aquinas' effort in the mid-thirteenth century is the best known and was the most influential, though his philosophy did not come to dominate Roman Catholic philosophy until later.

In the end, then, Aristotle's philosophy, modified according to various Christian sensibilities, became the core of the university curriculum, lasting from the middle of the thirteenth century until the seventeenth century. Using Aristotle's method, scholars critically examined the various positions on a topic taken in the available texts and then arrived at their own conclusion on the matter. In many cases, this meant reaffirming Aristotle's views. Aristotelian natural philosophy, for instance, had no serious competitor until the late Renaissance period. Given the wealth and depth of Aristotle's insights, it made more sense for philosophers to master and dispute what he had written than to strike out in ignorance on their own in an attempt to gain first-hand knowledge of nature. In addition, given the classical deductive model of science, reasoned disputation was all-important in natural philosophy, not observation or experimentation. Logic was thought to be the main tool of science, not mathematics.

Mathematical and scientific writings were translated into Latin during this time, first from Arabic manuscripts (twelfth century) and then from Greek (thirteenth century). For the most part, European scholars welcomed these works without reservation. Here, it seemed, was an abundance of new material that could be accepted without any worry about its implications for the Christian faith. Technical mathematics made no explicit demands upon biblical doctrines. Translators began to learn the content of the texts and develop Latin terminology for the new concepts. Besides Euclid's *Elements*, texts became available in elementary algebra (al-Khwârizmî), astronomy (Ptolemy), and optics (Alhazen). However, since this was the time in which Aristotle's writings were being assimilated, these materials were mainly appropriated to the extent that they fit the purposes of scholastic philosophy and Christian theology. Archimedes' scientific work, for example, remained largely unknown at this time.

In the area of philosophy of mathematics, the influence of Platonic and Neoplatonic ideas, usually in Augustinian dress, remained strong. Mathematics was praised for its ability to give rational insight into the structure of the world. The Creator was seen as the Supreme Mathematician; humans thought God's thoughts after him in discovering the mathematical structure of the world. This outlook was in conflict with Aristotelian notions about the limited value of mathematics and the nature of mathematical abstraction; nevertheless, it continued to coexist with them, occasionally gaining the upper hand. At Oxford University during the first and last halves of the thirteenth century respectively, Robert Grosseteste and his disciple Roger Bacon strongly promoted the mathematization of science, while also asserting the importance of direct experience with natural phenomena. Mathematics makes experiential knowledge precise and certain through its subject matter and method. Change is studied by science; all change is quantitative in some way and so requires mathematics. In Bacon's view, "the gate and key of [all] sciences is mathematics. . . . [H]e who is ignorant of it cannot know the other sciences or the things of this world. . . . [He who is able to apply it] correctly to knowledge of other sciences and things . . . will be able to know all things that follow, without error or doubt, easily and powerfully."[11] Mathematics confers its absolute certainty upon those fields of thought that make use of it.

Bacon made no original contribution to mathematical science himself, though he did popularize an excellent mathematical treatment of optics by the Islamic scientist Alhazen. A more developed outworking of mathematical natural philosophy arose a generation or so later (1325-1350), also at Oxford, among the so-called Oxford Calculators Thomas Bradwardine, William Heytesbury, and others. Repudiating some of Aristotle's ideas on motion and asserting the central value of mathematics for understanding physics, they defined the concept of instantaneous velocity and then analyzed the behavior of uniformly accelerated motion mathematically in terms of uniform motion. Their discovery of the *Mean Speed Theorem,* which states that a uniformly accelerated body traverses the same distance as it would if it were to travel uniformly at its average speed, provided the centerpiece for a new mathematical theory of motion, the science of kinematics. This result gave a quantitative basis for comparing speeds and formed the logical foundation for the time-squared law, which Galileo finally asserted and proved three centuries later in his analysis of freely falling bodies.

By around 1350, natural philosophers at Paris had generalized the work of the Oxford Calculators. The mathematician Oresme investigated not only motion in this way, but insisted on a similar treatment of any quality of a body capable of undergoing quantitative change or having different degrees of intensity, such as heat or even virtue. He represented the various intensities of these qualities geometrically by proportional lines erected on a base line. Thus, in the case of locomotion, line segments represent velocities at different times. This representation enabled Oresme to provide a geometric proof of the *Mean Speed Theorem* and its consequences. His approach introduced graphical representation long before Descartes and Fermat systematically developed it into analytic geometry using algebra in the seventeenth century.

All of this looks fairly modern, setting out a mathematical approach

11. David C. Lindberg, "Science as Handmaiden: Roger Bacon and the Patristic Tradition," *Isis* 78, no. 294 (December 1987): 531.

to motion that involves calculation and a sort of coordinate geometry, distance being identified with the area of a velocity diagram. It certainly contradicts the disparaging view held by pre–twentieth-century historians who believed medieval mathematicians did nothing new. However, there are still some critical features that separate fourteenth-century scholastic theories of change from seventeenth-century kinematics. For one thing, neither Oresme nor the Oxford Calculators based their kinematic theory on how bodies actually move. Experimentation was still off in the distant future; theirs was an abstract mathematical theory built on precise definitions and logical reasoning. No attempt was made to measure speed using real-life units of time and distance. Thus, paradoxical as it may seem, fourteenth-century kinematics involved quantification but lacked all grounding in measurement! It merely assumed that degrees of speed could be compared by means of a ratio and proved its results on that basis. This is no different from traditional geometry itself, where concrete units also play no role, but it is quite different from the way seventeenth-century physicists were to develop their science of motion. Fourteenth-century kinematics remained closely allied with the concerns of scholastic philosophy; it did not give rise to a new mathematical science. A closer link between mathematical concepts and techniques and those of physical science was still missing. The practical arts and mixed sciences that developed during late medieval and Renaissance times supplied this, connecting theory and practice with actual quantitative measurements.

Renaissance Mathematization Trends

By the fourteenth century, the political power of the Roman Catholic Church over Europe was beginning to wane under the challenge of nation-states asserting their own sovereignty. Various thinkers were also starting to question the scope of the Church's authority and the pronouncements of its clergy and theologians. As time went on, merchants and guilds of artisans gained political and economic power in the cities. A form of capitalism rooted in a lucrative international trade was developing. Renaissance humanism also arose and flourished in this time period (1350-1600), particularly in Italy. This was a many-faceted movement that self-consciously rejected medieval scholasticism with its domination by theological concerns. Renaissance humanists sought to restore the glories of ancient Greek and

Roman civilization and gave higher priority to secular matters, including a new emphasis on the importance of Man as the measure of life.

In this context of secularization and societal differentiation, mathematics found fresh patronage, uses, and stimuli for growth, strengthening alliances with other spheres of human activity as well as reviving old ties. In this section, we will consider some significant ways mathematization advanced in daily life, in various arts, and in the mathematical sciences themselves. We will also look at a few developments in philosophy during the late Renaissance that stressed the importance of mathematics.

Mathematization of Renaissance Life

The theoretical mathematics taught in late medieval universities was of little value for everyday life. The new technical, mercantile, and military projects being pursued throughout Europe called for a more applied mathematics involving calculation and proportionality and based upon measurements and the use of mechanical instruments. In many different ways mathematics helped to rationally organize and control a variety of phenomena and so improve human life. It was especially the mathematical frame of mind stirring in economic, civic, and artistic affairs that proved beneficial for linking mathematics more intimately with natural science.

While there were relatively few important developments in theoretical mathematics during the late medieval period or even the Renaissance, these periods witnessed a profound shift toward a quantitative way of experiencing the world. Time and space were both being quantified in ways that put a strong imprint on the fabric of ordinary life. Fourteenth-century artisans built large mechanical clocks and set them up in town squares. Measured, homogeneous, physical time thus began to regulate the affairs of human life. Elaborately designed clocks not only told time, they also exhibited the motions of the heavenly bodies that governed time. Such clocks were admired to such an extent that the clockwork image came to be used to describe the intricacy and beauty of the way the world worked. The use of measured time or meter was also central to changes taking place in music. Memorized Gregorian chants with their fluid, text-based rhythms gave way to more complex polyphonic music notated with a graph of sorts (the musical score) that indicated both pitch and uniform rhythms. Such a score was needed so that the various parts could be properly coordinated to give the desired effect. Besides these newer links between mathematics and music, the older intrinsic connection between harmony and ratios continued to be emphasized.

Spatial magnitudes also became more prominent in this period. Measured distances and angles became integral parts of navigation, surveying, gunnery, military fortification, machine construction, sculpture, architecture, painting, and even garden design. Renaissance thinkers were fond of pointing out the utter practicality of mathematics and emphasized the foundational relevance of basic geometry and measurement for various manual arts even while they also noted its sublime character.

Business practices spread the fussy habits of quantification and precise calculation into all corners of life. While money had been used somewhat in Europe before the late medieval and Renaissance periods, it now became nearly universal. This required numerical values to be assigned to everyday goods and services. The world of financial transactions stimulated the further development of arithmetic, and it in turn facilitated commerce and trade. People with mathematical talent were able to earn their living as reckon-masters, teaching the necessary computational and commercial algorithms for organizing the chaos of different weights, measures, and monetary systems. God may have ordered everything in number, measure, and weight, but he evidently did not supply the standards, and the Babel of different systems in existence throughout Europe and elsewhere gave ready employment to those who could subject them to some order. New accounting methods put into practice during the fourteenth century helped thriving merchants keep better track of their expenses and revenues over time and so figure their financial status. Balancing the company books brought clarity and understanding to what might otherwise have been an unmanageable confusion of economic activity.

Commercial interests of merchants and nations were aided by developments in navigation. Fourteenth-century maps known as portolan charts gave sailors compass headings to use in sailing from port to port. These were of little use for journeys to far-off lands or new worlds, however. For that, more accurate methods of navigating and of mapping new discoveries were needed. As we mentioned above, the recovery of Ptolemy's *Geography* in the early fifteenth century greatly aided global navigation. Columbus and later explorers to the New World would never have undertaken their voyages without the newer (albeit flawed) maps provided by Italian geographers.

Mathematization and Renaissance Art

Mathematics was important not only in the affairs of public life, but also in the arts. Early Renaissance artists sought to model their sculpture and architecture on classical Greek and Roman masterpieces. In order to emulate the realism of these works and determine their compositional secrets, Italian artists such as the architect Brunelleschi made copious exact measurements, calculations, and sketches. To achieve a realistic representation of an object, he and others thought, one must pay close attention to the proportionality of the parts. Brunelleschi believed that making the principal dimensions proportional to ratios of small whole numbers was as important for sound architectural design as it was for creating musical harmonies — such ratios are found everywhere in nature and thus also belong in art, which imitates nature, for they are divine in origin.

Brunelleschi also discovered how to make realistic and quantitatively precise renderings of architectural monuments, using the principles of linear or one point perspective. This occurred sometime before 1415. Interestingly, Ptolemy's *Geography* was the catalyst for this development. Ptolemy's method of coordinatizing space involved projecting a surface onto a plane using a visual cone whose vertex was located at the perceiving eye. A similar optical procedure can be used to create a geometrically accurate map of any three dimensional scene. The light rays passing from the source to the eye could be thought of as intersecting the plane of the picture at an appropriate distance from the eye.

Various architects and painters quickly adopted Brunelleschi's technique of perspective drawing and used it to carve and paint realistic three-dimensional scenes on flat surfaces. The amazing visual reality of this work, conceived in abstract geometric terms, was hailed as an important advance over medieval art. Brunelleschi's compatriot Alberti, who considered himself a follower of Ptolemy, gave a detailed mathematical explanation of the technique in a treatise on perspective painting (1436). The method of placing a veil or rectangular network of threads in front of a scene could be used to help an artist place objects properly within a picture, but geometrical principles regarding vanishing points and the appearance of parallel lines were also invaluable for organizing the painting.

The entire process of perspective drawing was depicted about a century later in a series of famous woodcuts by the German artist Dürer (1525) in a manual written for prospective artists and draftsmen in his country. About the same time, Leonardo da Vinci was making extensive use of perspective and mathematical figures in his painting and drawing, including anatomical illustrations of a more scientific nature. It is clear from the writings of Alberti, da Vinci, Dürer, and others that exact measurement, proper proportion, and mathematical principles were paramount in good painting, though they did not advocate following mathematical rules mechanically or slavishly. The use of mathematical principles transformed painting and other arts such as architecture from an empirical craft into a more precise science. A rational artist was deemed one who was well versed in mensurational geometry and the theory of proportions as well as the techniques of his craft. Naturally, Renaissance art was much more than mathematics, but it had a strong quantitative foundation. It was the result of precise planning and design based upon measurements and calculations.

Italian artistic techniques had a lasting influence on the practice of painting, but their impact went far beyond the realm of art. Arising about the same time as printing, the technique of realistic illustration was incorporated into technical books of various kinds. The ideal of representational painting provided mechanics, medicine (anatomy), engineering, and many other areas of thought with graphic ways to draft things as they are seen. This geometric mode of visual communication made possible a precision that could not be captured with words. In natural science, technical drawings became an instructional aid.

Already in late medieval and Renaissance times, then, we note the appearance of a robust trend toward increased measurement and quantification, a tendency that became even more pronounced in the modern period. The foundational significance of mathematics for all areas of human activity was no longer a mere slogan based on a mystical point of view and a limited number of successes. Now it began to be actualized in down-to-earth concrete ways. Intentional quantification brought about a measure of rational control over both social and material phenomena, though possibly at the expense of other aspects of a situation. Quantities gave the merchant a measure of control over his transactions, and they gave the artist a new command of his medium. The habits of careful observation, measurement, and mathematical analysis were present in various activities of daily life and certain arts, but they also began to show up in the physical sciences. We will now turn to look at the progress made by mathematical science toward the end of the Renaissance period.

Mathematization of Renaissance Science

The Italian Renaissance is often identified with trends in art, literature, and philosophy. What is less well known is that it included a program of revival in mathematics as well.[12] Many humanists were interested in locating, restoring, and translating uncorrupted versions of Greek mathematical and scientific manuscripts. Earlier medieval translations naturally formed the baseline for this work, but there were reasons for going beyond them. As the Turks began to threaten the eastern Byzantine Empire during the first half of the fifteenth century, many scholars fled to the West, bringing ancient Greek manuscripts with them, some of them completely new or more accurate than previously known versions. Fifteenth- and sixteenth-century humanists thus had an opportunity to recover Greek mathematical thought to a greater extent than was possible earlier. Their work here supplemented that done in other fields.

As humanist mathematicians worked to restore various sciences, using classical sources, they were forced to master the ideas and techniques of the works they were translating. In some instances they had to reconstruct missing reasoning or corrupted passages, and in so doing went beyond the original sources. In this way the seemingly conservative project of recovering and restoring the glories of ancient science and mathematics led to innovation and progress. Humanists did not stop with establishing and translating the purest texts; they wanted to revive the entire field to which these works belonged, to use the knowledge they learned to further its ends. Their motivation, in many cases, was to present mathematics in a form that once again did justice to the certainty of its knowledge.

The development of mathematical science was greatly accelerated in the middle of the fifteenth century by an invention that completely changed the nature of research and scholarship, namely, the invention of movable-type printing. Those involved in restoration programs undertook the publication and dissemination of their translations as an important and integral part of their task. Printed books became quite affordable around 1500, so it was then possible for scholars all over Europe to be engaged in a common scientific enterprise on the same foundations and to enter into critical dia-

12. For a scholarly treatment of this topic, see Paul Lawrence Rose, *The Italian Renaissance of Mathematics: Studies on Humanists and Mathematicians from Petrarch to Galileo* (Paris: Librairie Droz, 1976).

logue with one another. A Latin translation of Euclid's *Elements* was first printed from a Greek source in 1500; other Greek-to-Latin versions were also printed, as well as an Italian version and an English version, all by 1570. Archimedes' works in mathematics and mechanics were translated and first became available in 1558. Proclus' Neoplatonist commentary on Euclid was printed in 1560, and the work of Pappus on analysis was printed in 1588, though some scholars knew both earlier from summaries made at the beginning of the century. Apollonius' treatment of conic sections was printed in 1566, and Diophantus' algebra was done in 1575. By the last quarter of the sixteenth century, therefore, new and important Greek mathematical works were available in book form and exerted a weighty influence on the development of late Renaissance and early modern mathematical science.

Astronomy was one area in which scholars mounted a strong restoration program. This began already in the middle of the fifteenth century. Especially Regiomontanus attempted to upgrade what had been made available in medieval times. To further the development of astronomy, he gave a new and more accurate translation of Ptolemy's astronomical work, the *Almagest.* In his accompanying commentary, however, he also pointed out a couple of difficulties with Ptolemy's approach.

Regiomontanus thought that a revived astronomy could be built upon a corrected version of Ptolemy, but Copernicus decided otherwise. According to popular histories of science, this was because Ptolemy's geocentric system was at its core too common-sensical, while in its detailed outworking it had become both too complicated and too far out of line with the data. The Copernican system (1543) is widely supposed to have corrected these problems. However, Ptolemy's system was not supplanted because of the Copernican system's technical simplicity or greater accuracy. Nor did Copernicus advance his system because it agreed better with the physics of the time. Copernicus took his approach largely for reasons of mathematical simplicity, systematic coherence, and philosophical perspective. He developed an alternative astronomy because he found the greater geometric simplicity and harmonious arrangement of the resulting planetary and stellar motions philosophically attractive and because its organization of the world fit better with then-current Neoplatonic views of the importance of the Sun.

Rather than trying to adjust Ptolemy's methods, which Copernicus considered too *ad hoc,* he decided to take a completely new approach. After a great deal of thought, Copernicus developed a heliocentric system of the world, with the Earth rotating daily on its axis and revolving yearly about a stationary Sun. Planetary paths could be easily explained in a qualitative fashion now, without having recourse to retrograde motion along an epicycle. Copernicus also did away with certain technical devices of Ptolemy, making his astronomical calculations both easier and more uniform. Moreover, his system incorporated the ideal of uniform circular motion for the planets better than Ptolemy's and so seemed more natural from a mathematical point of view. Simplicity, aesthetic sensibility, and philosophical preferences compelled Copernicus to choose his system over Ptolemy's, not closer congruence with astronomical data.

Copernicus' text, published as he was dying, set forth a new system of the world. However, a preface added by the printer's proofreader, Osiander, as the work went to press, presented it in the traditional way as an alternative astronomy. Consequently, while Copernicus thought he had captured how the universe actually works with his mathematics, the preface disarmed the theological and physical opposition this would have generated by claiming that it was only a mathematical theory for computing planetary positions. Astronomers could thus use his system as they had used Ptolemy earlier. In the minds of many, Copernicus had restored mathematical astronomy to its previous high position. Sixty years passed before his system was seriously entertained as a candidate to replace both Ptolemy and Aristotle. The seventeenth-century mathematicians Kepler and Galileo were among the first to think of it as combining astronomy and physics. Both were instrumental in promoting the Copernican system as the true system of the world, claiming in essence that mathematics was sufficient for treating this part of natural philosophy.

The field of mechanics was another area that benefited from the restoration efforts of Renaissance mathematicians, and it, too, underwent further mathematization in the process. There was first of all a keen interest at this time in practical mechanics, including automata (machines that imitated life-like actions) and other technology. These were topics that certain late Greek mathematicians and Roman engineers had considered. This tradition embodied a more empirical approach to physical problems, a trend that would bear fruit in the seventeenth century.

Renaissance Italy also saw an increased interest in theoretical mechanics. This had both a mathematical side, which derived from Archimedes' work in statics, and a more philosophical side, which owed its existence to Aristotle's discussion of dynamics in his *Mechanics*. Renaissance mathematicians never integrated these two parts of theoretical mechanics. An Archimedean approach, involving mathematical proof, was used to treat centers of gravity, but no one before Galileo was able to use it to give a satisfactory mathematical treatment of dynamic phenomena. Nevertheless, Archimedes' influence seems to have been the deciding factor in Galileo's work. Renaissance restoration and extension of Archimedes' work was therefore important to the rise of modern physical science, even if it fell short of developing a quantitative science of motion.

Besides astronomy and mechanics, there is one more branch of mathematics that was affected by Renaissance culture and worldview: the field of algebra. This was an area that benefited in the end both from the practical concerns and the theoretical preferences of the time period. Algebra was not of Greek origin, however; it had its roots in Babylonian and Hindu computational mathematics. Medieval Islamic mathematicians took over these two legacies and combined them with the Greek tradition by illustrating and justifying their algebraic techniques geometrically. It was through their works that medieval Europeans became familiar with the field. They also learned of the Hindu-Arabic system of numeration and reckoning from Islamic sources. Merchants found this new Arabic way of calculating helpful and began to adopt it as early as the thirteenth century, though it became more widespread during the Renaissance. The Hindu-Arabic system of written computation finally replaced Roman numerals and the abacus in the sixteenth century.

Algebra was considered an extension of computational arithmetic, providing problem-solving techniques that went beyond routine calculations. Reckon-masters often taught their pupils both arithmetic and algebra, even though the more advanced parts of the latter had no immediate practical application. During the fifteenth century, scholars began to introduce abbreviations for certain algebraic operations and quantities and to develop procedures for manipulating algebraic expressions. Before this time, algebra had made very little use of special notation. Regarding algebraic methods, there was a technique for solving quadratic equations (completing the square) and a geometric procedure for solving various types of cubic equations, but that was about it, except for specialized methods that worked only for certain cases. Pacioli's mathematical compendium of 1494 summarized what was then known in the various branches of mathematics, and he pointed out the limitations of algebra with respect to solving cubics.

All this changed drastically during the sixteenth century. In the first half of the century, Italian mathematicians learned how to solve various cubic and quartic equations algebraically. Cardano's *Ars Magna* (1545) presented formulas and procedures, due to him and others, for solving such equations in general and gave geometric demonstrations of their validity. Bombelli (1572) extended Cardano's work and introduced the arithmetic of complex numbers to help simplify intermediate expressions in the process of solving cubics. Near the end of the century (1591), Viète put algebra forward as a general theory of equations of any degree and set it on the path to becoming totally symbolic. Using letters to stand for both knowns (consonants) and unknowns (vowels), he was able to prescribe certain techniques for solving problems and so arrive at formulas that embodied the general solution to standard forms of equations. His express goal was to construct a method so thorough and powerful that no problem could resist solution.

The works of Bombelli and Viète are interesting because both of them were influenced more by the Renaissance program of restoring ancient mathematics than by practical computational concerns. Each was stimulated by the recently discovered work of the late Greek mathematician Diophantus (AD 250). Viète was also deeply influenced by the ideas of Pappus (AD 325) on analysis. Pappus' ideas seem to have been the inspiration for his *Analytic Art.* Viète claimed that his version of algebra revived a hidden ancient Greek tradition and purified the barbaric algebra of his predecessors. Viète's algebra succeeded in attracting various followers in the seventeenth century. For example, in the 1630s Fermat used it in connection with his study of Apollonius' *Conics* (225 BC) to develop his version of analytic geometry.

Progress in symbolic algebra, therefore, was due both to the practical interests of Renaissance merchants and to the theoretical interests of Renaissance humanists. Several other computational advances around the start of the seventeenth century were triggered by the needs of business and science. Various people proposed an arithmetic of decimal fractions as a natural extension of Hindu-Arabic operations with whole numbers. This would make calculations with fractional quantities simpler and more efficient. Stevin wrote the most famous pamphlet on this topic in 1585. In it, he also suggested the adoption of a decimal system of measures and weights. Stevin's goal was finally accomplished two hundred years later, when a metric system was instituted (with some difficulty) in France as the result of a concerted effort to make measurement more rational.

With increasingly precise measuring instruments, astronomical calculations became more laborious. At the same time, more accurate astronomical tables were needed for such things as calendar reform. Improvements made in trigonometry during the sixteenth century helped, but methods were still needed to shorten the computations. Some headway was made using trigonometric identities, but the real breakthrough came about early in the seventeenth century. In 1614, Napier invented logarithms so that complex operations on numbers could be reduced to easier operations on their logarithms. Mathematicians then developed logarithm tables that could be used by astronomers to simplify their trigonometric calculations.

Overall, the Renaissance period saw a great increase in the growth and use of mathematics. Quantities played a larger role in the affairs of daily life, and this stimulated new mathematical theories and applications. Mathematization was also promoted and supported by the various restoration projects that were undertaken by Renaissance humanists. Precise measurement, proportionality, and geometric principles formed the basis of Renaissance art in both theory and practice. Archimedes' ideas were revived and extended in the field of mechanics, which also saw a trend toward a more empirical approach to natural philosophy. And astronomy was permanently changed by Copernicus' work, though its most revolutionary aspect — making mathematical concerns and analysis the basis for a physical astronomy — was not realized until early in the seventeenth century. To complete this picture of mathematization in Renaissance times, we will look at a couple of philosophical developments that are germane to it.

Mathematization and Renaissance Philosophy

During the Renaissance, philosophical works other than Aristotle's became more widely available. By the end of the fifteenth century, the Italian philosopher Ficino had translated and published a number of Platonic and Neoplatonic writings with extensive commentary of his own. By the middle of the sixteenth century, atomistic philosophy was also being revived. None of these systems was strong enough to overthrow Aristotelian natural philosophy, particularly in the university, but they had a devoted following in certain arenas, and they contributed to a climate that considered mathematics essential for attaining certain knowledge about the world. In the early seventeenth century, thinkers who challenged Aristotelian philosophy found aid and comfort in these alternatives.

A number of mathematicians openly espoused Platonic ideas, though they often gave them a peculiarly Renaissance twist. Pacioli, for instance, affirmed Plato's injunction about the crucial importance of studying mathematics, but he did so on account of its practical utility, a notion that Plato would have found abhorrent. Some university educators toward the end of the sixteenth century, such as the well-known Jesuit mathematician Clavius, combined ideas from Plato and Proclus with those of Aristotle on subordinate or mixed sciences in order to argue the value of mathematics for philosophy and science. Mathematics, they suggested, should be accorded primacy of place in the curriculum due to the absolute certainty of its demonstrations.

Hermeticism was another philosophical trend in this period that took an elevated view of mathematics.[13] This development was part of Renaissance humanism's obsession for returning to the purity of ancient ideas in order to progress beyond scholastic medieval thought. It contributed to the notion of mathematization in ways that are both peculiar and difficult to assess, but it had an impact that needs to be mentioned.

Hermeticism originates with the eclectic occult philosophy of the mythical Greek, Hermes Trismegistus. This work was translated in the mid-1460s by Ficino, who interrupted his translation project on Plato to do so. These writings, which date back to about AD 200, were largely inspired by Pythagorean and Neoplatonic thought, but this fact was only established in an exposé of 1614. Renaissance thinkers convinced themselves that they embodied mystical Egyptian thought going back to the time of Moses. Hermetic ideas thus carried the authority of a very ancient tradition reaching back toward Adam, who was believed to have had perfect knowledge of the world prior to the Fall.

The salient point for our interests here is the Hermetic view of mathematics and its relation to natural philosophy. Their world was literally an enchanted one. Material reality merged with spiritual forces that could be grasped and commanded through occult means, such as astrology, numerology, alchemy, magical incantations, and so on. Hermetic thought was frequently combined with cabalistic practices. This arose from a mystical Jewish tradition that associated Hebrew letters and words with numbers

13. For a groundbreaking but controversial treatment of occult philosophy in the Renaissance and early modern periods, see Frances A. Yates, *Giordano Bruno and the Hermetic Tradition* (Chicago: University of Chicago Press, 1964), and also *The Rosicrucian Enlightenment* (London: Routledge, 1972).

and that attributed special importance to certain numbers, yielding numerological interpretations of Scripture and other writing. Hermetic thinkers greatly admired mathematics for its ability to penetrate the mysteries of the universe and give humans a measure of control over their world. The mathematical simplicity and harmony of nature was emphasized, as was the practical value of mathematics for the arts and the design of mechanical inventions, but so was mathematics' contributions to magic through the use of esoteric numerological properties and potent magical diagrams.

A number of thinkers, including Christians, adopted this outlook, though sometimes with great caution. Some found the notion of a primitive mystical religion behind both Christianity and Judaism attractive in an age that saw different Christian traditions warring with one another across Europe. Many sympathized with the idea of mathematics as the privileged route to natural knowledge. Some, like the sixteenth-century English mathematician John Dee, swallowed a large dose of Hermetic thought. Others, while they were uncomfortable with the mystical mindset of its occult philosophy, were nevertheless affected by its outlook. A few aspects of Copernicus' philosophical outlook show signs of being influenced by this background, as do certain ideas of Kepler in the early seventeenth century, although he adamantly resisted the non-empirical tendency of Hermetic thought. Newton himself had more than a passing interest in alchemy and apocalyptic speculations.

In the end, the supernatural role of mathematics in gaining knowledge of the world largely died out. Scientists who found its magical approach at odds with their religious and philosophical outlooks rejected Hermetic thought with its otherworldly bearing. Mersenne, a seminal Catholic thinker in the first half of the seventeenth century and a one-man clearinghouse for the latest scientific ideas in Europe, championed a completely mechanistic viewpoint of the material world in order to combat the pantheistic tendency of Hermetic and animistic natural philosophies. A century later, deists would take this mechanistic approach as an indication of the self-sufficiency of the physical universe. However, as the scientific revolution was beginning, a mechanical worldview was developed and promoted as a Christian response to a perceived threat. The interesting point to note here with respect to our topic is that while a mechanical worldview was diametrically opposed to Hermeticism, both viewpoints remained committed to the belief that mathematics was the key to understanding the universe.

Summary

Let us briefly summarize what we have found so far in tracing the process of mathematization. The roots of the modernist perspective on mathematics that we will address in the next two chapters are ancient. Mathematization is anchored both in Greek philosophical views on the nature of reality and in their scientific practice. Over the two thousand years following Pythagoras, Euclid, Plato, Aristotle, and Archimedes, mathematics became increasingly embedded in Western science, commerce, and other dimensions of culture. This advance in mathematization was partially pragmatic — it was built on mathematics' superb success in dealing with various practical affairs. However, it was also the result of the compelling philosophical attractiveness of a mathematical approach to knowledge about the world. As epitomized by Euclid's *Elements,* mathematics provided science with a sure method. That is, through deduction from axioms, mathematics provided both the highest certainty possible and a systematic way to organize knowledge. But mathematics achieved this certainty in part through the exactness of its subject matter, and this was also taken over by other fields of thought. Focusing on the quantitative aspects of social or physical phenomena or of artistic matters enabled people to gain a deeper understanding of those aspects of the world and helped them to exert control over them.

From a Christian perspective, many aspects of this mathematization are commendable, but some are quite problematic. For some people, mathematical knowledge represents a form of certainty that can be found independently of God and, in their thinking, may even replace God. Furthermore, the Neoplatonic notions that mathematics gives us knowledge of God's nature and that mathematical knowledge is thereby superior to all other forms of knowledge about the world are not only dubious but have had unpleasant consequences. We will begin to see some of these consequences in the next chapter, where we focus directly on the modern period. However, Chapter 6 will still be primarily historical. We will save a more philosophical analysis of these matters for Chapter 7.

CHAPTER 6

Mathematization and Modern Science

Introduction

Let us take stock of where we are in the process of tracing Western mathematization. Some ancient Greek philosophers had abstractly asserted the preeminence of mathematics for understanding the nature and structure of reality. Other Greeks promoted mathematics more concretely by developing mathematical theories of natural phenomena such as Archimedes' treatment of the law of the lever. However, much of what we now call natural science remained unaltered in any essential way by mathematics. Various writers in the medieval period continued the rhetoric of mathematization, but they provided little solid evidence to back it up. Later thinkers instead followed Aristotle, emphasizing the importance of logical argumentation for natural philosophy and viewing each science as having its own (non-mathematical) subject matter. Mathematics was a field to imitate on account of its deductive method, not one to apply because of its subject matter. An exception to this occurred near the end of the Middle Ages, when the Oxford Calculators developed a mathematical theory of motion, but this remained tied to abstract philosophy.

Renaissance thinkers valued mathematics highly on account of its logical structure and the certainty of its conclusions, but they also put it to work in practical and artistic affairs. While they still thought largely in Aristotelian terms about natural phenomena, habits of measurement and quantification were becoming ingrained in everyday life and provided a platform for seventeenth-century developments in science. As Greek knowledge was being reassimilated during the Renaissance, old ideas were also being challenged, in part due to discoveries made on voyages to new lands. New ideas and techniques were arising in astronomy, algebra, and computational arithmetic, often in connection with projects that sought to restore mathematics to its ancient glory. Activities that lay at the juncture of practical affairs and mathematical science, such as architecture, engineering, and gunnery, also made advances in the sixteenth century. Furthermore, the Renaissance saw the rebirth and spread of philosophies that asserted the supremacy of mathematics. Since mathematics had proved itself so well on a human scale in Renaissance culture, expanding its applicability to cosmic proportions seemed reasonable. While neither Platonism nor Hermetic philosophy was overly disposed toward an experimental or mechanical approach to natural philosophy, they did contribute to an intellectual climate that challenged Aristotle and promoted the mathematization of science.

Aristotelian natural philosophy came under fire on a philosophical level in the seventeenth century, but that was not the most important showdown. It was the successful outworking of a quantitative outlook within astronomy and mechanics that finally ended Aristotle's domination of natural philosophy. The immense success of mathematized science throughout the century encouraged people to push the program into other areas of natural science and even into the human and social sciences.

Seventeenth-Century Mathematization of Natural Science

The seventeenth century produced unprecedented advances in mathematization. During this period mathematics was harnessed to a mechanistic program of natural philosophy. Science began to be mathematized in a way that anchored it in experienced measurable behavior of natural phenomena.[1] This development began in the mathematical sciences of astronomy and mechanics, where quantification already held a strong position and was even sanctioned by a modern version of Aristotelian natural philosophy that up-

1. Peter Dear traces various sixteenth- and seventeenth-century developments, linking quantification with different notions of experience and experimentation. See his *Discipline and Experience: The Mathematical Way in the Scientific Revolution* (Chicago: University of Chicago Press, 1995).

held the legitimacy of mixing science and mathematics. However, the newly discovered work of Archimedes in mathematical physics was the crucial factor driving this development. It gave scientists an important model and some essential tools for pursuing the project.

The two most important mathematicians who battled the traditional physics and astronomy of Aristotle and Ptolemy were Kepler (1571-1630) and Galileo (1564-1642). In different ways and independently of one another, they attempted to establish Copernicus' view of the universe, both with their scientific findings and their defense of its reasonableness. In Galileo's case, this was part of a broader attack on Aristotelian natural philosophy. He also developed a new mathematical science of motion that contradicted Aristotle's views. Kepler's main work was in astronomy, but he also made an important contribution to the field of ocular optics. Both of them were strong proponents of a mathematical viewpoint on nature; both of them made lasting contributions to mathematical physics. Thus we begin our story of the greatest advances of mathematization with their work.

Kepler's Mathematical Vision of the World

Kepler studied nature to gain a deeper understanding of its Creator. Kepler's success in revealing the secrets of God's magnificent work was offered back to him with exuberant praise and devotion. In an era of religious discord, Kepler was a Lutheran with sympathies toward both Calvinist and Roman Catholic positions on certain theological issues. What was most important for his scientific work, however, was his view of God as Creator. Here his ideas derive primarily from Neoplatonic/Neopythagorean philosophy. For Kepler, God is the Supreme Architect, the one who created the universe according to eternal geometric patterns located in his mind. As Kepler notes:

> Geometry, being part of the divine mind from time immemorial, from before the origin of things, being God Himself (for what is in God that is not God Himself?), has supplied God with the models for the creation of the world.[2]

2. Max Caspar, *Kepler,* revised edition (New York: Dover, 1993), p. 271. This quote appears in Kepler's work of 1619, but it represents his viewpoint throughout his life.

Thus, Kepler believed that God had embodied some of his essential mathematical nature in the creation. Being created in God's image, we humans can think his thoughts after him, using the ideas of number and magnitude he has implanted in our minds. True knowledge of natural phenomena can be attained when the geometric schemes in our minds correspond to those prototypes in the Divine mind that have been copied into the world. Scientific knowledge results from the use of human reason stimulated by experience. Because scriptural revelation does not aim to impart knowledge of nature, it is thus irrelevant to this task. To understand how the universe is regulated and even to learn more about the being of God, we must search out the spatial structure of the world and look for proportion and harmonious relations among its geometric magnitudes.

Kepler's mathematized natural philosophy is clearly seen in his astronomical work. Already as a university student, he became convinced of the correctness of Copernicus' system, being attracted by its overall simplicity and its emphasis on symmetry and harmony. After he became a professor of astronomy, Kepler developed his ideas further. He believed that the five Platonic solids, whose treatment had provided the grand finale to Euclid's *Elements,* could explain the planetary orbits. Using a series of inscribed and circumscribed spheres separated by the five nested regular solids, Kepler accounted for the number of the planets and the spacing of their paths along these spheres. God ordered the planets, he claims, to accord with this beautiful geometric arrangement. He explained this theory in his first astronomical work, *Mysterium Cosmographicum* (1596). Although existing astronomical data failed to support his hypothesis completely, Kepler gave reasons, including possible inaccuracies of the data, for why his theory might not quite agree with tabulated values. Kepler came back to this Platonic vision of the celestial array throughout his life, unwilling to toss it aside. He supplemented it in his third astronomical work, *Harmonice Mundi* (1619), with various other arcane mathematical ideas, such as the identification of musical harmonies with planetary motions. This gave wondrous mathematical detail to the Pythagorean "music of the spheres." Kepler discovered one of these harmonious mathematical relations while trying to find a numerical relationship to demonstrate the Sun's role in propelling planets about their orbits. This result is now known as Kepler's third law: the squares of the periods of the revolutions of the planets are proportional to the cubes of their mean distances from the Sun.

Aspects of Kepler's thought seem to mesh quite well with the blend of

Platonic and Hermetic philosophies of his time. Kepler was initially attracted to an animistic view of planetary motion — that is, that planets were either beings that had souls or were guided by such beings. However, he soon abandoned it for a purely mechanical outlook, comparing astronomical motions to a well-regulated clock. Kepler's use of mathematics also differed from his more mystical counterparts. Kepler pointed this out in a critique he gave of a leading Hermetic thinker who had accused his astronomy of not delving into the inner realities of nature. One must always subject mathematical conjectures to empirical testing, Kepler asserted, and not create mathematical fantasies. We are bound to the world God made and are not free to create one of our own. Kepler thus disassociated genuine mathematical science from numerological speculations and delineated proper uses for mathematics in opposition to the excesses around him. Astronomy can use mathematical hypotheses that accord with appearances, but these hypotheses must do even more; they must demonstrate the way things actually work. The harmonious geometrical structure of reality was an *a priori* given for Kepler, but the exact form of the mathematical regularities it exhibited must be determined from the facts of experience.

This attitude is aptly demonstrated in Kepler's second astronomical work of 1609, *Astronomia Nova.* Ten years earlier he had been hired as Tycho Brahe's mathematical assistant, doing the astronomical calculations associated with Tycho's program of observations. Kepler took on the task of calculating Mars' orbit. Given the precise data for which Tycho was known, Kepler discovered that the accepted ideas about Mars' motion failed to generate accurate values. Continuing to work on the problem after Tycho's death in 1601, Kepler eventually found a solution. It was a solution unlike any he or anyone else had anticipated, however. After years of struggling with computational and conceptual difficulties, he finally concluded that the orbit of Mars was an ellipse. This is an instance of Kepler's first law. The compound circular motions that had been postulated by others to describe each planet's apparent motion were now replaced by a single, simple, elliptical motion around the Sun, which was located at one focus. In addition to relinquishing circular motion, Kepler found that he also had to give up uniform velocity. Mars moves with uniform velocity only in the sense that its radius from the Sun sweeps out equal areas in equal times. This is Kepler's second law. Finding these results took about eight years of patient calculation with empirical data and involved numerous dead ends and mistakes. However, Kepler's new astronomy swept away two thousand years of false preconceptions and completed Copernicus' system of the world in a surprising and beautiful way.

In addition to discovering these two laws, which he generalized to all the planets, Kepler insisted that the motion of heavenly bodies required explanation. Traditional natural philosophy kept heaven and earth separate and saw no need to stipulate a cause for uniform circular celestial motion. Kepler challenged this. He proposed, for the first time, that physics be combined with mathematical astronomy. He was unable, however, to give a fully acceptable quantitative explanation for planetary motion (following Gilbert, he suggested that a rotating Sun pulls the planets around using some sort of magnetic force). The final outworking of this view had to await Newton's theory of universal gravitation toward the end of the century. Even so, Kepler's astronomy initiated the idea that the world is a mechanical universe in which the planets move due to the physical action of the Sun and in obedience to mathematical laws; they no longer move through the action and will of quasi-divine beings inhabiting the planets.

Kepler's astronomical writings contained technical mathematical discourse, overfull descriptions of the meandering process of his scientific discoveries, Neoplatonic philosophizing, and ecstatic religious utterances. This unique combination did not attract many readers. The full importance of his astronomical viewpoint and results were only gradually recognized. Nevertheless, his *Epitome Astronomiae Copernicanae* (1621) provided the main source for later scientists, such as Halley and Newton, to learn the basic details of the Copernican system he had completed.

Galileo's Mathematical View of Nature

Galileo's *Dialogue Concerning the Two Chief World Systems* (1632) was intended as a popular account of the Copernican system, and it was well known and accepted by many, even though the Roman Catholic Church quickly moved to ban it. In this book, Galileo presented a witty and engaging discussion of the merits of the Copernican system over against the entrenched coalition of Aristotelian cosmology and Ptolemaic astronomy. He showed that the usual physical arguments against Copernican astronomy, arguments directed against the motion of the Earth, are not decisive. The reason objects do not fly off a whirling world, for instance, is because they also participate in the Earth's motion, just as a weight dropped from the mast of a boat will fall at its base, even if the boat is moving. Galileo defused the standard criticisms with striking and sometimes humorous counter-examples dealing with terrestrial motion, a field in which Galileo was very much at home. In addition to defending Copernicus against his critics, Galileo presented his account of the tides as strong positive evidence for the correctness of the Copernican system. Consequently, he thought Copernicus' astronomy approached the requirements set up by Aristotle for being a demonstrative science. This posture got Galileo in trouble with the Church. He had been given permission to present Copernicus' system of astronomy in the conventional way, as a hypothetical mathematical theory, not as the true system of the world.

Galileo had adopted a Copernican outlook already in 1609. Investigating the heavens with a newly invented telescope of his own construction, he noted the following: the Moon's surface appears to be irregularly shaped, not perfectly spherical; the Sun shows spots that change; Venus goes through phases like the Moon, indicating its revolution about the Sun; and Jupiter has four moons that circle it, rather than the Earth. These discoveries were consistent with and lent support to the Copernican viewpoint, while they presented real problems for the traditional approach. Galileo's report of his findings and of the instrument he used generated far more excitement about astronomical possibilities than Kepler's work and cast him in the role of Copernicus' defender.

In the *Dialogue,* Galileo took a non-technical qualitative account of the Copernican system as the basis for his discussion — circles, uniform motion, and all. He did not advance the Copernican viewpoint using quantitative means, except indirectly by advocating combining physics and astronomy, thereby making physics more mathematical. In fact, while Galileo was aware of Kepler's earlier astronomical work, there is no indication that he ever adopted its results. His interest was more in the physical aspects of the situation and with what he could contribute to the discussion from his knowledge of mechanics. Galileo did promote a highly mathematical approach to physics in this latter field, however.

Like Kepler, Galileo's mathematical viewpoint of the world was grounded in his religious and philosophical orientation. Galileo took a traditional Augustinian viewpoint on the source of natural and scriptural revelation — God is the acknowledged Author of both the book of Scripture and the book of Nature. Therefore, divine truths revealed by either one cannot contradict those of the other, though human interpretations of each might give rise to conflicts. However, he proceeded to elevate knowledge of nature. Aristotle had said that demonstrative scientific reasoning based on sensory experience provides necessary and certain knowledge of the world and not mere opinion. Galileo agreed. Hence for him, results in this realm should be given primacy and accepted by everyone, including those engaged in interpreting Scripture. The natural light of reason must be granted a higher priority than faith in the realm of natural knowledge. Given the Church's attitude toward lay interpretations of Scripture in the wake of the Protestant Reformation, however, Galileo knew that he had to tread very carefully in this matter.

What makes the knowledge of nature so certain? Here Galileo, like Kepler, proffered a mathematized view of reality whose philosophical roots go back to Pythagoras, Democritus, and Plato. The deductive structure of geometry guarantees the certainty of its results and the use of quantity gives it precision. Nevertheless, Galileo, like Kepler, was critical of the mystical side of Neoplatonism. Natural philosophy is not fiction. In a work of 1623, Galileo explained his overarching viewpoint on science with the following well-known words:

> Philosophy is written in this grand book, the universe, which stands continually open to our gaze. But the book cannot be understood unless one first learns to comprehend the language and read the letters in which it is composed. It is written in the language of mathematics, and its characters are triangles, circles, and other geometric figures without which it is humanly impossible to understand a single word of it; without these, one wanders about in a dark labyrinth.[3]

Following Plato and the atomists and in opposition to Aristotle and scholastic philosophers, Galileo asserted that the essential and necessary properties of material things are the primary mathematical qualities of number, size, shape, and speed. Sensory properties such as color or sound are secondary qualities and reside in the sensing subject, not in the object itself. Galileo thus recognized the foundational role of mathematics for science, but he went further and reduced the central content of physics to geometry. His views on primary and secondary qualities were echoed else-

3. Stillman Drake, *Discoveries and Opinions of Galileo* (New York: Doubleday Anchor Books, 1957), pp. 237-38.

where in Europe, though they likely arose independently, from the same ancient sources that fed Galileo's viewpoint.

Galileo accepted Aristotle's deductive methodology of science,[4] but he gave larger roles to experience and especially to mathematics. Galileo was especially indebted to Archimedes for his approach to mathematical physics. Galileo's outlook in this respect is not unique, for as we noted above, earlier natural philosophers were moving in this direction, similarly taking inspiration from Archimedes' practice. Yet, Galileo was more thoroughgoing in his mathematization and ingenious in his empirical and mathematical exploration of phenomena. His familiarity with the Renaissance tradition of applying mathematics to practical technology may have helped him introduce measurements into physics. He rejected the ancient tendency to see qualities as pairs of opposites, such as hot and cold or being at rest and being in motion, and instead treated them as measurable quantities lying along a continuum. They are thus quantities that can be represented by line segments. In Galileo's hands, natural science was being transformed into a field of thought conducted mainly by means of the ideas and methods of mathematics rather than those of syllogistic logic. Galileo's mathematical notions were not mystical ideas floating high above the field of physics, but were linked concretely to the way things actually function. Experimental exploration was used by Galileo to suggest appropriate first principles and to eliminate or confirm hypotheses, but the core of any true science was down-to-earth mathematics suited for the task. An Aristotelian concept of science was combined with an Archimedean viewpoint on the importance of geometry and proportion, subject to experimental verification.

Galileo is best known in scientific circles for his mathematical analysis of motion. This was published in 1638 as the central part of his *Two New Sciences*, though his initial thoughts on motion go back nearly 50 years earlier. In this, his last work, Galileo presented a science of motion in three parts, first dealing with uniform motion, then with naturally accelerated motion, and finally with projectile motion. Each section is organized in standard Euclidean style, opening with the relevant definitions, axioms, and postulates, and then proceeding to various propositions. For uniform motion, Galileo proves several results relating speeds, times, and distances traversed, using

4. See William A. Wallace, *Galileo's Logic of Discovery and Proof*, Boston Studies in the Philosophy of Science, vol. 137 (Kluwer Academic Publishers, 1992).

the Archimedean tools of ratio and proportionality, ending with the equivalent of our result that distance equals speed times time.

Regarding naturally accelerated motion due to free fall, Galileo first notes that free fall yields uniformly accelerated motion since the results derived on that basis match those gotten by his experimentation with inclined planes. However, he also says this is to be expected since Nature always acts in the simplest way, and the simplest type of accelerated motion is that in which equal increments of speed are added on in equal increments of time. The conclusions he obtains from this agree with those gotten earlier by the Oxford Calculators (whose work he seems not to have depended upon), but they contradicted the commonly held notion that *speed* is proportional to the *distance* fallen. Instead, Galileo shows that *distance* is traversed as the square of the *time* elapsed.

In the last part of the treatise, Galileo determined the path of a projectile. Mathematicians earlier conjectured that the path consisted of three parts, a first and last straight path connected by a curved path, possibly circular. Building on his results about uniform horizontal motion and naturally accelerated vertical motion and assuming that these motions maintained their independence when combined, Galileo was able to establish the exact shape of the path as parabolic. Galileo understood the benefit this gave to military art, for it enabled gunners to compile charts relating the range of their cannon shots to the elevation chosen. Galileo was also able to prove that the maximum range of a projectile would be achieved at half a right angle.

Galileo proudly and rightly advertised his work on motion as an important new science. He had individually done for motion what others had done for optical phenomena and statics — he had created a mathematical science. Traditional natural philosophy, drawing on Aristotle's ideas of motion, treated motion in a mostly qualitative fashion. On those points where it made quantitative assertions, it was often spectacularly wrong. The work of the Oxford Calculators was an exception to this, but their work was abstract mathematics, not physics. Galileo's was both physics and mathematics, treating natural (if idealized) motion. He did not focus on the cause of motion or try to explain how motion could continue, as earlier philosophers had. Instead, he explored its mathematical features and established precise functional dependencies, using the language of proportionality. The question addressed, using mathematics, was not *why* but *how*. This mathematical model was to inspire other scientists in the seventeenth century and was to become the *modus operandi* of physical science. The significance of Galileo's specific accomplishment is further put into perspective by considering the importance assigned to motion in seventeenth-century mechanistic thought: along with the mathematical features of number, size, and shape, motion was thought to lie at the base of all other natural phenomena. Galileo provided the mathematical foundation and some of the tools for developing this outlook. Newton was to take Galileo's work and incorporate it into his own later in the century, using a methodology that owed much to Galileo's approach.

Descartes' Mechanistic Mathematical Universe

Mechanistic explanations were becoming the new wave of natural philosophy. Aristotelian natural philosophy had been modeled on living organisms and emphasized teleology, a notion of purposeful development. Now, scientists were beginning to analyze phenomena largely in terms of their mathematical relations and mechanical behavior. Complex automata and intricate machines became the new model. Kepler and Galileo were far from being the only scientists to adopt such an approach. Early in the century, Isaac Beeckman, drawing upon a practical mechanical tradition in the Netherlands as well as his knowledge of Archimedes' theoretical works, was one of the first to formulate a mechanistic outlook. Since quantitative features of natural reality are all-important, he says, mathematics supplies the hands of physics. These ideas exerted a lasting influence on seventeenth-century thought through the medium of Descartes (1596-1650), who learned this mechanistic approach first-hand from Beeckman in 1618-19.

Descartes was neither an astronomer like Kepler nor a physicist like Galileo, though he contributed to both fields. He was primarily a systematic philosopher, whose work in mathematics was also of the first magnitude. The methodology and subject matter of mathematics, Descartes believed, holds the key to natural philosophy. His goal was to argue this for all natural phenomena, not merely astronomy, mechanics, or optics. Focusing on shape, size, and motion advances the study of nature, for the core meaning of matter and its primary quality is extension. Physics is nothing more than geometry. Thus natural philosophy can be placed upon an indubitable foundation, for mathematics is the paradigm of irrefutable knowledge. Galileo, Descartes notes in a 1638 letter to Mersenne,

> philosophizes much better than the usual lot, for he . . . strives to examine physical matters with mathematical reasons. In this I am completely in agreement with him, and I hold that there is no other way of finding the truth. But I see a serious deficiency in his . . . [not] considering the first causes of nature. . . .[5]

According to Descartes, Galileo should have been more systematic and built his science on a firmer metaphysical foundation. Descartes aimed to develop a method that would solve specific scientific problems but that would also give absolutely certain demonstrations of the true structure of all of reality. In this respect his overall object and view of science was much the same as that of Aristotle, though his mathematized mechanistic approach was very different.

Descartes had hoped to demonstrate the veracity of the Copernican system of the world. However, when the Roman Catholic Church condemned Galileo's ideas, he sought to present his own system in a way that would not collide with the censors. This could be done, he thought, by grounding it in a metaphysics that was consistent with Catholic theology. Beginning with an attitude of radical doubt (in order to counteract even the extreme skeptic), Descartes notes that doubting requires a thinking subject. This in turn requires a trustworthy Supreme Being who can guarantee our existence and the truth of what we know. God alone can legitimize the nature and certainty of human reasoning. We humans have been given the capacity by God to generate clear and distinct ideas of things by the operation of our minds. These innate ideas provide us with mathematical and other notions for understanding our world.

Descartes thus proposed a strong rationalist philosophy. There are two irreducible types of reality, according to Descartes: mental and corporeal. Matter behaves in a completely mechanical fashion; mind comes to know how matter acts by means of its clear and distinct ideas. Humans alone have reason and are essentially thinking beings. All other living things are really nothing more than automata. We can apply our reason to attain true knowledge of the natural world from first principles. At times, however, especially given the incomplete state of our knowledge, we may have to resort to hypothetical reasoning, as was done earlier in astronomy. Then we must check the

5. William R. Shea, *The Magic of Numbers and Motion: The Scientific Career of René Descartes* (Cambridge: Science History Publications, 1991), p. 312.

deductive consequences of our hypotheses with experience to verify or falsify them. Experience provides the data requiring explanation, and it helps us to decide how to generate an explanation for them from *a priori* principles. While Descartes did quite a bit of experimentation in connection with optics and physiology, its role in his thought was mainly that of determining what Reason needs to render an explanation for and of suggesting some possible connections. True science is ultimately generated, however, by drawing necessary conclusions from self-evident axioms known *a priori* by our minds.

Descartes stressed the essential importance of method in acquiring knowledge. The general approach that Descartes saw embodied in the mathematical method of analysis can also be called the method of philosophical analysis. He exhorted anyone who wanted to advance in natural philosophy to consider all relevant phenomena, analyzing and clarifying the various notions involved into more basic constituent concepts, until arriving at the simplest ideas. These ideas can then be taken as foundational, more complex ideas being built up from them, until a theoretical explanation of the phenomena is obtained. It was this rationalist approach of breaking reality down, as one might a machine, and then reconstructing it from its abstracted elements, building knowledge upon a basis of clear and distinct ideas that attracted his followers. However, the metaphysical underpinnings for this approach (first doubt, then God, then clear ideas), which Descartes saw as necessary, was of less interest to many. They were captivated by Descartes' vision of a mechanical universe susceptible to mathematical analysis, but not always by his philosophical rationale.

The value of method was certainly not a new theme in philosophy, but Descartes was the one who made it fashionable for the modern age, drawing upon his work in mathematics. Mathematics contains not only the time-honored deductive mode for communicating known truths, but even more importantly, an analytic method for discovering truths. He published his ideas on this in *Discourse on Method* (1637), to which he attached three appendices as proof of the value of his approach: an essay on optics, another on meteorology, and a concluding one on geometry. Like Viète, Descartes found vestiges of the analytic method in Arabic and Renaissance algebra and in the geometry of Pappus and other ancient Greeks. Unlike Viète, however, he maintained that his algebra and analytic geometry were more advanced than those of the ancients, for he was able to solve problems that had defied their best efforts. In his essay on geometry, we can clearly see our modern approach to symbolic algebra and analytic geometry taking shape. Descartes' influence in this area is apparent from the similarity of his work to what we know these fields to be, though the presentation is still somewhat obscure in spots. In his appendix on optics, Descartes presented his law of refraction. Others in the same period had also come to a correct understanding of this law, but it was Descartes who publicized it and attempted mechanical explanations of it. He also offered an exemplary analysis of the appearance of the rainbow, explaining it in terms of the Sun's multiple refraction through raindrops. Descartes' *Discourse,* with its famous appendices, rightly established his reputation in mathematics, physics, and philosophy as an intellectual giant.

Descartes pursued his mechanistic natural philosophy in all realms, including mechanics, astronomy, optics, and physiology. He proposed a universe completely full of particles, whose collisions with one another produced motion in accord with a number of quantitative rules, some of which anticipate Newton's laws of motion but most of which we no longer accept as valid. The motion of the planets about the Sun was explained in terms of whirlpool or vortex motion of particles in the intervening space. Natural philosophers after Descartes tried to determine the precise mathematics of this motion, without success. Newton later discredited Cartesian physical astronomy by showing that the planets' motions would require contradictory behaviors from the vortex, but his refutation was not accepted by everyone as sufficient reason for rejecting the general idea. In optics, Descartes explained how light travels, using various mechanical analogies, one of them being the instantaneous transmission of pressure through the visual medium. These are then used to explain reflection and refraction. Following the lead of Kepler in his mathematical analysis of retinal images on the back of the eye early in the century, Descartes explicitly separated the physiological aspects of vision from perceptual cognition and treated them completely mechanically.

Descartes' system of natural philosophy may have lacked mathematical specificity in some areas, but on many fronts it initiated a very attractive program to be developed. Descartes showed for the first time the potential of taking a mechanistic approach to all of nature, of treating the material universe as a machine. In the last half of the seventeenth century and on into the eighteenth, Descartes' system of thought commanded the highest respect and formed the basis of further scientific work. After Descartes, Aristotelian natural philosophy was no longer a viable system; mechanistic natural philosophy had taken its place, with Descartes' ideas in the forefront. It was thought by some that Descartes had laid down the basic outlines of the true system of the world and that all that remained for others to do was to fill in the details.

Newton's Mathematical Treatment of Physics

Like every other natural philosopher of the late seventeenth century, Newton (1643-1727) was weaned on Descartes. He came, however, to reject Descartes' thought because of its speculative rationalistic character. For Newton was also nurtured on the ideas of his countryman Francis Bacon, who emphasized the need to generalize from firmly established facts.

Bacon, who wrote in the first quarter of the seventeenth century, stressed detailed empirical investigation of reality, leading to classification and determination of the true natures of things. His approach was not aimed at formulating mathematical laws. Unlike his earlier namesake, Bacon did not recognize the value of mathematics for science. He thought that mathematics' habit of abstraction was dangerous to physics, which needed to remain in close contact with reality. Robert Boyle, Bacon's disciple living in the Cartesian second half of the century, had a much stronger appreciation for mathematics as an aid to science and as a necessary part of mechanical philosophy than his mentor did. However, he still failed to connect mathematics very closely with experimental practice. He believed that mathematics was too independent of physical reality and that its fastidious precision was out of place in the laboratory.

For many at this time, contributions of mathematics to natural philosophy were largely of two sorts, either sublime speculations or mundane observations, but these were not often linked in any intimate way. Mathematics was admitted to be an essential component of mechanical philosophy, since the primary subvisible properties of things that determined their structure and behavior were taken to be quantitative. However, exactly how these microscopic mathematical features determined macroscopic behavior was difficult, if not impossible, to ascertain. Mechanistic natural philosophy thus led to imaginative non-verifiable conjectures. At the same time, more aspects of natural phenomena were being quantified and measured more precisely in this period (time, distance, speed, weight, volume, air pressure, temperature, volume, intensity of light, etc.). This gave a potential foundation for mathematizing various fields of thought and activity. Yet, these were rarely connected with any mathematical theory about how things worked. The habit of looking for quantitative functional dependencies was still weak; most scientists continued to look for underlying causes in the presumed nature of things. Moreover, many seventeenth-century scientists worked largely in the qualitative experimental areas of natural philosophy, such as chemistry and natural history.[6]

All of this changed with the arrival of Isaac Newton, whose work in the classical sciences of optics, mechanics, and astronomy made deep connections between quantitative features of phenomena and the way the universe worked.[7] Newton, more than anyone else in the seventeenth century, was able to forge an alliance between experimental research and mathematical analysis and so realize the enormous benefits of mathematizing natural philosophy. He synthesized earlier work in the mixed sciences and laid a foundation for eighteenth-century developments in various other physical sciences.

Newton first demonstrated his brilliance with his investigation of colors. Soon after receiving his Bachelor's degree from Cambridge in 1665, Newton undertook a series of optical experiments to determine the nature and behavior of colored light. Rather than holding that colors are mixtures of light and dark, modifications of white light, as most held, Newton asserted the opposite, that white light is a mixture of different colors. Light can be spread out into a spectrum by a prism because the rays of each color are refracted according to its own characteristic degree of refrangibility. Colors are primary for Newton; white light is the mixture. He established this revolutionary viewpoint by a series of carefully controlled experiments using various prisms, making the appropriate precise measurements and calculations to test his ideas against those of others, such as Boyle and Hooke, and to forestall their criticisms.

He summarized his work on colors in his first paper presented to the Royal Society (1672). In addition to describing the experimental basis for his results and organizing his conclusions in the deductive fashion of mathematics, Newton advanced a new corpuscular theory of light. It was this aspect of his work that received the strongest criticism by others, particularly

6. At the time, the fields that we today call biology, geology, anthropology, etc. did not exist. "Natural history" was their predecessor. Its approach was primarily descriptive.

7. Literature on Newton's scientific work is an entire industry. See, for instance, Richard S. Westfall, *Never at Rest: A Biography of Isaac Newton* (Cambridge: Cambridge University Press, 1980).

Hooke. The unpleasantness of this experience made Newton wary of including hypothetical elements in his scientific theories, and it forced him to try to develop a method of philosophical investigation that would prompt rational acceptance rather than dissent. In response to his critics, Newton asserted that in the main, he had investigated and positively established various properties of light by means of experiments, mathematically deducing their consequences. Only once this was accomplished, did he put forward a conjecture to explain the behavior. This, he noted, was the proper scientific method for doing physics.

Newton elaborated his method of philosophical inquiry in various later works as well. Experiments must be designed to elicit answers to specific scientific questions; thus regularities of observed phenomena are identified and measured. Once general principles are arrived at by induction, their consequences can be deduced by rigorous mathematical demonstrations. Additional experiments verify these consequences as a check on the principles. While this process may not generate absolutely certain knowledge, it gets as close to it as the subject matter permits. Newton's belief on this score conflicted with many of his peers who had come to believe that empirically based knowledge must remain fairly tentative, and also with those who were content to argue from hypothetical mechanical causes. Newton believed that true knowledge could be generated in physical science by applying the method and ideas of mathematics to the data of experience.

After one ascertained the mathematical behavior of phenomena, one might then attempt to discover the causes that produce such phenomena. Newton, like both Descartes and Aristotle, desired to penetrate to the true causes underlying the behavior of reality. Unlike Aristotle, however, he remained far longer on the level of phenomenological behavior and gave priority to quantitative features of the situation. Unlike Descartes and his followers, he was unwilling to use hypotheses to evade empirical exploration and induction. And, unlike both, he refused to build a grand philosophical system based on ultimate causes to encompass his results. He was content to leave this for posterity to work out, if possible. His mathematical and experimental exploration of more limited areas of thought put Newton more in the tradition of Galileo than Descartes or Aristotle.

Mathematics, for Newton, as for Kepler, Galileo, and Descartes, held the key to natural philosophy. Yet, for Newton it seemed to be more an operational stance than an ontological or epistemological position. Whatever the actual nature of the physical world or the precise character of human knowing, he was concerned to concentrate first on the indubitable mathematical regularities nature exhibits. This limited focus was emphasized in the title of his magnum opus, *Philosophiae Naturalis Principia Mathematica* (1687). As Newton noted in the preface: "the whole burden of [natural] philosophy seems to consist in this — from the phenomena of motions to investigate the forces of nature, and then from these forces to demonstrate the other phenomena." Since motions and forces are quantitative matters, and since demonstrative knowledge of these was sought, Newton used geometry throughout. His book was a masterful exposition devoted solely to mechanics and astronomy, but he held out the possibility that a similar exploration in other realms of physical behavior, such as optics or magnetism or chemistry, would likewise uncover relevant forces operating between particles. This suggestion, given in the context of the marvelous achievements of his work, was to encourage similar work in other fields and foster an even stronger viewpoint on the importance of mathematics than he himself may have entertained. Newton wanted to emphasize a dynamic mathematical approach to natural science in contrast to Descartes' more speculative mechanistic natural philosophy, but he did not think that this exhausted everything that could be said about how the world functioned. Nevertheless, the character of his work did lead to a more constricted view of the methods and scope of natural science.

The *Principia* resulted from about two and a half years of intense intellectual work in which little else engaged Newton's attention, including food and sleep. Like a man possessed, he threw himself into developing a force-based mathematical theory of mechanics and astronomy. To ground his astronomical conclusions, Newton first created both a version of geometry that drew from his earlier work in calculus and a new deductive science of mathematical dynamics that built on and corrected Descartes' mechanics. Newton elaborated these ideas in detail, refined or defined new quantitative concepts such as mass and inertia, and gave precise quantitative formulation to the laws that govern motion. Newton showed in the abstract, for example, using only mathematics, that Kepler's laws imply an inverse square law for a centripetal force attracting two bodies that obey his laws of motion, and conversely that an inverse square law entails Kepler's laws. Consequently, given the physical behavior of the planets summarized by Kepler's laws, the force of gravitational attraction between the planets and the Sun must obey an inverse square law. He also proved that the power of gravity must be proportional to the masses of the bodies involved. While Newton offered no explanation of the nature or cause of gravity, he asserted that the force of gravitational attraction acts uniformly throughout the universe, being able to account for and therefore being responsible for both the movement of heavenly bodies and that of naturally accelerated objects on Earth. It was this mathematical encapsulation of the cause of all natural motion, the law of universal gravitation, that most people found so utterly amazing — that the most diverse movements throughout the universe could be explained by a single rather simple principle was astounding! This achievement convinced even the most skeptical that absolutely certain knowledge of the world might be possible using the tools of mathematics.

Newton's ingenious efforts established him as the leading mathematician and the foremost natural philosopher of Europe. No other work in the history of science compares to the *Principia* in scope or importance. Newton combined, revised, and completed Galileo's science of motion, Descartes' mechanics, and Kepler's mathematical analysis of celestial motion, treating them all in his own new mathematical science of dynamics and astronomy. Notwithstanding this accomplishment, not everyone was ready to accept the intrusion of what seemed to be an occult force acting at a distance (gravity) into natural philosophy. In fact, resistance by Huygens and Leibniz on this point kept Cartesian natural philosophy alive for some time into the eighteenth century. However, the coherence, simplicity, scope, and depth of Newton's system were universally admired and eventually helped to overcome the opposition. Newton had shown how to unify various aspects of physical science in one simple harmonious mathematical system, and his success gave others hope of making additional conquests by the same mathematical and mechanical method. Newton's system of the world set the future course of Western thought and helped to define the modern worldview. More narrowly, it established a new mode of mathematizing natural science. Before Newton there were expert experimentalists and accomplished mathematicians, but few top-rank mathematical experimentalists. Newton's approach gave both experiment and mathematics their due. Now analysis (quantitative experimentation, involving measurements, leading to appropriate inductive generalizations) and synthesis (logical deduction from accepted principles, using mathematical theories, concepts, and techniques designed for the purpose) became full partners in the scientific process. Quantitative measurements were tied to theoretical analysis and related by means of functional dependencies, and mathematical tools were developed to help analyze experiential phenomena. Mathematics and physics were now fused on a deep level, making it possible to predict behavior of phenomena with great precision and to communicate these results in a universal language of clear and distinct ideas. Newton overestimated the ability of experiment to determine scientific principles, and he underestimated the hypothetical element involved in this process, but his emphasis on grounding scientific work in experimentation and placing restrictions on hypotheses was a necessary antidote to the practices and outlooks of many of his contemporaries.

Newton is also well known as one of the main founders of calculus. This is not the place to discuss his role in the history of calculus, but we will comment briefly on how his calculus fits into the mathematization of science. Newton's notion of fluxion, which he began working with already in 1665, is essentially a time derivative. Treating all variable quantities as magnitudes changing over time (fluents), his method of fluxions enabled him to conceptualize velocity and acceleration as mathematical notions and produce techniques for calculating rates of change when the rule for this change was known. He was also able to reverse the process and find fluents (functions) when their fluxions (derivatives) were known. Having discovered and demonstrated the *Fundamental Theorem of Calculus,* Newton could use these reverse methods for calculating areas, volumes, and arc lengths. None of these things appear in Newton's *Principia,* which was written using the classical language of geometry, but their relevance to the topics it covers should be quite apparent, and later mathematicians in the eighteenth century recast and extended his mechanics using the apparatus of calculus. Natural science deals with dependency and change; calculus is the premier theory of functional change. In this respect, too, Newton contributed in an essential way to the mathematization of physical science.

Leibniz's Mathematization of Thought

The other principal founder of calculus was Gottfried Leibniz (1646-1716). Though he arrived at his ideas about a decade after Newton, he was first to publish (1684) and attracted the two Bernoulli brothers to help him develop them further. Using notions of the sum and difference of infinitesimally small quantities, Leibniz made the differential the centerpiece of his brand of calculus. The first calculus textbook, written by L'Hôpital in 1696, followed this approach. Leibniz's ideas and notation soon became standard in

Germany, France, and Holland, while Newton's were largely stranded on the British Isles. After some initial acrimonious exchanges regarding priority of discovery, each side maintained the superiority of its approach without very much positive interaction with the other side for over a century.

Leibniz's ideas on calculus contributed more to the actual process of mathematization than Newton's, since his version became the tool of choice for eighteenth-century mathematical science, which was principally developed by Continental scientists. However, there is another side to Leibniz's thought that is even more strongly devoted to mathematization. That is his scheme for developing a universal calculus of thought.

Leibniz, like Kepler, Galileo, and Descartes, emphasized the ontological and epistemological value of quantification. Like many others at the end of the seventeenth century, Leibniz found a mechanistic viewpoint of science very appealing and strongly promoted it in his works. The essential properties of things are those which can be quantified: size, shape, position, motion, and force. Without the use of quantitative explanations, no understanding of the world is possible; with it, one can determine why things are the way they are and why they cannot be otherwise. For the philosopher Leibniz, as for his predecessors Aristotle and Descartes, the goal of science was obtaining necessary, demonstrative knowledge about the world. Like Descartes, Leibniz sought to ground natural philosophy in the nature of God through metaphysical argumentation. Leibniz found Newton's work rather deficient on this score; his system of the world would run quite as well without God as with Him.

In order to extend mathematics' success in generating absolutely certain knowledge to other fields of thought, Leibniz advocated the dual process of analysis and synthesis. This theme had its proximate source in the late sixteenth century and echoed throughout the seventeenth century, Descartes being the most important instance. On the surface, Leibniz's description of the dual process sounds quite traditional. Analysis means breaking down each concept/proposition into its prerequisite concepts/antecedent propositions and these into theirs, until finally arriving at the most primitive concepts/first principles. Synthesis means reversing the process and using these foundational results to define/prove the given concepts/propositions. Leibniz believed that this process could be applied in any area of human thought, not just mathematical science. By means of rational analysis, the basic concepts and principles of a given field can be determined, and these can then be combined to yield truths to which all rational beings will have to give assent. This emphasis on analysis would continue to gain strength into the eighteenth century, especially since analysis was associated with the field of algebra, which had been successfully widened to produce both analytic geometry and, with the work of Leibniz, calculus.

So far, Leibniz has merely amplified Descartes' approach, but he next gave the whole procedure a further twist that revealed his extreme commitment to mathematization. The process of analysis and synthesis, he believed, can be effected in a mathematical way by choosing appropriate symbols for the basic concepts and their combinations and then calculating with them to obtain the consequences of the principles, much as is done in ordinary algebra. Already in his *Dissertation on the Combinatorial Art* (1666), Leibniz gave voice to this vision: "there would be no more need of disputation between two philosophers than between two accountants. For it would suffice to take their pencils in their hands, and say to each other: Let us calculate." Eleven years later, after he had begun developing his differential and integral calculus, he reiterated the point, in the following words: "All inquiries that depend on reasoning would be performed . . . by a kind of calculus. . . . And if someone would doubt what I advanced I should say to him: Let us count, sir; and thus by taking to pen and ink, we should soon settle the question."[8] Leibniz's notion of a universal rational calculus that can be used to formulate and solve problems in all areas of human activity may seem an extremely naive form of utopian rationalism, but a similar hope has motivated others in Western culture since the time of Leibniz. Boole's algebraic logic of the mid-nineteenth century, for instance, matched some of Leibniz's ideas almost exactly, though it was developed independently. Aspects of computer science in our own era — the development of expert systems and the strong program of Artificial Intelligence (we will discuss this in detail in Chapter 9) — can also be viewed as intellectual descendants of Leibniz's brainchild.

It is clear that Leibniz is on Descartes' side of the philosophical seesaw pitting reason against experience. Leibniz stresses the value of *a priori* reasoning over experience. He thought that British scientists, such as Boyle, overdid it with all their emphasis on experimentation. The senses are unable to give rise to certain knowledge. Natural science demands more, which only demonstrative reasoning can satisfy. Leibniz's attitude toward a rational

8. Alistair C. Crombie, *Styles of Scientific Thinking in the European Tradition: The History of Argument and Explanation Especially in the Mathematical and Biomedical Sciences and Arts*, 3 vols. (London: Duckworth, 1994), p. 1009.

physics, combined with Descartes' similar sentiments, overpowered Newton's more balanced approach in subsequent developments. Eighteenth-century thinkers expanded the fields of physics opened up by Newton and went into related fields as well, but they did so using the rationalistic approach of Descartes and Leibniz. The lure of being able to attain absolute truth through human reason and mathematics was too strong to resist.

The Ongoing Mathematization of Natural Science

Eighteenth-century mathematics was primarily mixed mathematics. Mathematical research in calculus and differential equations went hand in hand with work in the mathematical sciences of mechanics, astronomy, fluid mechanics, acoustics, and others. There was no such thing at the time as separate fields of pure and applied mathematics, although distinct areas of mathematics did start to separate from physics as the century wore on.

Eighteenth-century mathematics' being tied to science meant that the mixed sciences were still as much a part of mathematics as they were of physical science. This fit the classical approach going back to the Greeks, but in the eighteenth century it was part of the Cartesian legacy, augmented by Leibniz's viewpoint. As we noted above, Newton's own approach, which emphasized experimentation as well as mathematics, was neglected by many eighteenth-century scientists, especially, ironically enough, in the field of mechanics, where a strong *a priori* rationalist approach predominated.

Leading scientists such as d'Alembert (1743) and Lagrange (1788) treated mechanics as a closed mathematical system whose results could be rigorously demonstrated from necessary first principles. This was accompanied by banishing geometry from mechanics and reformulating mechanics in purely analytical terms, sometimes without any diagrams or reference to spatial content. Such a treatment of mechanics was usually coupled with a deterministic philosophy of nature. Mechanics was thus deemed capable of generating absolutely certain knowledge about the ultimate structure of the world. Given the initial conditions of the world system, one could in principle predict the future state of the entire material world at any time. This sort of mathematical determinism was present in Laplace's analytic treatment of astronomy (1799), which showed that Newton's system of the world was even more stable than had been previously thought. Eighteenth-century mathematical scientists also advanced other areas of physical science worked on by Newton, such as fluid mechanics and acoustics. They attempted to place them on a firmer mathematical basis of self-evident mechanical principles and explore their deductive consequences using techniques of algebra and calculus.

The areas of electricity, magnetism, heat, and chemistry remained largely experimental during the first half of the eighteenth century as scientists were becoming more familiar with the basic phenomena they exhibited. By the end of the century, however, scientists had learned how to define and measure different quantities associated with them, using instruments designed for the task, and in some cases had begun to formulate mathematical theories for them. In 1787 Coulomb published the results of his carefully controlled experiments in static electricity and magnetism, establishing inverse square laws for both types of attraction, just as for gravity. In the first half of the nineteenth century, a deeper analysis of electromagnetic phenomena by Faraday and others led to an impressive unified mathematical treatment of all these fields by Maxwell in 1865, using notions of vector analysis.

Joseph Black was responsible for initiating a quantitative approach to heat. He first distinguished quantity of heat from temperature in 1760, relating the two by means of the notion of specific heat or capacity for heat. He also introduced the notion of latent heat associated with change of state. A fully mathematical treatment of heat, however, had to wait until early in the nineteenth century, when Fourier studied heat diffusion (1822), investigating it with what we now call Fourier series.

The related area of chemistry became more mathematical by quantifying heat, but also through the systematic use of weight, measured by improved balances. The need to make precise metallurgical analyses of ores for mining provided a strong economic impulse for quantitative chemistry by mid-century. Lavoisier's reform of chemistry (1789) made it more mathematical in a couple of senses. It introduced a more systematic way of naming and symbolizing chemical substances, and it attempted (though only with partial success) to use weights and equations to explain chemical reactions. A more thorough and deeper mathematization of chemistry came about when Dalton put forward his atomic theory in 1808. Using his notion of atomic weight and the law of definite proportions formulated by Proust a decade earlier, scientists could now explain chemical reactions quantitatively in a systematic manner.

Developments during the eighteenth and nineteenth centuries thus made it abundantly clear that the influence of mathematics in natural science was not about to dry up any time soon. Rather than assisting with a few matters in mechanics and then retiring to its own corner, mathematics continued to find significant and essential employment in the physical sciences. The stream of scientific mathematization continued to swell exponentially over time and has grown unabated into our own era, with no sign of let-up. Obviously the mathematical approach of modern science has uncovered genuine and intrinsic connections between natural phenomena and mathematical concepts and techniques. This conclusion seems undeniable. The question that this success often leaves unasked, however, is whether there may be important aspects that have been overlooked due to taking a rather narrow quantitative perspective. And, more importantly, whether the achievements of mathematical science and the elevated view of mathematical truth have given false encouragement to other areas of human life to pursue mathematization where it is less appropriate. We will address some of these matters in the next chapter.

Mathematization of Science and the Modern Worldview

Developments in seventeenth- and eighteenth-century natural science brought about a major change in people's perception of the world. At the beginning of this period the world was still a universe full of inherent purpose and Christian mystery, though a number of people were becoming skeptical about the validity of knowledge provided by Aristotelian philosophy and traditional religion. By the middle of the eighteenth century, however, the world envisioned by Europe's leading thinkers was one of matter in motion operating mechanically according to universal mathematical laws. Mathematical science began to replace scriptural revelation as the acknowledged authority about the nature of the world. Guided by this mechanistic and mathematical outlook, natural philosophy had made remarkable progress in producing certain and reliable knowledge about the physical world.

The role of human agents in such a mechanical universe was unclear and problematic. For some, humans transcended nature on account of their rational faculties. Humans also had free will that enabled them to be more than passive lumps of matter obeying deterministic laws of Nature. Others felt that the methodology of mathematical science should be pushed as far as possible. Given its grand successes in astronomy, optics, and mechanics, why not adopt a similar approach in analyzing human nature and social behavior? This path seemed to them to hold the potential for finally arriving at incontrovertible objective truths in these areas as well.

Our next chapter examines some of these more radical developments; we will conclude this chapter by first summarizing what we have learned about the mathematization of natural science and then evaluating these developments from a Christian perspective.

Historical Summary: The Mathematization of Modern Science

The dominant role played by mathematics in natural philosophy in the seventeenth and eighteenth centuries is closely connected with its philosophical heritage. The original Pythagorean viewpoint, adopted and modified by Platonic and Neoplatonic philosophies, deified mathematics, raising it to a position of absolute importance for scientific knowledge. Mathematics alone was seen as capable of penetrating the secrets of the universe, of tracking all things back to their lair. This pagan viewpoint went virtually unchallenged by medieval thinkers. It also influenced a number of Renaissance developments. A revival of Neoplatonism along with Pythagorean-like Hermeticism in the late Renaissance encouraged early modern thinkers to elevate mathematics far above the place given it by traditional natural philosophy.

However, more than philosophy was responsible for the exalted position of mathematics. Mathematics also made itself indispensable in the arts and in the arena of practical affairs during the Renaissance. Here a number of very down-to-earth connections between mathematics and reality were established. This assumed a vastly different role for mathematics than that envisioned by mystical mathematical philosophers, but it was one that nevertheless emphasized the essential necessity of mathematics. Recovery of a number of ancient Greek works, both in mathematics proper as well as mathematical science, also gave a boost to mathematization. Toward the beginning of the modern era, the works of Archimedes exerted a strong positive influence by demonstrating just what could be accomplished in mathematical science.

As Kepler, Galileo, and Descartes worked out their philosophical perspectives in mathematical science, their impressive achievements established a closer working relationship between mathematics and natural philosophy and seemed to validate their mathematized approach. Newton seems to have entertained a more moderate view of mathematics' place in the universe, but his success in determining the mathematical principles of natural philosophy only gave additional momentum to mathematization in science. Advances in mathematics itself also made a strong impression on scientists. Tools were now available for tackling problems that the ancients could not even formulate, much less solve. The magnificent success of mathematics seemed to feed the philosophical perspective that gave it birth, producing a spiral of scientific progress and mathematized scientific philosophy.

By the start of the eighteenth century, mathematical science had transformed natural philosophy in a revolutionary way. The enterprise was no longer what it was at the start of the modern era. The goal of mathematical science was simultaneously more modest and more ambitious than that of natural philosophy. It was more modest in that now the goal of science was restricted to describing mathematically the way natural phenomena function. This involved a twofold reduction: only mathematical features were thought relevant now, and the ultimate nature of things need not be determined, only its behavior. On the other hand, the goal of mathematical science was more ambitious, for it aimed to plumb the very depths of reality with detailed mathematical precision and logical certainty. It was not content to settle for postulating occult causes or constructing a deductive system of hypothetical knowledge, but intended to specify how reality actually works on a deep level.

The success of mathematical science in achieving what was judged to be universally true knowledge of the world made it the envy of all other areas of thought. As we will see in the next chapter, various social sciences and humanities followed suit, using the rational methodology and techniques of mathematical science. In that way they hoped to push back ignorance and rise to the level of science themselves. Uncovering basic laws of human nature and society, they would then be better able to master it and so gain control over human destiny.

Looking back on these developments with historical hindsight, we can see fruitful connections between philosophical perspectives and positive scientific work, but we can also see the limitations and aberrations that this collaboration produced on a broader scale. Science, powered by mathematics, was installed as *the* source of absolutely certain, objective knowledge, or at least, as the source of the very best knowledge that was humanly possible.

Human Reason muscled out Divine Revelation in the end, though it did not start that way. The development of modern science was closely associated with Christianity, and Christians were deeply involved in developing it,[9] but this collaboration was pursued with an insufficiently critical testing of the philosophy that came along with it. As time went on, scientific philosophy first moved to a deistic viewpoint in which God no longer had any lasting role to play in his mechanistic universe. It finally became a naturalistic viewpoint in which God was completely irrelevant, if he even existed.

The Role of Mathematics in the Modern Scientistic Worldview

What part has mathematics played in the development of this secular outlook? Mathematics has contributed centrally to the modern scientistic worldview, through both its content and its methodology. We will look at each of these briefly in turn.

The aspects of reality that are considered important for scientific work in the modern era are the primary qualities associated with quantity: number, size, shape, position, motion, and force. Other aspects of reality are ruled out as irrelevant or reducible to those of mathematics. Scientific measurements generated by experiments provide the numbers upon which the techniques of mathematics operate. Scientific laws stipulate functional dependencies holding between such magnitudes. Mathematical theories explain the lawful regularities of observed phenomena and predict behavior. This role for mathematics holds in the physical sciences; it was extrapolated into other natural, social, and human sciences as well. Today numbers and graphs are used for quantifying anything and everything. Quantifying gives knowledge, and knowledge is power. At the very least, statistics have now invaded every part of our life, presumably giving us an objective basis for rational decision-making.

The methodology of mathematics has also had a great impact upon the development of the modern scientistic worldview. We can distinguish

9. This theme has been developed by a number of works on science and religion, such as Nancy R. Pearcey and Charles B. Thaxton, *The Soul of Science: Christian Faith and Natural Philosophy* (Wheaton, Ill.: Crossway Books, 1994). Our treatment of mathematization in Western culture looks at the relation between Christian thought and science from quite a different perspective.

two main aspects here. On the one hand, the traditional axiomatic method of mathematics contributed its view of truth and consequences to science. The ultimate goal of physical science, according to the modern view, was to determine absolutely certain first principles and then to logically derive all other results in the field from them using the tools of mathematics. In this way, one will arrive at the most certain knowledge possible, given the empirical nature of the subject matter. Different people evaluated the attainability of this goal and the need for experimentation differently, but arriving at certain knowledge remained the goal of science. Over time, it was thought, scientific progress would generate closer and closer approximations to the truth.

On the other hand, the analytic tradition in mathematics revived a more general notion of analysis that became the paradigm for how scientific discovery should proceed in all areas. The period from Descartes through the Enlightenment can aptly be called the Age of Analysis. Mathematics also provided a well-developed science of analysis, which in the end included algebra, analytic geometry, calculus, and differential equations. These areas provided the ideas and techniques that helped physics analyze basic concepts, solve problems, and derive results from given principles or conditions using algorithmic procedures. Mathematical analysis provided the motive power for the scientific revolution.

A Christian Response to Western Mathematization of Science

What might a Christian perspective on mathematics have to say about all of this? First, a Christian perspective on mathematics can acknowledge with appreciation the positive contributions that mathematics has made to the development of natural science. The process of mathematization has uncovered intimate connections between mathematics and science that reveal the marvelous coherence of creation — something for which we can glorify God. Furthermore, pursuing such knowledge is consonant with a Christian vision of human beings as God's stewards of his creation. That is, an enriched mathematical understanding of how the natural world behaves can help us serve God's purposes in various areas of life. Mathematical knowledge helps us fulfill the cultural mandate given to humanity by God in the garden of Eden.

As Christians, we can also give assent to the reality-oriented stance of the modern scientific outlook: the subject matter of mathematics is certainly relevant to our experience of created reality, and vice versa. Mathematics is not a purely human mental construction, even though mathematics obviously involves the rational operations of our own minds — abstraction, generalization, comparison, deduction, etc. A Christian outlook on the nature of mathematical objects agrees with Platonism to this extent: mathematicians discover lawful regularities in conceptual entities whose existence and properties are largely independent of human intellectual activity.

On the other hand, while affirming mathematics as a good gift of God for understanding quantitative aspects of our world, a Christian perspective on mathematics will take exception to those aspects of the modern worldview that arise from the absolutization of mathematics. Mathematization frequently rejects non-mathematical aspects of life as unimportant or nonexistent and so promotes a lopsided vision of reality. Such a reductionistic program denies the validity of the rich variety of aspects within creation that go beyond quantitative properties. Radical mathematization only works when people are willing to narrow down their perspective of what is real, when they accommodate reality to mathematics as well as conversely.[10] As Christians, we should instead hold a more modest view of the nature of mathematics and its accomplishments, and we should welcome other dimensions of reality as complementary to those studied by mathematics. Thus the perspectives of the craftsman fashioning a beautiful object, the dramatist using words to create powerful portrayals of emotions, and the historian who interprets the meaning of past developments, ought to be respected equally with that of mathematics. A Christian perspective will deny the modern claim that mathematics has a corner on the truth about the world, that it is the final arbiter of all meaning. Mathematics cannot penetrate to the very essence of the universe; God is more/other than a supreme mathematician. The world has non-mathematical as well as mathematical structure.

Rejecting mathematical imperialism, a Christian perspective on mathematics allows us to consider alternative visions of how God, humans, and the world are interrelated with respect to mathematics. In the Neoplatonic

10. This point is argued by a number of case studies in Theodore M. Porter, *Trust in Numbers: The Pursuit of Objectivity in Science and Public Life* (Princeton: Princeton University Press, 1995).

outlook, adopted by various seventeenth-century thinkers, humans thought God's mathematical thoughts after him in order to understand the structure of the world, and these thoughts were seen as eternal and essential to God's nature. Such a view tends to deify mathematics, as we have seen. If we do not equate mathematics with necessary knowledge that is true of all possible worlds, we can entertain other ideas about how mathematical knowledge arises in human experience. We may not accept the postmodern alternative to modernism on this point, but we are certainly freed to explore other options than Neoplatonic rationalism.

Finally, a Christian perspective will view human beings more as integral parts of creation than what is envisioned by dualistic, rationalistic philosophy. Humans are not rational agents set over against the rest of reality; we are not thinking beings situated in a material world that possesses only primary quantitative features. Nor are we masters of our destiny or in control of science and culture solely because of our mathematical scientific abilities. We are creatures of the Lord, meant to exercise our analytical and quantitative abilities in the service of other people and the rest of creation, not to further our own ends or challenge God's sovereignty. As Christians, we must take responsibility for what mathematics we develop and how it is applied in the world around us. Our overall motivation should be service, not mastery and control. In the chapters that follow, we will develop these ideas in further detail and present a positive alternative to both the modern and the postmodern outlooks on mathematics.

CHAPTER 7

The Mathematization of Culture

Introduction

In the previous two chapters we traced the mathematization of science from ancient Greek times into the modern era. Our main focus was the mathematization of natural philosophy, though we also noted the importance of mathematization in other areas of life for developments in science. The striking accomplishments of mathematical physics during the seventeenth century provided a strong incentive and a clear model for extending its methods and outlook to other parts of human culture. In this chapter we will complete our discussion of Western mathematization by exploring the role of mathematics outside science. This is necessary, for while mathematization made its greatest inroads and achieved its greatest success in the area of natural science, its impact upon our modern worldview cannot be fully ascertained without noting how it penetrated other areas of human endeavor as well.

Our approach at first, then, will be historical, tracing the ways in which mathematization took place in the social sciences and humanities during the seventeenth and eighteenth centuries. We will also look at the beginnings of a modern reaction to mathematization and scientism by Romantic thinkers. Much more can be done with these topics than we have room for here,[1] but

1. Some books that discuss this period in a general way are Ernst Cassirer, *The Philosophy of the Enlightenment* (Princeton: Princeton University Press, 1951); Charles Coulston Gillispie, *The Edge of Objectivity: An Essay in the History of Scientific Ideas* (Princeton: Princeton University Press, 1960); Thomas Hankins, *Science and the Enlightenment* (Cambridge: Cambridge University Press, 1985); and Richard Olson, *Science Deified and Science*

our survey will be sufficient to show that mathematization has contributed in an essential way to the rise and character of the modern scientistic worldview.

Having established the nature and depth of mathematization in modern Western thought and culture, we will then look at more contemporary forms of mathematization. We will exchange a historical presentation for a more topical approach. We will demonstrate how pervasive mathematization is in today's world by looking at ways in which mathematization has contributed to our culture even when little or only elementary mathematical content is involved. We will see that mathematization has yielded both benefits and problems. In order to help us understand these mixed consequences, we will carefully reflect on some rather profound and difficult issues associated with mathematization. The chapter concludes with an evaluation of mathematization and its consequences from a Christian perspective.

Western Mathematization Outside the Natural Sciences

Mathematization of the Social Sciences

The desire to reproduce the success of mathematical physics in the realm of the human and social sciences goes back as far as 1630 when the political philosopher Thomas Hobbes first came across Euclid's *Elements.* He was particularly impressed by how one could begin with the simplest propositions and end, after a lengthy train of conclusive reasoning, with results that were far from obvious. Galileo's use of this same methodology in developing his new science of motion provided Hobbes both with a model of how to proceed and with ideas he could use to attempt a fully mechanical explanation of sensation. His views on the nature of man in society were published in the 1650s. Hobbes' approach was too radical for nearly all of his contemporaries, but his views had a lasting impact on later treatments in these and related fields.

Hobbes claimed that it is in the nature of human beings to act in ways

Defied: The Historical Significance of Science in Western Culture, vol. 2: *From the Early Modern Age Through the Early Romantic Era, ca. 1640 to ca. 1820* (Berkeley: University of California Press, 1995). Morris Kline, *Mathematics in Western Culture* (Oxford: Oxford University Press, 1964), addresses this topic in a more mathematically focused way than the others, but his assessments must be read critically, for Kline occasionally sacrifices historical accuracy for the sake of telling a good story.

that seek to maximize pleasure and minimize pain. This quasi-quantitative axiom underlies his science of humans. Seeking their own self-interest in society, individuals interact with one another much like colliding particles. A competitive mechanistic outlook on society thus came to replace the earlier organic view of society as an ordered community and became axiomatic for later social theorists and political economists.

Political economy, also known as political arithmetic, was the first self-consciously mathematized social science, both in approach and content. In England about 1660, William Petty and his friend John Graunt began to analyze quantitatively a wide range of economic and demographic data as background for forming rational public policy. They believed that statistical regularities could be observed in aggregate data and that general laws relating various socially significant quantities might be discovered and developed into a science. The importance of this sort of analysis was soon recognized by those involved in selling life insurance and by governments desiring vital statistics on their subjects. A bright future was opening up for those individuals and agencies that collected and analyzed economic data, for they would be able to use this information to predict and shape economic phenomena.

Petty and Graunt did their work before physical science began to receive its definitive formulation at the hands of Newton. Nevertheless, they drew from earlier strong tendencies toward mathematization, both in commerce and natural science, and in return made their own contribution to this outlook.

Mathematical economics continued to develop in the work of various seventeenth- and eighteenth-century thinkers. Particularly in France and England there was a move to develop a mathematical theory of society and political economy. Social theorists asserted that while rational economic agents act in their own self-interest, the marketplace has its own natural mechanism of supply and demand that invariably converts the collective pursuit of individual gain into maximum public wealth. Advocacy of a live-and-let-live free-market economy received its most definitive formulation in Adam Smith's *Wealth of Nations* (1776), the Bible of all Western industrial economies. There are natural laws at work in society — pursuit of individual self-interest and a self-regulating market — that make economics as much a deterministic science as Newtonian mechanics. Using these basic principles, social theorists explained a wide variety of economic phenomena, thus providing a measure of justification of these principles. In the nineteenth and twentieth centuries the connections between mathematics and economics became even tighter, especially once the ideas and procedures of calculus were imported into economics.

Naturally, economics was not just a theoretical science. Its outlook motivated both public policy and individual enterprise. Tabulating various sorts of statistics gained momentum in the eighteenth century under the encouragement of governments such as France, who wished to assess and better manage their resources. As a result, statistical methods of analysis were further developed and statistics was applied to an ever-wider circle of phenomena.[2]

Closely related to the use of statistics was the development of a mathematical theory of probability.[3] Mathematicians made use of probability to make statistical inferences in natural science, but they also discussed its application to areas of social interaction in which strict causality was difficult to uncover or collective human judgment played a decisive role.

Around the end of the eighteenth century, the program of quantification was advanced in all fields when weights and measures were intentionally standardized in France. Scientists and government officials associated with the French Revolution created a decimal metric system in order to gain better access to and control of data and to compare and communicate numerical results more easily. Uniform measures and quantification brought order, objectivity, and a degree of certainty into many areas of human activity.

The functions of organization and rational control made quantitative analysis attractive for individuals and businesses as well as governments and social analysts. Toward the end of the eighteenth century, the entrepreneur Josiah Wedgwood, Charles Darwin's maternal grandfather, undertook a scientific analysis of the entire process of manufacturing pottery. He carefully experimented with different temperatures and materials to determine just how to create the glazes for which his family business became famous. In the process he invented a pyrometer for measuring the extreme temperatures produced in a kiln. He used the same experimental approach to analyze the production process itself, instituting a number of scientific management practices, such as division of labor, worker specialization, and synchronized assembly line production. Workers were literally viewed as part of the overall

2. Various articles in Tore Frängsmyr, John L. Heilbron, and Robin E. Rider, eds., *The Quantifying Spirit in the Eighteenth Century* (Berkeley: University of California Press, 1990), elaborate these points.

3. See Lorraine J. Daston, *Classical Probability in the Enlightenment* (Princeton: Princeton University Press, 1988), for the history of probability in the seventeenth and eighteenth centuries.

machinery of the process, subject to quantitative measurement and manipulation. This controlled approach to labor became popular with other employers, and it soon became standard operating procedure during the spread of the Industrial Revolution in the early nineteenth century. Quantification and scientific analysis made possible various advances in industrial practices, increasing profits through decreasing production costs, but they simultaneously enabled employers to inflict inhumane conditions on the workforce.

Here, to many, was the dark side of an overly robust mathematization, of an overblown mechanistic approach to life: humans were reduced to cogs in the wheels of industry. Needless to say, this prompted a strong reaction from some. Before we describe this response, however, we will once again go back in time to look briefly at how mathematization affected art, ethics, and religion.

Mathematization of the Humanities

By the end of the seventeenth century, mathematics was firmly entrenched in physical science, and it was starting to show its potential for social science. But what about its relation to those areas that seem more human, more central to the creative human spirit? What about artistic expression, morality, and religious belief? Even there the effects of mathematization were increasingly being felt. We have already seen the importance of earlier mathematics for perspective in art and for harmony in music theory; in the seventeenth and eighteenth centuries, mathematics established and intensified contacts outside these more traditional areas.

A number of thinkers were quite exuberant about the attitudes and goals of mathematical science and sought to extend them into all areas of human culture. Fontenelle, a French scientist whose hundred-year life spanned the last half of the seventeenth century and the first half of the eighteenth, expressed this outlook in his famous essay, *On the Usefulness of Mathematics and Physics* (1699). Other things being equal, he asserted, a work in any field whatsoever will be better the more it exhibits the habits and approach of mathematical physics, the more it adopts the orderliness and precision found in mathematics.

This perspective had a fairly wide following during the Enlightenment with detrimental consequences for some of the humanities. Literature, for example, was forced into a more rational mold. Poetry with its imaginative excesses was scorned and fell out of favor. Those who continued to write poetry did so in a very controlled manner and with the most serious intent. In prose writing, the plain factual style of the scientist was held up for emulation by others. Natural language was perceived as being too fluid and ambiguous for analytical purposes. Eighteenth-century thinkers sought to regulate it better by constructing dictionaries to fix the form and meaning of words and by writing grammars to prescribe the proper linguistic structure for rational discourse. Leading language theorists found mathematics, and particularly algebra, with its univocal use of symbols and its strict rules for manipulating expressions, the ideal model for a universal language.

Mathematics has close ties to symbolism and language, so a connection between them is not totally unexpected. But what about morality and religion? The more progressive Enlightenment thinkers were convinced that all areas of life and thought should be pursued with the method of mathematical science. This included morality and religion. For instance, the utilitarian philosopher Jeremy Bentham developed a system of ethics and law (1789) that built upon Hobbes' theory of human nature. In order to determine what course of action among several was most correct, he said, one should assign quantitative values to the pain and pleasure of each action for all persons involved and then calculate the aggregate pain and pleasure, subtracting the one from the other to determine an overall value for the action. Comparing the alternatives, one must choose the action that gives the greatest good for the greatest number; that is, one should maximize net pleasure, quantitatively measured.

The Enlightenment Legacy: Mathematization and the Modern Scientific Worldview

It seems quite obvious and natural to those of us nurtured on Western culture that mathematics provides the ideas and tools that make real progress possible in science and technology, and through them, in the quality of our lives. We live in a quantified world whose mathematical and scientific infrastructure extends into all its nooks and crannies.

However, we need to keep in mind that our high expectation of mathematics and science is a direct legacy of the Enlightenment. This should be apparent from our survey of mathematization in the modern era: mathematics continued to make strong contributions to physics and other natural sciences; it provided the tools to jump-start the social sciences; and its assistance was eagerly sought for developing scientific theories of human nature. As time went on, some applications of mathematics passed out of vogue while others arose to take their place. But what remained constant throughout was the Enlightenment belief that our world, with its natural, social, and personal realms, is the sort of reality that is amenable to rational scientific and mathematical treatment. As one historian has observed,

> . . . the essence of the Enlightenment was the belief that the world could now be seen in mechanical terms and defined in mathematical language . . . all reality . . . acts by natural laws, if we had eyes to see them. The world of human affairs, in sum, is the same as the natural world because the same laws that govern each govern all. The task for Enlightenment thinkers, therefore, was to ascertain those general laws that governed reality and then apply them to the various cases that came up, whether political, economic, social, or religious.[4]

The typical perspective of the Enlightenment was thus highly rationalistic. Leading thinkers affirmed the idea that human reason and empirical observation were the only instruments available for solving problems connected with human beings and their societies. Furthermore, they saw these instruments as completely adequate for addressing these problems. This was an era of great confidence in human capacities. According to the perspective of the time, reason should be unhampered by belief in revelation, submission to authority, or deference to established customs and attitudes.[5]

The "Enlightenment project" of using mathematical and scientific investigation to discover knowledge and establish non-controvertible bases on which all of culture can be built has held sway over the majority of European and North American scholars from that time until fairly recently. The appropriateness, even the desirability, of engaging mathematized science to address all the problems of our world has become an undoubted article of faith among us and remains particularly strong in the mathematics and scientific communities.

4. Ronald A. Wells, *History Through the Eyes of Faith: Western Civilization and the Kingdom of God* (San Francisco: Harper and Row, 1989).

5. See Frederick Copleston, S.J., *A History of Philosophy* (London: Burns, Oates and Washbourne, Ltd., 1958), vol. 4, ch. 1.

Romantic Backlash Against Mathematized Science and Culture

By the second half of the eighteenth century, however, there were a number of prominent thinkers who were beginning to dissent from this new orthodoxy. Their reaction against a mechanistic worldview and its underlying mathematical science ideal is known as Romanticism.[6] Such people saw the universal application of the content and methods of mathematical science as a real threat to human freedom and personality. In opposition to mathematical physics, the more scientific among them allied themselves with the sciences of natural history, such as botany, zoology, and geology, where the mechanistic perspective did not predominate. Romantic thinkers espoused other, more participatory, ways of getting in touch with Nature.

Much of what is important in the world, the Romantics claimed, cannot be captured by clear and distinct ideas or constrained by measure, number, and mechanical contrivances. This holds in art and literature; it is true in ethics and religion; and it is the case in politics and economics. Some critics, such as Edmund Burke and William Blake, became particularly agitated by the modern approach when they saw its ramifications being worked out in the French Revolution and the Industrial Revolution, both of which entailed major social upheaval. They believed that a mathematized approach yields an arrogant and impoverished reduction of life to its quantitative and mechanical aspects, ignoring important social and cultural dimensions of reality.

This Romantic backlash had its supporters, but it failed to gain cultural power. The Enlightenment belief in the power of mathematical science to create progress was too strongly entrenched to be denied. Romanticist ideas were rooted out of natural science, including biology, and they fared no better in the social sciences. They eventually found a more congenial residence in literature, the arts, and history, where individuality, creativity, and a sympathetic oneness with Nature were celebrated and could not so easily be converted into fixed rules and equations. In recent years, however, they have made a transformed reappearance in various postmodern reactions to science and the modern scientistic worldview. A stress on the importance of culture and human subjectivity is central to the postmodern critique of sci-

6. We are encompassing several diverse trends under this term. A more detailed analysis would have to make finer distinctions than we have room to develop here.

entific theorizing, as we saw in Chapter 1. This would seem to be largely irrelevant to fields that study objective necessary truths, but its contemporary critics no longer grant this status to science and mathematics. Finding aid and comfort in various twentieth-century developments in the foundations of mathematics and physics, postmodern thinkers now boldly challenge the appearance of objectivity, universality, and certainty long thought to reside in natural science and mathematics. Their critique is not quite ours, but before we turn to evaluate the mathematization of culture from our perspective, we will first discuss some ways in which mathematics is used in contemporary culture.

Mathematization in Today's World

Levels of Mathematization

There are two different levels on which mathematization takes place in our modern world. In the first place, mathematization obviously occurs every time some field of knowledge or practice is explicitly rendered more mathematical. This involves quantification, measurement, application of mathematical theories, structures, concepts, and techniques, and the use of graphical and symbolic representations, such as graphs, charts, and equations, to express knowledge. Mathematization in this sense includes simple applications of mathematical ideas as well as more complex cases of mathematical modeling. Our historical survey of mathematization has already indicated a few of these developments. In those applications, the products of science and technology largely mediated the impact of mathematics on culture. This is the popular conception of the role mathematics plays in contemporary life.

This image isn't wrong, but it is incomplete. Calculus and other branches of modern mathematics have certainly had an enormous and quite visible impact on Western culture via science and technology. However, there are other influences of mathematization on our culture that are even deeper and more pervasive. This broader impact is not primarily the result of advanced mathematics; it frequently uses elementary mathematical content outside of traditional applications. Our thesis is this: While it may be less evident, the ideals and values implicit in the practice of mathematics and mathematized science have had a substantial impact on the human and social sciences, on government, on management, and on popular culture. We have already pointed this out somewhat in the preceding section; now we want to elaborate this thesis in a more systematic way.

What are some of the traditional ideals and values of mathematics? Here is a list of the main ones; they express many features of the modernist spirit:[7]

- impersonality (objectivity, in the sense of intersubjectivity — universal truths that are independent of any particular individual's or group's perspective or experience),
- abstraction from particulars (generalization),
- the precise use of language in defining and relating basic concepts,
- open knowledge (explicit and unambiguous presentation of all of one's assumptions and analysis for anyone to critique),
- the standardization and formalization of ideas and procedures,
- the orderly organization of knowledge into definitions and fundamental principles (axioms) from which other knowledge (theorems) is derived,
- rigorous thinking,
- deductive reasoning (proof) as a standard for making claims of certainty and truthfulness.

It should be clear from this list that the values and ideals of mathematics have played an important role in mathematizing empirical science. For instance, quantification and measurement are motivated in part by values of impersonality, abstraction, and precision. However, these ideals also function outside conventional uses of mathematics. We will support this thesis by looking at several examples from contemporary life.

Everyday Examples of Mathematization

Our first example of the influence of mathematization is time. Quantification of time is ancient — even the New Testament account of Good Friday speaks of the sixth and ninth hours. However, the advent of rigid calendars

7. One very readable discussion of some of these values and ideals is Alfred Renyi, *Dialogues on Mathematics* (San Francisco: Holden-Day, 1967), pp. 3-25.

and the invention of the clock in the Middle Ages made precise quantification of time an integral part of everyone's life. The popular conception of time thus moved away from an association with natural phenomena such as sunrises and the change of seasons to a physically and mathematically based social convention. Anyone who has spent a few weeks in a culture (for instance, in Africa or South America) where precise quantification of time is not so highly valued can testify how significant an impact this precision has had on Western culture in general; it's not just a matter affecting mathematical physics. Note the value judgments implicit in this cultural development: precise is better than imprecise, quantitative is better than qualitative, an impersonal, abstract measure of time is better than one linked to everyday human experience of natural phenomena.

Another example is space. We are not referring here to "outer space," but rather to the locations in which we live. For instance, in Grand Rapids, Michigan, most of the urban and rural streets are laid out in a rectangular grid and run either north-south or east-west. The north-south road at the center of the city is called "Division Avenue." Many east-west county roads have names like "Three Mile Road" and "Four Mile Road" and they are literally that distance from the east-west line running through the center of the city. This is quite different from cities like Philadelphia or Boston that developed before 1800. Cities like Grand Rapids are often criticized for lacking the charm of cities laid out less simply, but finding a location in Grand Rapids is typically much easier than in Boston or Philadelphia. Nevertheless, even in places that are not laid out using a rectangular grid, property boundaries are precisely defined and maps of them are included with deeds that reside in a government office. Sale of property normally requires a fresh survey to clarify boundaries. In fact, the entire globe has been laid out using coordinate geometry and (at least within North America) every property boundary is located within that grid. While this does not eliminate disputes over property boundaries, it does provide a means that can be used to resolve them peacefully.

Standardized units have been developed for length, weight, volume, and many other measures. Today we take this largely for granted. However, as we noted above, two hundred years ago uniform standards did not exist. For instance, there were separate units for measuring linen and silk. In France, each local geographical area had its own bushel. Standards varied (and disputes arose) as to how to handle the heap at the top of the bushel. Most mathematicians in Renaissance Europe were employed changing units for trade outside local geographical areas. Decimal units for currency provide another example of the impact of mathematics, as does the rise (in the nineteenth century) of the accounting profession. The mathematics here is not difficult, but decimal units for currency, accounting, and standardized measures incorporate mathematical values of impersonality, abstraction, and precision.

Economic expansion was responsible for most of these developments in the history of measurement. In order to do business with strangers, whose character was unknown, a basis for trust was needed. Standardized measures, quantification, and means to enforce honest measures provided such a basis. A medium for communication across languages and cultures was also needed; standardized measures and quantification provided this as well. Thus the distance that quantification bridged was both geographic and interpersonal.

The capacity of mathematization to enhance communication applies to science and industry as well as to business. The language of precise measurements, algebraic notation, and equations provided a means for scientists across Europe to express their work in a common format that could be widely understood. In the nineteenth and twentieth centuries, industrialization depended on standardization. Mathematization of this sort continues to play a major role today — for instance, the standard format in which corporations present their annual financial reports makes it easy for investors to assess a company's financial health. A global movement toward the metric system has facilitated international trade. In short, in a large society where personal trust is hard to develop or where uniformity is needed for comparison purposes or cooperative ventures, mathematization provides a common impersonal basis for public decision-making.[8]

Another area in which the ideals of mathematization have been felt is that of politics and government. Note that the ideals of mathematics — openness, independence of knowledge from social class or positions of authority, impartial impersonality — comport very well with those of democracy and the rule of law. Numbers have substantial rhetorical power. They appear independent of the passions so prominent in political debate, and they represent an ideal of diligence and careful analysis. Thus they are very attractive to decision-makers in a democracy; they provide a means for leaders to protect themselves from charges of bias, self-interest, or incompe-

8. See Theodore Porter, *Trust in Numbers: The Pursuit of Objectivity in Science and Public Life* (Princeton: Princeton University Press, 1995), for a detailed discussion of this matter.

tence. Moreover, the impact of numbers has extended beyond rhetoric; they have advanced the very application of democratic principles. For instance, advances in accounting have helped root out corruption. Careful reporting of employment and housing statistics have called attention to systematic differences in treatment of women and minorities and have led to major changes. Thus, mathematization has contributed to a shift in the social basis for authority — away from powerful elites and toward abstract concepts of equity. Thus, mathematization and democracy have mutually advanced each other.

Mathematization is also important for the growth of government bureaucracies. Bureaucracies typically gather data and enforce rules in the course of providing a public service. The one charge that is anathema to a bureaucracy is inconsistency. "True," "best," and "efficient" are typically far less important than "consistent." Bureaucracies have a great need to provide uniform treatment. Hence, they tend to develop highly standardized categories for data collection and highly standardized procedures. Note how the ideals of mathematics — precise, open, abstract, impersonal, standardized knowledge — are embraced by bureaucracies. Such knowledge is very useful for administration; for example, the precise geometric layout of land parcels in North America makes it possible for bureaucrats to manage the transfer of land ownership without ever leaving an office that may be hundreds of miles from the land being bought and sold. Furthermore, the fruit of the bureaucratic data-gathering process is precisely the kind of information that the political decision-makers spoken of earlier need. Of course, administration by the numbers can lead to anomalies as well — for instance, a congressional mandate permits the United States Forest Service to cut no more lumber than is renewed by annual growth. Since that law was put into effect, growth rates have been greatly enhanced, at least in the Forest Service accounts, by using new herbicides and pesticides and by planting different varieties of faster growing trees to replace the ones cut down.[9]

Governments and organizations now depend on enormous amounts of data for making informed decisions. Collection and analysis of such data has become a major concern of our modern Western world. The Inter-University Consortium for Political and Social Research is one group that provides an on-line archive for social science data. This is how it describes its holdings.

9. Porter, *Trust in Numbers*, p. 44.

> Beginning with a few major surveys of the American electorate, the holdings of the archive have now broadened to include comparable information from diverse settings and for extended time periods. Data ranging from nineteenth-century French census materials to recent sessions of the United Nations, from American elections in the 1790s to the socioeconomic structure of Polish poviats, from the characteristics of Knights of Labor assemblies to the expectations of American consumers are included in the archive. Surveys, aggregate data, and computer-based teaching packages in various substantive areas are continually deposited in the archive by leading scholars around the world. The content of the archive extends across economic, sociological, historical, organizational, social, psychological, and political concerns. Topical expansion is taking place to include urban studies, education, electoral behavior, socialization, foreign policy, community studies, judicial behavior, legislators (national, state, and local), race relations, and organizational behavior.[10]

Note that all of these areas are described by *data* and the collection of data minimally requires the precise definition of categories. It may also require quantification and a fair amount of abstraction.

Another area where quantification has taken deep root is that of mental measurements. IQ testing is the most familiar example, but the 11th edition of *Buros' Mental Measurement Handbook*[11] lists 477 commercially available tests designed to measure an extraordinary range of human qualities. The tests are classified into the following categories: personality, vocations, developmental, English, education, achievement, intelligence and scholastic aptitude, reading, speech and hearing, mathematics, neurophysiology, behavior assessment, sensory-motor skills, science, social studies, fine arts, and multi-aptitude. Yet another area of human statistics is personnel data maintained by employers. Again, these are not sophisticated mathematically, but such data are typically stored in computerized databases, so they have to be mathematized in the sense that they have to be recorded in precise, abstract categories. Beyond personnel information, businesses typically collect data

10. ICPSR, *Guide to resources and services,* 1993-1994, p. vii. P.O. Box 1248, Ann Arbor, MI.

11. J. J. Kramer and J. C. Conoley, eds., *Buros' Mental Measurement Handbook* (Lincoln: University of Nebraska Press, 1992).

on market research, the economic climate, competitors' sales, and various aspects of productivity. In fact, the bulk of information stored in computerized databases is highly impersonal since its content and format are expected to be independent of the perspective of the data gatherer.

We could continue. Political scientists and polling agencies have become very sophisticated in their ability to predict election outcomes based on data. Much of the management of our economy is done via econometric models. Major business decisions are made with the aid of operations research. In fact, major engineering decisions are frequently made with the assistance of formal decision models. On a more popular level, graphs, charts, and equations have become standard communication tools across cultures and languages. For example, in a recent edition of *USA Today,*[12] seventeen articles in the first section alone depended in some essential way on data, tables, or graphs.

Furthermore, Jacques Ellul, an influential critic of Western culture, has identified a significant way in which mathematization has altered the "inner life" of our culture, that is, has shaped the patterns in which people think.[13] He calls this pattern "technique," referring to a process of reducing human activities to routines that can be optimized for efficiency and productivity. (Ellul distinguishes "technique," as a way of thinking from "technology," the production of artifacts.) It is easy to cite examples to show how widespread the application of technique has become in Western culture. Assembly lines in manufacturing, systematic methods of accounting in business, standardized procedures for computing taxes, the use of bar codes to enhance grocery store check-out and keep track of inventory, the use of systems analysis to model the flow of information through a corporation, the use of Roberts' Rules of Order to govern a deliberative body, and the standardized procedures that large bureaucracies use to manage their activities are all examples. In fact, any computerized process is an application of technique. One of the clearest illustrations Ellul gives is how home economists at Cornell University in the early part of the twentieth century altered the design of American home kitchens. Their idea was to design kitchens that would optimize the process of food preparation, for instance, by minimizing the walking dis-

12. February 24, 1995.

13. Ellul's most well-known book is *The Technological Society* (New York: Knopf, 1964). Two more recent works are *The Technological System* (New York: Continuum, 1980) and *The Technological Bluff* (Grand Rapids: Eerdmans, 1990).

tance of a person preparing food. Thus, one of the principal recommendations of these home economists was to organize the kitchen around three distinct centers focused on the activities of preparation, cooking, and cleaning. The centers were placed to allow efficient flow of food and utensils from one to another. Nineteenth-century kitchens typically had a table in the center of the kitchen. Hence, another recommendation was to remove the table. Of course, technique is not explicitly mathematics. However, it does instantiate several of the values and ideals of mathematics — open knowledge, precision, abstraction, and impersonality — and thus is part of the process of mathematization.

In summary, then, it should be clear that mathematization has become deeply entrenched in Western culture over the past 500 years or so. It is a significant part of various dimensions of our daily lives as well as of technologically sophisticated devices. Our long list of examples in this section has demonstrated that at least some of the mathematical values cited earlier have become major dimensions of contemporary culture. In addition to the widespread application of mathematical content in science, technology, and everyday life, the method and values of mathematization have contributed to the way in which we think about issues and seek for solutions to problems. Not surprisingly, some deep and difficult issues accompany such an extensive cultural impact. In the next section, we will discuss four of them. We will then conclude our discussion of mathematization by sketching a Christian alternative to the modern outlook on this matter.

An Evaluation of Mathematization

Four Issues Accompanying Mathematization

Our first issue is that using mathematics to address a real-world problem shapes our perception of that problem. In other words, the use of mathematics in studying anything is neither epistemologically nor metaphysically neutral. Thus mathematization does not possess the neutrality and total objectivity for which Enlightenment thinkers had hoped.

Consider the distinction that some social scientists (especially economists) frequently make, between "positive" and "normative" approaches to their discipline. The positive approach in economics means data collection and mathematical modeling (frequently based on regression analysis) and analysis based on these techniques. That is, the positive approach seeks to merely *describe* "what is," using the tools of mathematization. The normative approach, on the other hand, seeks to *prescribe* "what ought to be." For instance, an economist in the positive tradition might approach the study of inflation by gathering large quantities of data and seeking to understand what factors influence inflation rates. Nevertheless, she would leave for the political process the normative question of what level of inflation in society is tolerable. Most social science research uses the positive approach.

Now imagine a social scientist who decides to use the positive approach to explore a research question. Typically, the researcher takes this approach to avail herself of the credibility it possesses in contemporary culture. In the process of carrying out her research, she exhibits respect for the values of open knowledge, careful definition, precise measurement, abstraction from particulars, and rigor. She follows a formalized process for selecting a random sample and carrying out the analysis of her data. Her personal opinions and hunches must be, if not set aside, severely disciplined. If she writes or speaks by presenting an interpretation from her own perspective, she is expected to acknowledge that she has stepped out of the positive paradigm. Such an approach has advantages and disadvantages. The advantages are that it is widely understood; it lends itself well to attempts to replicate results; and it provides a measure of protection against bias. Some disadvantages are that for scholars (and policy makers!), aspects of a situation that are hard to quantify are easy to neglect. Also, this process is unable to deal well with the uniqueness of individuals and communities and with other non-replicable matters. The fact that most results about human beings involve elements of interpretation may be obscured. In addition, the element of intentionality, that is, individual free choice, is usually ignored. Thus, the positive approach aims to restrict the kind of knowledge accessible to its users to observable regularities that exclude the observer's unique perspective. For this very reason, however, it is not epistemologically neutral. The researcher has made a choice to focus on aspects that can be quantified and replicated and to ignore those that cannot.

Measuring something numerically assumes that a meaningful correspondence can be made between the entity being measured and the set of real or natural numbers. However, the decision to make such measurements and the development or selection of a means by which to do it must occur before data are gathered and before experiments or observations are performed. The positive approach also involves implicit metaphysical assumptions, such as the uniformity of nature, the quantitative character of the phenomena being investigated, etc. Taking all of this into account, we see that the positive approach is not metaphysically neutral. Since the positive approach to research is part of the broader project of mathematization, all of these conclusions about positivism apply to mathematization in general.

Note also that use of the positive approach is often accompanied by the value judgment that only knowledge accessible by this method is legitimate. When such a judgment is made, the postmodern criticism that claims of objective knowledge are often accompanied by exercises of power is clearly applicable.

Our second issue is that mathematization often implicitly introduces norms into situations. The word "norms" typically refers to principles or standards that guide human behavior. In democratic societies, norms are among the most powerful forms of social control that societies exert over individuals.[14] Quantification of human qualities inevitably introduces norms. The intelligence quotient or "IQ" is a good example; however, our comments will apply to all of the mental measurements discussed earlier as well as to other measures of the characteristics of individuals and groups. Alfred Binet, a psychologist, first introduced the notion of IQ in France around 1904. The French government commissioned Binet to develop a standardized test to identify children who would not benefit from ordinary classrooms and who needed special education. Thus, IQ was intended to provide an important social service. Binet created a list of questions that tested everyday intellectual skills. He varied the level of difficulty of the questions so that children of different ages could answer different numbers of questions. He then administered the test to large numbers of children and found the average number that could be answered correctly at various ages. This list provided a means to translate a child's raw score into his or her "intellectual age." Dividing this number by the child's chronological age gave an "intelligence quotient." In short, then, the IQ test provides a means to map the intelligence of individuals (operationally defined by the test questions) into the positive real numbers. This mapping automatically gives us a norm for intelligence: if a is greater than b, then a is better than b. Measuring IQ by a single number implicitly assumes that intelligence is best represented by a linearly ordered set of numbers rather than by a more complex structure such as a partial order. In contrast, Howard Gardner, a leading psychologist, asserts that intelligence

14. See Porter, *Trust in Numbers.*

is multi-dimensional. However, even his approach still yields linear norms for each dimension.

It is not hard to come up with many other examples of how mathematization introduces norms. For example, North American cities are regularly rated for their quality of life. Faculty grade students. Students evaluate faculty. *US News and World Report* ranks colleges and universities. Each of these procedures reduces complex phenomena to quantified aspects of the full reality and sets up quantitative norms for evaluating them.

Such norms are often helpful — for instance, a low grade may motivate a student to study more effectively. A low quality of life indicator may stimulate a city to improve the environment for its citizens. However, those making such assessments are assuming a great deal of responsibility. It is clear that measurements prioritize aspects of a situation and alter people. It is not clear, however, what meta-normative principles ought to be applied to assess these mathematically generated norms. Nor is it clear whether or not the mathematics community ought to assume some form of responsibility for these norms — for instance, by providing a critical analysis of them.

A third issue is that mathematization has limits that are often forgotten. Suppose, as we have argued above, that mathematization is indeed the principal method that large democratic societies use to establish a basis of trust enabling transactions to occur among strangers. Mathematics' limits affect the extent to which it can serve in such a critical role. Here are several examples.

- Models typically idealize and simplify complex situations. In many cases, this is helpful as it enables the modeler to focus on aspects of the situation that are of central importance and ignore those that are irrelevant. However, most social situations are very complex, so it is often impossible to include all features one knows (or suspects) to be important. Thus, models necessarily misrepresent such situations through oversimplification.
- Social entities are often difficult to define precisely enough to quantify them, so proxies represent them. For example, gross domestic product, the market value of final sale of products, is a proxy for economic well-being. Most natural science concepts are regarded as having at least a modicum of realism. For instance, we measure the mass of an object by measuring its weight, then dividing by the acceleration of gravity. So mass is not measured directly but indirectly. Even so, we still regard

it as having real meaning. However, many social science models do not claim this sort of realism. They are typically regarded as instrumental means to make predictions but built out of proxies that do not necessarily correspond to the underlying reality one is trying to study.[15] Therefore, they are very different from natural science models and do not render the same kind of understanding Enlightenment thinkers anticipated based on their experience with the natural sciences. However, because they seem so similar to natural science models, it is easy to regard them as having the same type of realism.

- Mathematization usually (but not always) requires the collection of data. However, in recent years, scholars have increasingly recognized that sense data are an inconsistent and not fully reliable guide to knowledge. One's prior understandings affect even perception itself. This situation is commonly described by saying that perception is "theory-laden." For example, if we look at the object on which we have been sitting, we perceive a chair. Note we do not say that we perceived certain colors, shapes, or materials. We report our perception as being a chair. However, imagine a person from a culture that did not use chairs and had never seen one. In spite of the fact that the object being observed has the same colors, shapes, and materials for him as for us, his *perception* would be very different. The reason is that the concept of chair is culturally formed. So even our *perceptions* of social entities are culturally shaped. (Some commentators have captured this idea with the phrase, "There are no immaculate perceptions.") Thus, empirical observations of social entities can never be totally free of cultural influences, and thus neither can our models.
- Many important social concepts are value-laden. As such, they do not lend themselves well to approaches that require impersonality, precision, and abstraction. Consider important concepts such as mental health, friendship, hostility, aggression, and even religion. It is hard to imagine how one could study mental health without categories that distinguish healthy and unhealthy characteristics of personalities and/or behaviors. However, such criteria necessarily entail value judgments. Consequently, in gathering data or "facts" about mental health,

15. For an extended discussion of this notion, see W. James Bradley and Kurt C. Schaefer, *The Uses and Misuses of Data and Models: The Mathematization of the Human Sciences* (Thousand Oaks, Calif.: Sage Publications, 1998), ch. 6.

such facts (as well as the models based on them) are inseparable from values.

- Axiomatic approaches to the social sciences have not been notably successful, in spite of the most stringent efforts of some of the best minds for over 125 years. As one historian of mathematics writes, ". . . no thinker has yet built up a quantitative, deductive approach to an entire social science that would enable us to direct, control, and predict phenomena in that field. Especially in economics has success been signally absent."[16]

In short, placing trust in a mathematical entity such as a measurement or a model is reasonable as long as that trust is qualified in ways that recognize the limits of that entity.

These are technical limitations. However, there is another type of limitation. For example, consider contemporary mathematical economics. Most economic models are built on the concept of "preferences," represented either as ordinal lists (a simple ranking) or as cardinal utility functions (numerical values indicating the strength of preferences). These notions have been made mathematically rigorous and powerful theorems have been proven about them. However, in the process of such formalization, the critical edge is usually lost. That is, preferences are taken as givens — they are not critiqued. When preferences are taken as givens, greed becomes undefinable. More generally, mathematization can easily be used to avoid consideration of issues of personal responsibility, ethics, virtue, and caring.

Our fourth issue is that technique (a dimension of the larger project of mathematization) often leads to the neglect of primary values. Ellul not only sees technique as widespread in Western culture, but also views it as harmful. This conclusion of Ellul's may seem surprising, as technique appears to be morally neutral. That is, it is used simply to identify and optimize routine processes. However, technique is non-neutral because of the ways it shapes how people formulate and solve problems. Ellul argues that its appearance of neutrality is precisely what makes technique such a serious problem. That is, because technique *explicitly* excludes consideration of moral and religious values, its *implicit* values — productivity and efficiency — become the only values of its users, and as the only values, they become elevated to absolutes.

16. Morris Kline, *Mathematics in Western Culture* (New York: Oxford University Press, 1953).

Ellul is not arguing that efficiency and productivity are of no value, nor that inefficiency is preferable to efficiency or unproductivity to productivity. Rather he is arguing that values that ought to be regarded as means to an end have been elevated to ends in themselves. In the process, values commonly held to be ends in themselves, such as human dignity and worth, are overlooked and debased. In his book *The Technological Society,* Ellul provides an enormous number of examples of how such human values have been ignored in the quest for efficiency and productivity.

A Christian Response to Western Mathematization of Culture

As we have seen, mathematization of culture has provided some extraordinary social benefits. Among these are that it has provided a basis for trust in large heterogeneous societies, has enhanced cross-cultural communication, has provided some objectivity in political debate, has provided a means for scholars to order and discipline their investigations, and has given (via the vehicle of mathematical proof) a degree of certainty and logical interconnectedness to some pieces of knowledge. On the other hand, it has often been accompanied by a neglect of ethical and normative issues and has been used in ways that have ignored limits intrinsic to the mathematical method and excluded the substantive issues of values, purposes, and interpretation. These actions have had significant consequences. Valuable sources of knowledge have been neglected. Dimensions of human thought and culture that are not primarily rational or empirical (such as the arts) have been treated with disrespect. Social sciences have been modeled on the natural sciences and in the process have tried to avoid the critical matters of interpretation and valuing of human qualities and behaviors.[17] Intrinsic human values have often been neglected for the sake of productivity and efficiency.

Can Christian concepts help us assess and respond to this complex situation? If we consider only the abstract, formal aspects of mathematics, the answer is "no" or at most "not very much." For example, the statement of the Pythagorean Theorem is the same for Christians and atheists, and its proofs are equally convincing to the one as to the other. While its content might

17. For an excellent discussion of this issue, see Stephen Evans, *Wisdom and Humanness in Psychology: Prospects for a Christian Approach* (Grand Rapids: Baker, 1989).

lead a Christian to praise God for how he has ordered his creation, the theorem itself seems only remotely connected to deeper religious matters. However, there is a sense in which the answer is "yes." Consider this metaphor. For a carpenter, there is no Christian way to cut a board (beyond the obvious approach of using the best tools available and doing one's best work). Christians and non-Christians alike use the same tools. However, it makes a huge difference whether that board will be used to build a school or a brothel. In the same sense, most of the *technical* content of mathematics is immune from Christian influence. Nevertheless, for a Christian, building the kingdom of God is an ultimate concern. For many Christians this entails a concern with social structures in this world and their effect upon people. Therefore, the larger question of what kind of a culture people are building and how they are using mathematics to do that is of extraordinary importance. The focus of this chapter has been on how mathematics has been used in building Western culture since the Enlightenment. Christian concepts have a great deal to contribute to a discussion of this issue. In what follows, we focus on possible answers to two questions. What are the roots of the weaknesses in the edifice that has been erected? What challenges face contemporary culture in attempting to build a more solid structure? We begin with the weaknesses.

Starting from the Christian presuppositions that man is created in the image of God but has rebelled against him, the roots of the unpleasant fruits cited above (sources of knowledge being neglected, etc.) are not hard to identify. As we saw earlier, since the time of Galileo there has been a movement in the direction of increased autonomy for mathematics. Many thinkers have regarded mathematics (and possibly empirical science) as the only certain source of truth. Relatively little attention has been paid to the limits of mathematical thinking.

According to the modern Enlightenment outlook, we humans now relate to the world not as an integral part of creation, subject to forces beyond our knowledge and control, but as masters of a reality that is primarily mathematical in character. Natural science and technology are the tools we can use to control our world. By means of the results and methodology of science, we can gain objective and certain knowledge of the world, true at all times and in all places. Furthermore, science can provide a rational basis on which to construct a human culture transcending subjective differences and irrational prejudice. Positive science has been freed both from divisive religious disputes and from speculative philosophy and so is able to provide a common foundation for all future developments in human history.

From a Christian framework, this Enlightenment perspective on mathematization is a form of idolatry. This may seem a strange and inappropriate word to contemporary ears. The popular image of an idol is a stone or wooden object that people worship. In Scripture, however, idols also included the sun, the moon, and animals — natural objects that people worshiped instead of God. From a Christian perspective, anything viewed as having an autonomous existence and on which other things depend has been put in a place that rightly belongs to God (who alone is autonomous) and hence is an idol.[18] Thus, the Enlightenment perspective of regarding rational analysis and empirical methods as the only valid sources of truth, as sufficient to solve all human problems, and as not subject to any standards, principles, or agents greater than themselves puts mathematics in the same position as ancient people awarded to various natural objects of worship. However, we want to emphasize two points. First, as we shall see in Chapter 8, from a Christian perspective, mathematics has intrinsic value. The fact that some people have used it as an idol does not diminish its goodness as part of God's creation. Second, we are not suggesting that all mathematicians or users of mathematics in the past few hundred years have been guilty of the hubris described above (although, certainly some have been). Rather, we see this attitude as inherent in the Enlightenment perspective that has tended to shape the way mathematicians and users of mathematics have conceptualized their discipline and its place in culture.

At this point, we hope we have demonstrated the need for a different perspective on the role of mathematics in human thought and culture than the predominant perspective since the Enlightenment. In fact, some significant changes in Western thought in the past few decades have already begun to change that perspective. A widespread agreement has developed among scholars that the Enlightenment project has failed. Given the intellectual arrogance that has surrounded rationalism, as awareness of this failure spreads, one would expect a broad cultural reaction. This appears to be happening in popular culture. Although postmodernism is not a well-defined movement with specific tenets of belief, it does have typical characteristics

18. See Roy A. Clouser, *The Myth of Religious Neutrality: An Essay on the Hidden Role of Religious Belief in Theories* (Notre Dame: University of Notre Dame Press, 1991), for a more extensive treatment of this idea.

that can be understood as a reaction to rationalism. We saw some of these characteristics in Chapter 1 — for example, rejection of the notion that truth is purely rational, rejection of the belief that unified all-encompassing explanations are desirable, emphasis on non-rational ways of knowing, heightened status for emotion and intuition, less concern for systematic and logical thought, and abandonment of the Enlightenment ideal of the dispassionate, autonomous, rational individual. While such trends are evident in popular culture, a postmodern approach to the natural and social sciences has not yet developed to the point that it provides a viable alternative paradigm to the positive approach. So in spite of the failure of the Enlightenment project, it seems likely that mathematization will continue at least until an alternative paradigm arises, if one ever does. Even so, scholars critical of the positive approach to knowledge have been strengthened and will probably grow even stronger. Thus, the "two cultures" gap that C. P. Snow identified a generation ago[19] will probably widen, and hostility between the two subcommunities will likely intensify. We see a serious danger of Western culture oscillating from overly optimistic, unrealistic expectations of reason in general and mathematics in particular to a point where significant subcommunities hold an equally inappropriate hostility toward reason and mathematics. Thus, those who care about rationality, mathematics, and human culture have some major challenges facing them.

We turn now to our second question — what challenges face contemporary culture in building a stronger structure for mathematization?

We listed above a number of social benefits and harms that mathematization has produced. One challenge that users of mathematics face is *to eliminate the harms without losing the benefits.* Furthermore, this challenge has to be met in a complex context where the unified approach that the Enlightenment vision formerly provided has been eroded. In general terms, it is clear what is needed to meet this challenge. Mathematization needs to be accompanied by respect for other modes of thinking and other sources of knowledge. In addition, it needs to incorporate reflection on values, norms, purposes, and interpretation, and its limitations need to be frequently addressed. Encouraging signs of some such changes are currently appearing. These suggest that some users of mathematics are already taking steps to meet this challenge. For example:

19. Charles Percy Snow, *The Two Cultures and the Scientific Revolution* (Cambridge: Cambridge University Press, 1959).

- Some works have recently appeared on the role of ethics in the applications of mathematics.[20]
- Some mathematicians have shown an increased concern with interfaces with other disciplines. For example, a new subdiscipline called environmental mathematics has recently arisen.
- There has been an increased recognition of the legitimacy of other modes of thought than the purely rational. For instance, some scholars in artificial intelligence now emphasize the use of metaphor and case study as essential human problem-solving tools. And some recent research in brain physiology demonstrates a much stronger linkage between reason and emotion than had been previously imagined.[21] The latter works have been publicized in widely read mathematics journals.

Furthermore, some technical changes have taken place in the content of mathematics. One that is well known is the advent of chaos theory, which has demonstrated that certain complex natural phenomena such as the weather are inherently resistant to long-term prediction. That is, mathematical analysis itself has demonstrated the existence of significant limits to its computational capabilities. While such a change might seem to be independent of cultural influences, chaos theory in particular has received a great deal of attention and hence has influenced research agendas and funding.

However, in spite of these encouraging signs, the bulk of work being done in the social sciences as well as the natural sciences continues to be done as if the Enlightenment project's failure had never become known. Thus, there is still a great deal to be done.

Another challenge that mathematicians face is *to find a meaningful social role in a postmodern culture.* Historically, mathematics has only flourished in situations where it has enjoyed a broad cultural support. However, the predominant tone of postmodernism is anti-rational. Furthermore, mathematicians set a high value on abstraction and see truth as precisely

20. A book edited by William A. Wallace, *Ethics in Modeling* (Tarrytown, N.Y: Elsevier Science, 1994), is one. The article "Toward Ethical Guidelines for Social Science Research in Public Policy" by D. P. Warwick and T. F. Pettigrew in the book *Ethics, the Social Sciences, and Policy Analysis* (New York: Plenum, 1983) is another. And a third is Bradley and Schaefer, *The Uses and Misuses of Data and Models.*

21. Antonio R. Damasio, *Descartes' Error: Emotion, Reason, and the Human Brain* (New York: Avon Books, 1994).

formulated propositions, deductively derived from axioms. The postmodern climate values particulars rather than abstractions and, as we saw earlier, sees truth very differently. In addition, mathematics from Pythagoras to the present has often possessed something of an "otherworldly" quality. For instance, during the first two thirds of the twentieth century, the focus of the American mathematical community was overwhelmingly on pure mathematics; applied mathematics was de-emphasized and issues of social responsibility were largely ignored. Thus, mathematicians have tended to isolate themselves from practical affairs. In contrast, the postmodern climate tends to emphasize connectedness, service, and cultural context. Therefore, the mathematics community faces cultural challenges that it did not face in the Enlightenment era.

To some extent, the mathematics community can depend on tradition to resist these adverse cultural forces and avoid change. Mathematical patterns of thought are ancient, easily traceable back more than four thousand years. In addition, mathematics programs are deeply entrenched in well-established curricula in schools, colleges, and universities. Even so, it seems to us that the mathematics community ought to make some adjustments to contribute more effectively in the current cultural climate. Specifically, we would like to see the mathematics community become more critically aware of the role mathematization has played in Western culture. We would also like to see that community come to believe that understanding this role and helping to guide the process of mathematization is part of its professional responsibility within whatever culture its members reside. Of course, such changes on the part of the mathematics community would require a major shift in its professional values. At present, this community sets a very high value on the discovery of new mathematical knowledge, as indeed it should. However, it sets relatively little value on understanding the social implications of a widespread adoption of its ideals and values or on community service. Nevertheless, these matters are important and far from elementary. One implication of Christian thinking for the discipline of mathematics is the idea that the mathematical community ought to develop a broader perspective on its own discipline, one that embraces an understanding of its social role and the contributions it can make to culture.

SECTION III

FAITH PERSPECTIVES IN MATHEMATICS

CHAPTER 8

Mathematics and Values

Introduction

Up to this point, we have focused on the nature of mathematics, its history and role in culture, and how it is developed and learned. However, these are only part of the "disciplinary matrix" of mathematics. Like other disciplines, mathematics possesses a collection of professional values that give direction to the discipline. What problems are important? Which approaches are most likely to be fruitful? In what style should papers be written? What constitutes good mathematics? The answers to these questions have changed over time; they are governed by the values of the mathematics community. While not solely the concern of mathematicians, additional questions arise in many contexts about how — and if — mathematics should be used; they are part of a broad understanding of the meaning of mathematics.

In this chapter, we discuss how we can assess the value of mathematical work. One reason that mathematics is valuable, of course, is that it is useful for so many things. Mathematical models provide helpful insight into environmental problems, management decision-making, and political decisions; and this barely scratches the surface. However, as we will see, current mathematics research is often far removed from these everyday concerns and even from scientific uses. It is frequently not clear to granting agencies, academic administrators, students considering a possible career in mathematics, and others that such research is valuable. Thus, we need to articulate reasons for doing mathematics apart from applications.

Different cultures have valued mathematics in dramatically different ways. As we saw in Chapters 5 through 7, before the modern era (early seventeenth century to mid-twentieth century), mathematics was valued as a model for how knowledge ought to be organized, but many prominent thinkers were skeptical of its practical value. During the modern era, neither the practical nor the theoretical value of mathematics was much questioned. In fact, for most mathematics done between 1600 and 1850, it was often difficult to distinguish mathematics from its application. Today, the influence of science and other societal needs on mathematical work is much less pronounced. Contemporary viewpoints tend to see mathematics as autonomous; it has its own subject matter and can progress without depending on the sciences or other applications for new questions or concepts. Some empiricist and naturalist philosophies point to applications as the ultimate ground of mathematical meaning, but much mathematical practice is, prima facie, without concern for application. Philip Kitcher calls it "the science of human physical and mental operations."[1] However, even Kitcher describes its autonomy:

> In an important sense, *mathematics generates its own content.* The new forms of mathematical notation that we introduce not only enable us to systematize and extend the mathematics that has already been achieved, but also to perform new operations or to appreciate the possibility that beings released from certain physical limitations could perform such operations. Such extensions of our repertoire seem to me to have occurred with the development of notation for representing morphisms on groups and for constructing sets of sets. In both cases, the notation is a vehicle for iterating operations that we would not be able to perform without it.[2]

Advances in the modern era such as the development of consistent alternative geometries, abstraction in algebra, and axiomization in set theory gave impetus to the perspective that mathematics is autonomous. Such thinking coincides with the view of mathematics as pure — the queen of science. Of course, much mathematical work is, to varying degrees, responsive

1. Philip Kitcher, "Mathematical Naturalism," in *History and Philosophy of Modern Mathematics,* ed. W. Aspray and P. Kitcher (Minneapolis: University of Minnesota Press, 1988), p. 313.

2. Kitcher, "Mathematical Naturalism," p. 314.

to a particular application — the servant of science. However, Kitcher considers it an open question whether applications are continuing to alter mathematical ideas *fundamentally* or whether such ideas are independent of applications.

We can pursue two paths in articulating the value of a piece of mathematics apart from scientific or practical goals. One rests on the premise that it may be applied in the future. That is, the work may contribute to other progress in mathematics, leading to a general understanding of fundamental questions or a broad corpus of knowledge. This knowledge may eventually be relevant to seemingly unrelated scientific or technological issues. Note that this argument appeals to potential, rather than current, applications of mathematics. Thus, we put it in a different category from those arguments that justify mathematical research on the basis of its contribution to known applications. Call this the *future value* argument. The other path, called the *intrinsic value* argument, aims to find intrinsic worth in mathematics itself. We will explore both of these paths.

To provide a context for our discussion of valuing mathematics, we begin broadly. First, we discuss mathematics as *technique*, summarizing both the role that mathematics has played recently in Western culture and a postmodern critique of this role. We then distinguish mathematics as a *conceptual framework* from its use as a tool. By drawing an analogy with *art*, we also examine aesthetic value and various internal criteria for valuing mathematics. Having examined these three modes of thinking about the social role of mathematics, we critique the future value argument and the traditional idea of intrinsic value within mathematics. In particular, we ask whether these ideas are an effective motivation for a career within mathematics. We then propose a more nuanced view of intrinsic value, emphasizing the actual practice of mathematics. Finally, we illustrate how Christian presuppositions can provide additional ways of valuing mathematics.

Mathematics as Technique

It is hard to overstate the influence of mathematics on our technological society. The mathematical sciences, and particularly computer technology, have captured the attention of even the usually disinterested spectator. Additionally, mathematics has an important role in shaping our culture's conceptual categories and values. As we have seen, the Enlightenment project elevated mathematics and empirical science to the role of providing a common, non-controversial basis for society. Mathematics was, as Galileo expressed it, the alphabet of the universe, as the standards of truth and rationality persuasive to any rational person were fundamentally mathematical and scientific. The dream of René Descartes that "the entire universe is a great, harmonious, and mathematically designed machine"[3] has become a symbol for the extravagant hopes placed in mathematics.[4] The Enlightenment gave mathematics a well-defined and valuable place in human culture as one of the primary means through which we can know truth and establish justice and order. It gave rise not only to modern natural science but "positivist" social science — that is, social science based on careful observation, data collection and analysis, and attempts to infer scientific laws. The application of scientific methods to various human problems, such as law and politics, required a belief that human society, as well as nature, could be understood and manipulated through these methods. These beliefs had profound implications for public policy. For example, in 1962, President John Kennedy said,

> The fact of the matter is that most of the problems, or at least many of them that we now face, are technical problems, are administrative problems. They are very sophisticated judgments which do not lend themselves to the great sort of "passionate movements" which have stirred this country so often in the past. Now they deal with questions which are beyond the comprehension of most men.[5]

This faith in technical solutions led to a burgeoning of policy studies by social scientists. Two philosophic outlooks, naturalism and empiricism, gave impetus to the success of the scientific method. It is easy to see why this was the case. Naturalism requires that explanations of physical phenomena refer back to the same physical or natural realm; supernatural entities and divine purpose are excluded. Empiricism, likewise, limits knowledge to that which rests on observed facts.

3. M. Kline, *Mathematics in Western Culture* (New York: Oxford University Press, 1953), p. 106.

4. See, for example, Philip J. Davis and Reuben Hersh, *Descartes' Dream* (San Diego: Harcourt Brace Jovanovich, 1986).

5. W. James Bradley and Kurt C. Schaefer, *The Uses and Misuses of Data and Models: The Mathematization of the Human Sciences* (Thousand Oaks, Calif.: Sage Publications, 1998), p. 181.

While the Enlightenment gave mathematics a well-defined place in human culture, many forms of postmodern thinking do not give it any explicitly acknowledged place at all. In fact, the primary characteristics of mathematics — precise expression, careful reasoning, and the formulation of universal truths abstractly expressed — are viewed with considerable suspicion. If the scientific community adopts theories for subjective reasons, and the decision of what data are worth collecting is also relative to one's cultural context, then scientific knowledge is much less secure. Other, non-mathematical ways of knowing offer legitimate competition. The role of mathematics in the social sciences is even more suspect. Post-noeticentrism (a part of the postmodern critique) rejects the notion that there exist abstract mathematical laws capable of describing human life analogous to the laws of physics. Relativism and the acknowledgment of diverse viewpoints very legitimately question whether a community can expect agreement about morals and public norms, regardless of what analyses are performed or data provided. The current gun control debate exemplifies the disagreement. Extensive research on the issue has done little to mediate the claims of the two sides. Gun control opponents welcome studies suggesting that laws allowing concealed weapons reduce certain violent crimes. However, they reject other data suggesting that many people accumulate more weapons than they could legitimately need. In rejecting or ignoring the latter data, gun control opponents allow their prior convictions about freedom to overrule the data. Analogous things could be said about gun control proponents.

The consequence of the advance of postmodernism is that at present the Western academic world is suffering from a kind of intellectual schizophrenia. In both the natural and social sciences, most scholarship is still "positive" — that is, it is based on observation, collection of data, and theory formation. However, other scholars are approaching issues from a position that gives a much greater role to subjectivity. The Western academic world is struggling with two competing modes — one that would make science and mathematics the foundation for culture and one that regards them as incapable of producing anything of the sort. Thus, one of the principal characteristics of the academic world in our time is the division between these two dramatically different views of the proper role of mathematics and science in culture. In addition, while it might seem that American culture is rapidly moving away from the Enlightenment perspective in the direction of postmodernism, an anchor is retarding that motion, namely technology. Despite an element of distrust, Western culture embraces technology at its very core. Materialism and consumerism feed off it. There are enormous economic and national security motives for developing new technology. Even voices that speak against the excesses of technology do not wish to avoid it; rather, they seek a balance.

Jacques Ellul, an influential French Christian writer, has studied technology not primarily as technological artifacts, but as a way of thinking.[6] He sees "technique," what we might call algorithmic thinking, as the predominant mode of thought in Western civilization, most notably in the United States. Technique seeks to identify routinizable aspects of processes so that these can be optimized in terms of productivity and efficiency. It shares with mathematics an emphasis on abstraction, the separation of idealized processes from context, careful representation, the study of patterns rather than unique phenomena, and optimization. His criticism of technique is that the only values it recognizes are efficiency and productivity. Things that are not quantifiable or reducible to formal logic are explicitly excluded. In such a society, careful thinking about norms, purposes, and ends is neglected and productivity and efficiency unconsciously assume the role of ultimate values.

In summary, one role for mathematics in culture is that its approach provides a basis for technological thinking. While this fact might seem to provide an obvious demonstration of the value of mathematics, many people have pointed out problems with technique, and thus are legitimately questioning the value of this mode of thought. We will return to this tension after looking at two alternative ways to look at the role of mathematics.

Mathematics as Conceptual Framework

When thinking about the influence of mathematics it is important to note that it is not simply a tool for computation. Raymond Wilder, a mathematician who has applied concepts from anthropology to the study of mathematics as a cultural system, describes another influence beyond the obvious uses of elementary and even modern "core" mathematics:

6. Jacques Ellul, *The Technological Society* (New York: Knopf, 1964); *The Technological System* (New York: Continuum, 1980); *The Technological Bluff* (Grand Rapids: Eerdmans, 1990).

> The uses of mathematics in the other sciences are generally of two kinds: (1) as a tool and (2) as a source of conceptual configurations. . . . Not so well known are the uses cited in (2) regarding mathematics as a source of conceptual configurations. These need to be more generally recognized, since they call attention to the importance of the core of mathematics. . . . For Einstein, the Ricci calculus and Riemannian geometry stood ready for his needs, so that even as tools the core mathematics proves itself in meeting future demands. But most theoretical physicists, for instance, know that when their imagination seems to fail them, they can turn to the core mathematics for further insights.[7]

That these conceptual configurations move beyond science into wider areas of thought particularly supports the notion that core mathematics has a role in culture. Here are three examples.

First, our understanding of genetic code is shaped by the mathematics of information theory. The idea that DNA is a digital blueprint that somehow represents an organism has shaped our view of life. Thinking digitally about DNA requires, in addition to biochemical knowledge, a digital age. Without the theories of digital information that flourished alongside computers, we would not be able to contemplate seriously the encoding of life. Questions about the relationship between the digital information and the organism are being pressed further by recent success in cloning.

Second, algorithmic thinking is pervasive. The notion that following a fixed set of rules can solve many problems is a large part of what is meant when we say our culture has become mathematized. Algorithms underlie computer "thought," limiting the role given to human creativity. Algorithmic thinking is evident, for example, in medicine when a doctor goes through the stages of reaching a diagnosis. We will explore the possibility of computer thinking in Chapter 9.

Finally, the power of recursive calculation has been a particularly fertile idea. Given a function, f, that represents a physical law governing the change of some physical system, the iterates

$$f(x), f(f(x)), f(f(f(x))), \ldots$$

7. Raymond L. Wilder, *Mathematics as a Cultural System* (Oxford: Pergamon Press, 1981), pp. 156-57.

of the function represent the trajectory of the system over time, starting from the state *x*. The long-run behavior of the system can be as simple as reaching a fixed point or settling into a cycle that repeats periodically. However, the trajectory may also approach an "attractor" that is bounded but non-repeating — a chaotic attractor. This behavior, known as chaos, is inherently more complex and has implications that have captured the imagination of the scientific community. One implication is unpredictability. Physical systems that behave chaotically — and apparently many systems do — are extremely sensitive to initial conditions. Exerting a small force on an orbiting body alters its orbit slightly so that future position gradually deviates in a linear fashion from what it would have been. The orbit remains elliptical. Applying a small disturbance to a chaotic system, however, leads to changes that grow super-linearly and to qualitatively different future behavior. To explain this significance, consider an iterated function that is also spatially recursive in the sense that the argument x represents the state of one localized subsystem, such as the regional weather, or perhaps several adjacent subsystems that influence each other, and $f(x)$ is the next state of one subsystem. The long-run behavior of these systems, known as cellular automata, can be self-organizing: patterns emerge that involve many nearby subsystems. The patterns are dynamic and sensitive to conditions. They might be compared to the flight of a flock of birds rather than the regular motion of a wave or the growth of a crystal. These discoveries have led to speculation that they might explain the emergence of complex structures in a collection of molecules or other particles.[8]

Extrapolating even further, the physicist Paul Davies has popularized the idea that the entire universe is one massive iterative computation based on fixed physical laws.[9] This view of the universe as computation differs from the "clockwork image" of some Enlightenment scientists in that there is no sense of function or unifying pattern or principle. It is a *distributed* computation in that calculations are performed for each particle. From Davies's perspective, it also allows innovation to occur truly, as the outcome of the calculation defies prediction, and even the qualitative nature or com-

8. See S. A. Kauffman, *The Origins of Order: Self-Organization and Selection in Evolution* (New York: Oxford University Press, 1993), for a detailed example. See also M. Waldrop, *Complexity: The Emerging Science at the Edge of Order and Chaos* (New York: Simon and Schuster, 1992), for a non-technical survey.

9. Paul Davies, *The Mind of God: The Scientific Basis for a Rational World* (New York: Simon and Schuster, 1992).

plexity of the outcome cannot be anticipated. In light of this innovation and the openness or incompleteness of some physical laws, theologian Arthur Peacocke describes the universe as both being and becoming.[10]

Just as we saw earlier with technique, mathematics plays a powerful role in culture by providing conceptual frameworks. This role is less widely recognized than that of technique. However, once again, the role can be problematic: it is not clear that these frameworks are always beneficial to society or contribute to knowledge. We now turn to a completely different approach.

Mathematics as Art

While mathematics has permeated society as technique, its autonomy has led to a different view within the field. The freedom not to consider applications makes mathematics in one sense a pure science. However, attempts to define mathematics as a science with a self-contained subject matter, without reference to empirical science, run into ontological and epistemological questions. What is mathematics about (an ontological question)? Moreover, if we abandon the notion of a priori mathematical knowledge, how do we obtain mathematical knowledge that is not empirically based (an epistemological question)? For these reasons, among others, many mathematicians find the comparison with art more appealing, siding with Bertrand Russell, who wrote:

> Mathematics possesses not only truth but supreme beauty — a beauty cold and austere, like that of sculpture, without appeal to any part of our weaker nature, sublimely pure and capable of a stern perfection such as only the greatest art can show.[11]

Mathematics may be valuable, then, for aesthetic reasons.

Art suggests invention, not discovery, so the art analogy avoids the ontological question (What is mathematics about?) and focuses on mathematical practice. This line of thinking asks how mathematicians select axioms in

10. Arthur Peacocke, *Theology for a Scientific Age: Being and Becoming — Natural and Divine* (Oxford: Basil Blackwell, 1990).

11. Bertrand Russell, *The Study of Mathematics,* New Quarterly, 1902.

set theory and logic, how they formulate problems, and why they take various approaches in solving these problems. At one extreme is the conventionalist answer that these developments are arbitrary choices, but this response does not do justice to the coherence, unity, and usefulness of mathematics. How do the values, or internal criteria, that guide mathematics lead to progress and objective knowledge?

The prominent twentieth-century mathematician Armand Borel takes the art analogy further, likening "mathematics for mathematics sake" to abstract painting. He compares the dangers of doing mathematics without the guidance of empirical questions to the dangers of painting without a subject — exhaustion of creativity or fracturing into insignificance. "Up until now," he asserts, "mathematics has been able to overcome such growth diseases."[12] When ideas are developed that do not stimulate further development, mathematics tends to correct itself, pruning the unfruitful branch. The extension of the number system to quaternions is an example of an idea that did not appear to lead to much and was soon neglected. Such ideas may be revived at a later time, however, and quaternions have recently been shown to have potentially powerful applications to airplane guidance systems and to virtual reality.[13] Borel describes mathematical practice as largely free from considerations of application and sees this freedom, where work is guided by a consensus of internal aesthetic considerations, as leading to valuable developments.

> Our so-called aesthetic judgments display a greater consensus than art, a consensus that goes far beyond geographical and chronological limitations. . . . There are differences of opinion and fluctuations in time in the evaluation of mathematical works, though not to such a strong degree [as in art].[14]

Although Borel's claim of consensus must be qualified by recent studies of cultural influences on mathematics, it is certainly more compelling than in disciplines where some postmodern thinking has cut a wider swath.

G. H. Hardy also emphasized the aesthetic qualities that guide mathe-

12. Armand Borel, "Mathematics: Art and Science," *The Mathematical Intelligencer* 5 (1983): 15.

13. See Jack B. Kuiper, *Quaternions and Rotation Sequences* (Princeton: Princeton University Press, 1999).

14. Borel, "Mathematics: Art and Science," p. 16.

matics. Reflecting on his career at the center of the mathematical world in the early 1900s, he states, "I have never done anything 'useful.'"[15] His view of what mathematics should be is seen in his disparaging remarks about proof by enumeration of cases and the dullness of ballistics. In good mathematics, Hardy writes, "there is a very high degree of *unexpectedness* combined with *inevitability* and *economy*."[16] It should also be significant in its connection to previous and new theories. All of Hardy's criteria are internal to mathematics. The first three are truly aesthetic judgments, while the last has to do with use within mathematics. Indeed, Borel suggests that "our aesthetics are not always so pure and esoteric but also include . . . meaning, consequences, applicability, usefulness — but within the mathematical sciences" (Hardy, p. 15). He lists four criteria, of which only the third is purely aesthetic: (1) solving open problems, (2) comprehensiveness and making connections between various areas, (3) elegance or beauty, economy of principles, and originality, and (4) lasting influence and opening new paths.

Most traditional philosophies of mathematics do not emphasize the role of such values. Instead, they see mathematical developments as the inevitable result of rational inquiry. However, the realist philosopher Penelope Maddy recognizes internal criteria for justifying mathematics. She identifies seven "non-demonstrative" methods that have been used to argue the validity of axioms in set theory:[17]

1. Verifiable consequences that have been proven without the axiom.
2. Power in solving open problems.
3. Simplification of previous results.
4. Proof of previous conjectures.
5. Implying of "natural" or intuitive results.
6. Connections (similarity) with previous results.
7. Insights and simpler proofs of previous results.

These criteria have much in common with "purist" or aesthetic criteria, but emphasize intuitive knowledge. Interestingly, in Maddy's set-theoretic

15. G. H. Hardy, *A Mathematician's Apology* (Cambridge: Cambridge University Press, 1940), p. 150.

16. Hardy, *A Mathematician's Apology*, p. 113.

17. Penelope Maddy, *Realism in Mathematics* (Oxford: Oxford University Press, 1990).

realism they are viewed as "rational methods in the pursuit of truth" (Maddy, p. 35).

In summary, we have seen two significant social roles for mathematics — source of technique and provider of conceptual frameworks. These provide *external* criteria for justifying mathematics. We have also seen that a significant number of mathematicians and philosophers have rejected the notion that mathematics ought to be valued for its social utility, asserting its aesthetic qualities instead. Some of them have provided lists of *internal* criteria for justifying mathematics. Our goal is to examine the intrinsic value of mathematics, and to present a *social concept* of intrinsic value. However, we need to address some difficulties first.

Difficulties with the Future Use Argument

A powerful historical argument has been used to counter the question "Of what use is mathematics?"

> It becomes more apparent, year by year, that the core of mathematics serves the general culture by producing new concepts whose future, in addition to their uses in mathematics proper, will be to move into the general culture, meeting needs unforeseen at present. . . . Who, during the latter part of the nineteenth century and the first decade of the twentieth, would have foreseen that the studies initiated by Boole, Frege, Russell and Whitehead, and Hilbert would soon generate notions such as recursiveness (Gödel, Turing, etc.) which would serve as one of the basic notions of computer theory? Or that matrix theory, initiated by Cayley (*ca.* 1858), would turn out to be the precise tool needed by Heisenberg in 1925 for mathematical description of quantum mechanics phenomena? Or that the theory of analytic functions, involving complex quantities, would turn out to have widespread applications in physics and especially in electrical phenomena? One could go on with the citing of such cases. The "moral" should, however, by now be clear: That the "pure" mathematics of today will be the applied mathematics of tomorrow.[18]

18. Wilder, *Mathematics as a Cultural System*, pp. 152-53.

Other famous examples include the use of Riemannian geometry in relativity and the many contributions of group theory to particle physics.[19] Recognizing that unanticipated applications have occurred and will continue, how does this phenomenon affect the value of a given project? Can *all* projects be said to have the possibility of leading to powerful new applications? Can this argument legitimately be used as a blanket justification for all mathematical undertakings?

The first difficulty with the future use argument is essentially statistical. Major breakthroughs are rare. Thus, as the volume of publications has increased in the last 50 years, the chance of any one project leading to a breakthrough has become more remote. On the other hand, the increased value of mathematical applications might provide an economic argument in favor of exploring every possible mathematical path. Perhaps these trends cancel each other out. Increased specialization, however, can tip the balance strongly toward a decreased value of potential applications. It is not sufficient for *a* physicist to turn to *a* mathematician, or for a person to receive training in both. Rather, the persons working on some application have to learn of just that mathematical research which is needed, and in a way that allows them to recognize the connection. Viewing information transfer as completely "blind," modeled as a random bipartite graph, illustrates the dilemma. Mathematical and application "specialties" comprise the two vertex groups of this graph, with a vertex representing each mathematical and application specialty available. Edges represent information transfer between a mathematical and application specialty and occur randomly with some mean number of edges incident to each vertex. Now, suppose that the number of specialties grows by some factor, in both mathematics and applications. Assuming the number of connections made *by each specialty* (the degree) stays constant, we see that the chance of the one fruitful connection being made drops off by this constant factor. Even worse, the chance that an individual connection is fruitful drops as the square of this factor, thus discouraging efforts at dissemination.

This simple model of dissemination is overly pessimistic, however, in at least two ways. First, the trend toward specialization is being countered by many recent examples of unification and cross-fertilization. Specialties

19. Freeman J. Dyson, "Mathematics in the Physical Sciences," *Scientific American* 211 (1964): 127-46; reprinted in *Mathematics in the Modern World* (San Francisco: W. H. Freeman, 1968), pp. 249-57.

within mathematics are finding more reasons to interact, and some of the interactions include an area that is close to applications. Second, information transfer is not completely random. Much mathematics is done with a specific application in mind; the randomness argument does not apply here at all. As for the rest, dissemination is some combination of randomness and intentional matching. For instance, classical dynamics uses methods of solution of differential equations; particle physics uses group theory, and so forth.

However, the ability to anticipate the fruitful matchings is exactly what the future use argument claims is impossible. John von Neumann, who was involved in some of the most successful twentieth-century applications of mathematics, emphasized this point:

> But still a large part of mathematics which became useful developed with absolutely no desire to be useful, and in a situation where nobody could possibly know in what area it would become useful and there were no general indications that it even would be so.[20]

If it is impossible to know which mathematics will be useful and in what area it will be useful, then every area of application would have to be informed of every development in mathematics to realize these potential applications. Such a complete information transfer is impractical. Not only can one individual not master all the relevant knowledge. In the sense discussed in Chapter 4, the total *effort* involved in a team approach would be prohibitive.

The notion of a partial matching — that scientists have some idea what mathematical specialty might be useful to them — leaves a large amount of mathematics that is not matched with applications. This situation allows better information transfer. The admission that some mathematical topics are more likely to be useful than others, however, undercuts the argument that all mathematical work is valuable because of potential future applications. The future use argument, then, faces a dilemma. To the degree that we *can* match mathematics with application areas, not every mathematically sound project will be justified on the basis of potential future use. To the degree that we *cannot* know such a matching, information transfer becomes problematic. Of course, these observations do not mean that a pro-

20. John von Neumann, "The Role of Mathematics in Science and in Society," address to Princeton Graduate Alumni, 1954, cf. *Collected Works,* vol. 6, pp. 477-90.

gram can be constructed for choosing among mathematical projects to pursue, but they do suggest that the future use argument needs to be refined and used with more care. For example, mathematicians may impose the requirement on themselves that their results be produced with an eye towards a larger edifice, evaluating first each field of mathematics in terms of its individual value to other fields and sciences, and then surveying the value of each field's internal results to the whole. Indeed, the practical constraints of hiring, funding, and publication standards have already made pure mathematicians conscious of whether their work might bear on more practical issues in the sense that their future use can at least be speculated.

Our simple model of information transfer assumed that only one mathematical idea would meet the needs of a particular application. Another model emerges if different mathematical topics are taken to be alternative theories or conceptual frameworks for a particular application. In the understanding of science introduced by the philosopher Karl Popper, these theories are *underdetermined* by the empirical data. That is, several theories fit the data and are not "falsified." If this is true of the mathematical content of a theory, then there may be a number of mathematical approaches that meet the needs of the application. The amount of information transfer needed for mathematics to support the applications would be much less, helping explain the steady stream of new applications that are occurring.

Experience with mathematical modeling in engineering suggests how this might occur. For example, there are three fundamentally different approaches to modeling a manufacturing system: discrete stochastic (queuing theory), continuous with discrete stochastics (jump Markov processes), and continuous with continuous stochastics (Brownian motion). Researchers' choices of models depend mostly on who their advisors were, not which is best in a particular situation. Thus, successful applications can occur without complete transfer of information. As long as engineers are trained in one of the three approaches, they are likely to make progress on their problem. Of course, awareness of all three approaches is preferable, since each one works best in different situations, but applications can occur nevertheless. Schrödinger's wave operator and Heisenberg's matrix operator formulation of quantum mechanics provide an example at a higher mathematical level. In this example, the two mathematical formulations are equivalent. If one of the mathematical techniques did not exist or was not applied to quantum mechanics, presumably some fruitfulness in solving problems would have been lost, but the same mathematical rigor and generality would have occurred.

Once again a dilemma arises: To the extent that alternative mathematical topics can provide the conceptual framework and even the tools needed for applications, the justification for pursuing all mathematical areas is undercut. For instance, suppose that if all mathematics research were pursued with the same vigor as it is today, ten ideas would be discovered in the next fifty years, any one of which would solve a particular applied problem. Now suppose research is cut back so that only half as many ideas are discovered in the next fifty years. The probability that all ten of our important ideas are missed, and our problem is not solved, is only 1 in 2^{10}. Interchangeability of mathematical ideas, then, reduces the need to pursue every topic. Given the limits of dissemination, the marginal benefit of pursuing additional topics would be even less as the lines of communication become clogged.

Of course, we have painted an overly gloomy picture. Many researchers have in view the larger edifice of mathematics and applications in the first place, which in turn drives their thinking. Perhaps it is the utility of their edifice that should be critiqued. Regardless, there are practical difficulties with information transfer, and specialization increases these difficulties. Consider the sequence of events that would have to occur for a typical "pure" research project to become applied mathematics. First, the theory would have to be put in a form where its potential use could be recognized. Results would need to be surveyed and organized. Introductions to the specialty would need to be written. Then someone cognizant of the area of application must learn about the new theory. The connection is most likely to be made by someone well versed in both fields, as the scientist and mathematician work within different frameworks of thought and speak different technical languages. Seeing the connection often requires an extraordinary leap in thinking, not just the mere fact that the fields are put in contact. This is not to say that potential applications are necessarily pre-dictated. Historically, abstraction sometimes occurs by generalizing an insight into, say, how a computation is achieved, with future applications not occurring until a situation arises where the specific fails (at least in some sense) but the general applies. For instance, the utility of Riemannian geometry in relativity theory comes to mind, as it gave a better explanation of observed phenomena than theories based on a strict Euclidean geometry, though the latter were still valid on small-distance scales.

Further Difficulties in Establishing Intrinsic Value

Thus far, we have considered the value of mathematics on the basis of its potential future applications, its contribution to broader strains of thought, and its aesthetic qualities. All of these arguments are strengthened by the objectivity of mathematics. In some sense, it is the language of precise thought and therefore essential to rational inquiry. However, what are the relative importance of tastes, cultural values, and historical developments on one hand and the inevitability of mathematical results on the other? Current understanding of how mathematics evolves makes it clear that the beliefs and values of the community — the social and historical context — are an important influence on the actual corpus of results.[21] The objectivity of mathematics must be balanced against the human role in mathematical practice. How one strikes the balance varies among mathematicians — it depends on their conception of mathematics, and how mathematics fits into their worldview. For instance, to Platonists, mathematics contains timeless truth and is often held to describe a realm of abstract objects. Empiricists also regard mathematics as true, but for them, it shares the contingent status of scientific knowledge, and unapplied mathematics is more questionable. Formalists view axioms as human inventions, but theorems as following objectively from them. Constructivists emphasize fallibility and the human role in axiom and proof, with psychological, historical, or sociological explanations of mathematical practice.

A strong belief in objectivity might lead to a liberal conception of intrinsic value in which any new mathematical knowledge is progress and inherently valuable. However, as Philip Kitcher points out, even the Platonist can critique mathematical practice as not fruitful because it might be uncovering only insignificant truths.[22] Thus, even with a high view of mathematics as truth, values are needed to guide practice.

The primary difficulty with the idea of intrinsic value is the difference between the value of mathematics as a whole and the value of individual projects. Kitcher seems sensitive to this great gap when he distinguishes between individual rationality and community rationality. We will argue that a more nuanced intrinsic value argument is needed, and that it applies in different ways to different individual projects. We reach this conclusion largely

21. See, for example, Raymond L. Wilder, *Mathematics as a Cultural System.*

22. Kitcher, "Mathematical Naturalism," p. 312.

based on observation of current mathematical practice, in the same spirit as many studies of mathematics in the naturalist school, but also in light of some societal changes.

Hardy's defense of mathematics includes the assertion that it is a "harmless and innocent occupation."[23] In the isolationist leisure class of 1930s England, such a negative justification was somewhat persuasive. In the intervening decades, society — and normative concepts of a good society — has shifted toward an economic perspective of interdependency and productivity. We no longer accept the idea that doctors should decide how much is spent on health care, that NASA should decide how much is spent on space, or that scientists should decide how much is spent on research. Symbolic of this change was the cancellation of the superconducting supercollider project, which clearly offered the potential for new knowledge. The productivity pressures being faced in other occupations are beginning to be felt in academia. In mathematics, teaching is being reformed and research funding is being scrutinized for tangible results such as educational impact or collaboration with industry. Corporate and foundation funding is generally even more focused on anticipated applications, rather than the intrinsic value of mathematics. These pressures do not simply reflect budget difficulties. They also demonstrate the shift toward a society in which professions are subject to external scrutiny and the resulting pressures for productivity and societal value. Thus, justifying mathematics today requires a much stronger argument than Hardy's.

While expectations have risen in the last half-century, the changing face of mathematics has also affected the argument for intrinsic value. In many subfields, autonomy has continued to increase, making Borel's concerns about painting without a subject loom larger. Another important change throughout the twentieth century has been the sheer volume of publication and increasing specialization. Some areas, such as computational mathematics, can be clearly recognized as growing subjects. In other areas, it is hard to see how the "playing field" of important problems could have expanded very much. Although it is dangerous to make judgments of this type in individual cases, the overall impression is of a more crowded playing field. With specialization and increased volume of publications, it becomes very difficult to argue that a particular piece of research contributes in any meaningful way to the pursuit of truth. On purely statistical grounds, one can ar-

23. Hardy, *A Mathematician's Apology*, p. 141.

gue that such a piece is unlikely to have a lasting impact on mathematics (one of Borel's criteria). While this may not be a critique of the intrinsic value of mathematics in and of itself, it at least points out the difficulties of an individual or group's ability to survey the general body of mathematics and recognize its various possible entailments.

Another difficulty is that many areas of mathematics have been so extensively studied that the remaining questions are highly specialized. Most mathematicians are studying problems that are less well known, even within mathematics, than in previous eras. Keith Devlin, for example, distinguishes between "enlarging our fundamental understanding of the universe" and "tidying up the loose ends."[24] Large extensions, Devlin claims, fueled by applying mathematics to new domains, cannot proceed indefinitely. He also makes an argument of diminishing returns in mathematics as in mature sciences such as chemistry. Most observers do not share his sentiment that "The Golden Age of mathematical expansion may well be over." Indeed, scientists have asserted before that knowledge of some field is nearly complete only to have the field revolutionized by a new discovery. Still, the prospects for an individual mathematician to make a historic contribution do not seem as bright today. Again, while this may not be a critique of the intrinsic value of mathematics itself, it at least serves as a comment on current human limitations.

As we suggested earlier, one argument for the intrinsic value of mathematics is the timeless nature of mathematical facts. Mathematics has immutability to it. Of course, the weight given to this argument depends on one's philosophical position, but when this reasoning is applied to current work, two problems appear. First, current work usually deals with the more esoteric mathematical objects and theories. Without the obviousness of elementary mathematics, these concepts do not demand as prominent a place in many philosophies of mathematics. They are in the upper tier of the two-tiered epistemologies of Gödel and Maddy. That is, they do not "force themselves upon us" through some intuitive or naturalist means.[25] A second problem is that these objects, and the questions being posed about them, of-

24. Keith Devlin, "The End of Mathematics," *Focus* 17 (Washington: The Mathematical Association of America, 1997), p. 2.

25. In the lower tier, axioms are justified by their simplicity and intuitiveness. Concepts in the upper tier, such as cardinalities of infinite sets or the axioms used to refine set theory, are justified extrinsically by their consequences. See Maddy, *Realism in Mathematics*, p. 33.

ten appear to be contingent on current developments in mathematics or applications. Mathematics seems more likely now to discard concepts as it evolves than in previous eras. Of course, even forgotten mathematics can be viewed as eternally true, but the proliferation of mathematical objects reduces the sense of timelessness in terms of their availability to be shared by a temporal community.

Because of the difficulties sketched above, a person seeking motivation for her work in pure mathematics may encounter a dilemma. On the one hand, she would like to believe that she is pursuing fundamental truth. But when she looks at her project, it seems much smaller than that. Typically, it involves making incremental progress on an idea that already existed when she started. How can the rhetoric that inspired her with the ideals of beauty, timelessness, and truth possibly apply to her work? How can her contribution ever be noticed and used outside of her specialty, much less outside of mathematics? In addition, if she is honest about these things, why should she dedicate her professional life, or even her higher education, to mathematics? Many people are still choosing to enter mathematics, particularly mathematics education. However, they may not view their own work as interesting to anyone other than themselves and instead point to more famous work, such Andrew Wiles's proof of Fermat's Last Theorem, when talking to those outside their specialty. They may teach but never discuss their research with their students. The usual defense of this practice is that the research is not accessible. A more honest reason may be that they do not think their students need to know about their research. The dilemma is likely to be more serious for a person who sees a career as a calling. While we do not pretend that what we have described will inevitably occur, it is an important consideration, and we will return to it in the last section.

Valuing Mathematical Practice

To summarize, traditional intrinsic value and future use arguments cannot be used as a blanket justification for all mathematics being done today. They can justify some projects, but not all, and must be balanced against other priorities. If we view these priorities as primarily social, we must make intrinsic value arguments that ultimately talk about benefit to society, though in the broadest sense, including artistic and other kinds of value. Hence, the ability to communicate the work beyond specialists takes on added importance. Our proposal for thinking about intrinsic value will include some criteria for evaluating different projects. We are not proposing guidelines with the intent of restricting the freedom of mathematical pursuits. Rather, our primary goal is to help individuals who are trying to reconcile their mathematical interests with their background beliefs and values. We recognize, of course, that when resources are scarce, allocating those resources requires difficult value judgments. We aim to contribute to a discussion of how to make such judgments as well. Before laying out our proposal, we address two possible objections to any effort to assign value in mathematics.

First, there is a prevalent attitude that one learns what is good mathematics by seeing and doing it, not by discussing values. The knowledge needed by the person entering the field will rub off on her. The classroom clearly reflects this attitude. Students are shown proofs but rarely told what constitutes a proof. They are given research problems but rarely told what makes a good research problem. According to this view, discussion of values, particularly by outsiders, is not helpful. It is seen as associated with "science policy" and budget setting, rather than with mathematics proper. In fact, given that there is so little discussion of what is valuable in mathematics, there is a surprising degree of agreement about it; Borel makes this point. More so than in most other disciplines, there seems to be a consensus about what is the best work being done. There is also an unmistakable unity to mathematics. However, the agreement is limited, partly because the principles or criteria by which specific pieces of mathematics could be evaluated are not articulated and debated. We believe that some distinctions are possible, even without the benefit of hindsight, between the value of different projects. Individual career decisions can also benefit from discussion of standards.

Another objection to discussion of values comes from the positive-normative distinction.[26] A positivist might argue that mathematics should be value-free in the sense that it would investigate abstract structures through dispassionate inquiry but not consider any other values or social goals. Recent critiques of positivist science have led to an increasing recognition of the role of various values and background beliefs. Given these critiques and the related analyses of mathematics, the positivist objection loses its force.

26. "Positive" statements simply say what is; "normative" statements say what ought to be. Recent critiques of positivism argue that it is impossible to make such a sharp distinction.

Underlying the uneasiness with value judgments is a related controversy about philosophy. Should mathematicians philosophize? Philosophy of mathematics is not a field to which a mathematician can easily contribute. Developments such as the "failure" of foundationalism, the retreat to formalism, and the professional ethic that mathematics is objective and impervious to philosophic scrutiny contributed to reduced interest in philosophy of mathematics starting around 1950. The recent rejuvenation of philosophy of mathematics has helped, but it is still avoided by many. Hopefully those who see mathematics as a calling will not share this view. Persons who want to contribute to mathematics as part of a purpose-filled life might very well not contribute to the field of philosophy, but they need to think about the significance of what they are doing. "Our inability to sustain public discussion on values betrays philosophical unawareness and incompetence [in the mathematics community]."[27] It is not enough to say, "I got into algebraic topology because my advisor was working on it."[28] There are bigger stories and larger callings. To work out one's own story and develop a sense of purpose in one's career, a broadly contextual framework is needed for thinking about mathematics. That this task is more difficult in mathematics than in other fields makes it all the more important. Having addressed these objections, we now turn to our proposal.

Our proposed concept of intrinsic value is essentially social. The pleasures that an individual or a small group of specialists derives from mathematical activity, while very real, are not taken to be a sufficient motivation for a career. Additionally, using one's God-given capabilities, a spiritual sense of encountering truth, and recreational mathematics are frequently a part of the mathematical experience. But they are *individual* experiences. We are not proposing a precise way of valuing mathematical projects. However, we are suggesting three characteristics of their value. First, value increases as the number of people influenced by the work increases — more is better. Second, in order for a project to be highly valuable, it must ultimately benefit either the field in which it is embedded or a community larger than the mathematical community alone. Third, as already suggested, we recognize four types of value: (1) purely aesthetic considerations, (2) epistemic ends

27. Reuben Hersh, *What Is Mathematics, Really?* (New York: Oxford University Press, 1997), p. 40.

28. Admittedly, often the reverse is the case; a person chooses an advisor because she likes the field of that advisor.

within mathematics, (3) usefulness within mathematics, and (4) potential applications outside of mathematics. Let us consider these four types of value in more detail.

The aesthetic aspect of mathematics seems as strong today as ever. This may be due to the inexhaustibility of creativity. One might say there are more elegant proofs than important theorems. Beautiful results can be obtained in all branches of mathematics, even if the motivation was quite pragmatic. A concern, however, might be the increased complexity, or "messiness," of the problems being solved today. Long proofs, and particularly computer-aided proofs, are less likely to command aesthetic respect. On the other hand, computer visualization has brought great attention to the beauty of mathematics. Computer visualization is very important to a social view of aesthetic value, because, as Benoit Mandelbrot has observed, it reveals a beauty that was previously hidden to all but a few.[29] For any piece of mathematics, aesthetic value increases with the number of people who can experience that aesthetic.

Epistemic ends within mathematics are achieved when fundamental new mathematical knowledge is obtained. In Kitcher's naturalism, such knowledge is relevant only if it is based, however remotely, on some empirical scientific question. We disagree. It seems to us that the mind can construct mathematics without direct ties to experience. That is, we allow for mathematical knowledge apart from, but not necessarily prior to, other scientific knowledge. However, mathematics only possesses this value to the degree that it is *fundamental.* Such claims should probably be rare and modest. Decisions about what is fundamental rely on the mathematics community and the hopes for consensus discussed earlier. One relevant criterion involves how problems are stated. Topics are more credible when they can be simply stated and recognized by non-specialists. Another is that some problems seem to be relevant because they are an unavoidable test of the machinery. Studying these problems helps build confidence in the consistency and realism of axioms. This type of work should not involve generalization or new definitions, but merely ask questions about objects that are considered important for other reasons. Yet another is that some mathematical topics

29. He is quoted in *For All Practical Purposes* [videorecording 1991], a project of COMAP, the Consortium for Mathematics and its Applications, Inc.; a production of the Chedd-Angier Production Company, South Burlington, Vermont: The Annenberg/CPB Collection. Mandelbrot's pioneering work in fractal geometry has also increased the connection between visual art and mathematics.

have a special role in understanding the nature of mathematics and its limitations, such as dependency on axioms or formal incompleteness. History, foundations, and philosophy of mathematics would be pertinent to this objective.

The potential of a project for usefulness within mathematics and application to science or other thought, we maintain, is not a complete unknown. Rather, thinking through such potential can be a useful exercise — particularly for a person trying to choose an area in which to work. Recall the criteria for usefulness within mathematics discussed earlier given by Borel and Maddy. Those pertinent for this evaluation seem to be (1) power in solving open problems and proving previous conjectures, (2) simplifying and unifying previously unconnected areas, and (3) insights into and simpler proofs of previous results.

Another obvious criterion for usefulness is *dissemination.* This is a very common expectation; however, we might be inclined to disregard its moral value precisely because it is an expectation placed on us by our deans. Other than training our minds, our work influences the world through other people. Telling them about it is just as important as doing it. In addition to communicating clearly, we should recognize that advocating for a good idea is an important job, remembering that models and paradigms compete. Balanced against this imperative is the recognition that most published material will not be used and will only be read by a few. Hence, we should seek the form of dissemination appropriate for our work, which may be the classroom, a departmental seminar, a presented paper, a conversation with a colleague, or a real application.

A Christian Vision for Mathematical Practice

We suggested earlier that a framework for valuing mathematics emanates from a set of guiding principles or "control beliefs."[30] We now make some additional observations that more narrowly constrain the view of value, and that are clearly faith-informed. We do not claim that all of these views are unique to the Christian perspective, but we will refer to our Christian beliefs in developing them. Our concluding chapter will take up some of the ideas we introduce here in more depth.

30. We will also develop this term more fully in Chapter 13.

From a Christian perspective, the question of value takes on added dimensions. The sincere believer seeks to construct a value system based on God's revelation of his purposes. Theological perspectives inform all aspects of life, including vocation. How can a career in mathematics be of service in God's kingdom, and participate in redemption of our culture? The believer seeks normative values that guide her work, not just individual tastes. The values of the profession should be carefully examined to see if they are compatible with her values and commitments.

Aspects of being a mathematician today tend to create a tension with these faith commitments. Along with all scientists, mathematicians may be led by their work into a *de facto* naturalism (thinking that only the natural world exists because we spend our lives studying it) and scientism (thinking that scientific knowledge is the only kind of knowledge). The rare and specialized skills of a mathematician make it easy to be aloof from the affairs of the world. A mathematician may feel that she is not quite as good as other people at doing things outside her profession. On the other hand, outsiders are dramatically less adept at doing mathematics. If she wants to excel, then she has a strong motive to stick to mathematics. The academic culture encourages aloofness through specialization, the solo nature of most research, and the limited rewards for interdisciplinary work.

At the same time, most academics in all disciplines have deep moral beliefs, such as equal treatment for disadvantaged groups, compassion for those who are suffering, and altruism, that suggest a moral responsibility. However, the cultural relativism that dominates academic thought prevents them from constructing an intellectual defense of these beliefs. The relativism they embrace undercuts the moral traditions they wish to maintain.[31] This lack of an intellectual justification that meets the standards of the liberal academy has a chilling effect on living out these convictions. Apart from the moral codes sanctioned by the entire community, such as academic honesty, discussion of morals tends to be privatized and moral action decoupled from one's career, if not neglected completely. Academic work seldom responds to these moral obligations and addresses a specific human need, though it can have a great effect when it does. Indeed, positivist social scientists argue for value-free inquiry that does not address normative issues of

31. Examples of moral stands taken by academics and the dilemma of justifying them are given in George M. Marsden, *The Outrageous Idea of Christian Scholarship* (New York: Oxford University Press, 1997).

value and social goals. The problem faced in reconciling one's sense of moral responsibility with academic life is particularly acute for the mathematician. It can be countered by the theological contexts of creation and God's compassion.

The doctrine of creation recasts many questions about the nature of mathematics. One concern is the holistic nature of truth. If we start with the premise that an omniscient God created reality, then the various ways of knowing or levels of meaning that we confront will possess an underlying coherence. Truth is holistic and God-given. Mathematics conforms to this truth and assists in understanding reality. God's truth and his creation transcend our mathematical knowledge; neither the physical world nor mathematics is autonomous. Belief in creation and a spiritual reality does not require one to ignore the study of the natural world; rather, it puts that study in context. Acknowledging transcendence also fosters humility regarding mathematical ways of knowing, a humility that reengages us with other disciplines and aspects of life. From this perspective, mathematical (and scientific) thought are not seen as complete or exhaustive. Other modes of thought may be as important, or even more important, in understanding broad questions about meaning. Coherence is expected between different modes of thought. Further, inquiry is not absolutized as an end because it is viewed as the study of God's creation, which will be regulated by divine purposes for our lives and balanced with other priorities. The Christian tradition also emphasizes that God has a character, or personhood, that is marked by a love for and valuing of all aspects of his creation, including mathematics. Thus, when studying mathematics, we know we are studying something God values.

Note that these principles do not exclude basic research and do not uniformly point in the direction of areas where applications are apparent. However, the contexts of creation and divine compassion suggest that a career in mathematics must affirm human value. Applications that might serve human needs are one expression of such an affirmation, but by no means are they the only one. Excellence can be pursued and God acknowledged as the source of one's talents and inspiration, people can be treated with dignity and compassion, and teaching can include an appropriate understanding of the nature and limitations of mathematical knowledge. Mathematics, in this context, is not to be pursued literally "for its own sake" but under an obligation to work for God's kingdom in one's scholarship. The framework of social value that we have proposed does not accept the blanket statement that any research, no matter how arcane, is justified if it seeks to uncover more of the truth that God has created. Rather, the value of mathematics must be balanced against other priorities. Mathematics is not done in a vacuum.

In conclusion, then, the challenge each mathematician faces is to develop a more conscious scheme for valuing mathematical work, based not only on what is interesting to the mathematician, but also on what is relevant to serving in our world.

CHAPTER 9

Creativity and Computer Reasoning

Introduction

Throughout this book we have made reference to Kurt Gödel's famous undecidability theorem. It is a landmark result that had tremendous implications on how mathematics was viewed at that time. In this chapter we take a look at a result in computer science that uses similar methods of Gödel. It was proved by the British computer scientist Alan Turing in 1936, and is known as the *halting problem.* We will describe what this problem entails and look at its bearing on an old question dealing with the possibility of computer intelligence: Can a digital computer simulate human thought in its fullest form? Many experts have answered this question strongly in the affirmative. Consider the following quotation from Herbert Simon, who won the Nobel Prize in Economics (1978) for his work in a topic known as "bounded rationality," and who made many contributions to the field of cognition.

> When considered as a physical symbol system, the human brain can be fruitfully studied by computer simulation of its processes. . . . We can formalize our theories for human thinking as difference equations, or, which is the same thing, as computer programs.[1]

1. Herbert Simon, "Studying Human Intelligence by Creating Artificial Intelligence," *American Scientist* 69 (May-June 1981): 300-309.

In more or less the same camp as Simon is Marvin Minsky, noted not only for his work in artificial intelligence, but also for his contributions to cognitive psychology, computational linguistics, mathematics, optics, and robotics. He is currently a research scientist at M.I.T., and has been quoted as saying, "There is no reason to suppose that machines have any limitations not shared by man."

Simon often collaborated with the late Allen Newell, who was a Professor of Computer Science at Carnegie Mellon University for 42 years. Here is a statement taken from one of their essays dealing with the human mind.

> A physical symbol system has the necessary and sufficient means for general intelligent action.[2]

What produced this optimism? In some sense, it represents an epitome of mathematical reductionism by claiming that humans are, at bottom, sophisticated automata. Fueled by modernity, perhaps, such thinking may have been reinforced in popular culture by a series of remarkable computer programs developed in the 1965-75 time frame. Let's briefly examine two of them. The first involves a student, seeking therapy, and a therapist, which in this case happens to be a computer. The program controlling this computer, named ELIZA, was written by Joseph Weizenbaum of M.I.T.[3] Considering that it was composed between 1964-66, it is quite extraordinary.

STUDENT: Men are all alike.
COMPUTER: IN WHAT WAY?
STUDENT: They're always bugging us about something or other.
COMPUTER: CAN YOU THINK OF A SPECIFIC EXAMPLE?
STUDENT: Well, my boyfriend made me come here.
COMPUTER: YOUR BOYFRIEND MADE YOU COME HERE . . .
STUDENT: He says I'm depressed much of the time.
COMPUTER: I AM SORRY TO HEAR YOU ARE DEPRESSED.
STUDENT: It's true. I am unhappy.
COMPUTER: DO YOU THINK COMING HERE WILL HELP YOU NOT TO BE UNHAPPY?

2. Allen Newell and Herbert Simon, "Computer Science as an Empirical Inquiry," in *Mind Design*, ed. John Haugeland (Cambridge, Mass.: MIT Press, 1981), p. 41.

3. Joseph Weizenbaum, *Computer Power and Human Reason* (San Francisco: W. H. Freeman, 1976), pp. 3-4.

STUDENT: I need some help, that much seems certain.
COMPUTER: WHAT WOULD IT MEAN TO YOU IF YOU GOT SOME HELP?
STUDENT: Perhaps I could learn to get along with my mother.
COMPUTER: TELL ME MORE ABOUT YOUR FAMILY.
STUDENT: My mother takes care of me.
COMPUTER: WHO ELSE IN YOUR FAMILY TAKES CARE OF YOU?
STUDENT: My father.
COMPUTER: YOUR FATHER . . .
STUDENT: You are like my father in some ways.
COMPUTER: WHAT RESEMBLANCE DO YOU SEE?
STUDENT: You are not very aggressive but I think you don't want me to notice you.
COMPUTER: WHAT MAKES YOU THINK I AM NOT VERY AGGRESSIVE?
STUDENT: You don't argue with me.
COMPUTER: WHY DO YOU THINK I DON'T ARGUE WITH YOU?
STUDENT: You are afraid of me.
COMPUTER: DOES IT PLEASE YOU TO BELIEVE I AM AFRAID OF YOU?
STUDENT: My father is afraid of everybody.
COMPUTER: WHAT ELSE COMES TO MIND WHEN YOU THINK OF YOUR FATHER?
STUDENT: Bullies.
COMPUTER: DOES THAT HAVE ANYTHING TO DO WITH THE FACT THAT YOUR BOYFRIEND MADE YOU COME HERE?

The computer's impressive inference in the last line, of course, comes from something that the student said earlier in the dialogue. It appears the computer is doing some reasoning, not unlike what a human therapist would do.

Our second program, chosen because we will use it for some arguments later in the chapter, is named SHRDLU, and was completed in 1972. For simplifying purposes imagine that a person is in dialogue with a robot that has a TV camera capable of sensing objects. The robot has been programmed so it can recognize colored blocks and pyramids that are resting on a table. To talk to the robot, a person types commands into a computer terminal. Any response the robot makes is also displayed on the terminal. Comments in parentheses are edited versions of those by the program's author, Terry Winograd, a well-known computer scientist and cognitive psychologist.[4]

PERSON: Find a block which is taller than the one you are holding and put it into the box.
ROBOT: BY 'IT,' I ASSUME YOU MEAN THE BLOCK WHICH IS TALLER THAN THE ONE I AM HOLDING.

(The hand sets down the block it is holding and picks up a taller one.)

PERSON: Can the table pick up blocks?
ROBOT: NO.

(The semantic rules in the computer program make this a nonsensical question which therefore must be false.)

. . . . .

PERSON: I own blocks which are not red, but I don't own anything which supports a pyramid.
ROBOT: I UNDERSTAND.
PERSON: Do I own the box?
ROBOT: NO.

As is the case with ELIZA, it looks as if the computer is able to understand human discourse. One can easily see why these programs may have led to the optimistic statements cited above. They certainly lend credibility to the belief that computers are capable of at least *simulating* human activity. Suffice it to say that many lay people accept uncritically this conclusion. In some circles, one is in danger of being labeled "ignorant" for thinking otherwise. At least, it is a sign of enlightenment to be aware of the capabilities of computers. The desire to be so enlightened perhaps carries with it the pressure to grant that computers can be programmed to duplicate human activities. This pressure comes not only from experts like the ones cited above, but also from statements made from the respected carriers of knowledge to the general culture. For example, the late astronomer and science popularizer

4. Terry Winograd, *Understanding Natural Language* (New York: Academic Press, 1972).

Carl Sagan has reacted to the idea of computer therapy machines in the following way:

> I can imagine the development of a network of computer psychotherapeutic terminals, something like the arrays of large telephone booths, in which, for a few dollars a session, we would be able to talk with an attentive, tested, and largely non-directive psychotherapist.[5]

Marilyn Vos Savant, a columnist celebrated for having the highest tested IQ in history, responded to a reader's query, "Will a computer ever be the world chess champion?" as follows:

> I don't think so. That is, not unless a race car wins the Boston Marathon. If a Computer were called an entrant, it would be no more significant than if a Hydraulic lift were allowed to compete in the Olympic weightlifting event.[6]

Evidently Ms. Savant thinks computers will eventually be far superior to humans in the game of chess. It is not clear whether she has this view for other cognitive tasks, but many readers could well infer that from her statement.

Ray Kurzweil, however, indeed sees computer superiority in a general sense:

> Over the next several decades, machine competence will rival — and ultimately surpass — any particular human skill one cares to cite, including our marvelous ability to place our ideas in a broad diversity of contexts.[7]

The Possibility of Computer Thought

The above remarks are perhaps typical of the optimism accompanying the advances we have seen in the computer revolution. In this section, we will examine beliefs like these with special attention to ideas in computer science

5. Carl Sagan, in *Natural History* 84, no. 1 (January 1975): 10.
6. M. Savant, in *Parade Magazine*, December 14, 1986.
7. Ray Kurzweil, *The Age of Spiritual Machines* (New York: Viking, 1999), p. 5.

and mathematics. Our aim is to cast doubt on the certitude with which such beliefs have been asserted, at least by some people. To do so we will have to examine some ideas very carefully. They may seem at first to be unnecessary diversions, but we hope to show eventually that they all tie together. We begin with some ideas from computer science, specifically with an analysis of exactly what we mean by a computer program, as programs form the basis of any computer activity. We shall learn, perhaps surprisingly, that there are theoretical limitations to computer capability.

The notion of an algorithm is basic to understanding what a computer program is, so it is important to understand the term. It is derived from *algorism,* the process of doing arithmetic with Arabic numerals. According to the internationally known computer scientist Donald Knuth, algorism came from the Persian textbook author Abu Jaʿfar Mohammed ibn Mûsâ al-Khwârizmî, who wrote the book *Rules of Restoration and Reduction* (c. 825). "Gradually, the form and meaning of 'algorism' became corrupted with 'arithmetic,'"[8] hence the word "algorithm."

Informally, an algorithm is an unambiguous, finite, effective procedure (or recipe, or specific set of rules) for performing a task. By "unambiguous" we mean there is a universally agreed-upon set of semantics that we can refer to when following the procedure. The finiteness requirement simply means that the list of instructions comprising the algorithm is finite and that, when following these instructions, we will eventually finish. To understand what we mean by "effective," let us examine an algorithm that lacks this property:

> If your paternal great great great great great great grandfather had a broken arm at any time in his life, then take a shower, otherwise take a bath.

This instruction is certainly finite and unambiguous (ignoring some possible ambiguity in how the term *broken arm* is defined, and perhaps, too, in identifying the person in question). If given this instruction, however, we would not be able to carry it out, since most of us have no way of telling whether or not such an ancestor indeed had that misfortune. Thus, by "effective," we mean not only that the instructions accomplish the task we have in mind, but also that we will be able to follow the given set of instructions.

8. D. E. Knuth, *Fundamental Algorithms* (Reading, Mass.: Addison-Wesley, 1969), p. 2.

In 1936 the British logician, Alan Turing, formalized in a mathematically precise way the notion of an algorithm.[9] We can think of this formalization as a machine (called, suitably enough, a "Turing Machine") consisting of several components, as the diagram below illustrates.

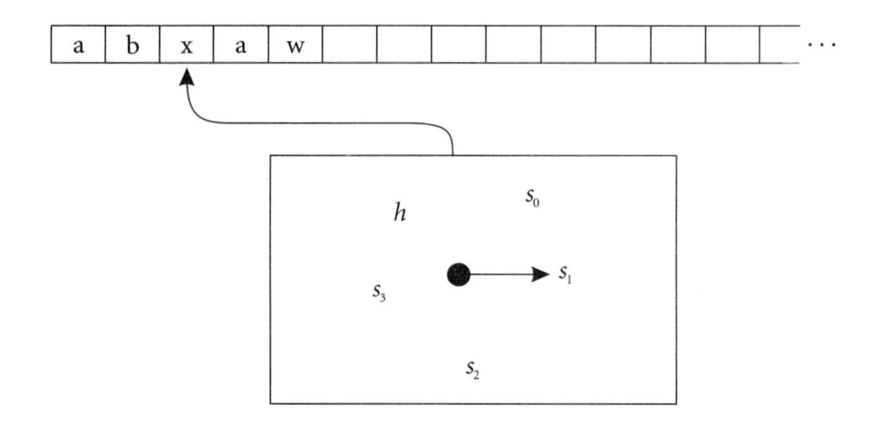

The first component is a tape that is infinitely long in one direction (say, to the right). Below that is a read/write head, capable of moving in both directions. The core of the machine, so to speak, is the box below the read/write head. It consists of several internal "states," labeled as s_i (in the illustration, s_0, s_1, s_2, and s_3) where the symbol h is used to indicate a special halting state, so that if the machine enters that state it stops. Depending on its state and the symbol currently scanned by its read/write head, the machine will enter a (possibly new) state and either write a symbol on the tape square currently being scanned (replacing the one there) or move its read/write head one square to the left or right. A *transition function* (not shown in the diagram, but think of it as the internal workings of the machine) is what controls this process. We can make this very precise mathematically, but the above description will be good enough for our purposes.[10]

9. A. M. Turing, "On Computable Numbers, with an Application to the Entscheidungproblem," *Proceedings, London Mathematical Society* 2, no. 42 (1936): 230-65, and no. 43, 544-46.

10. For those interested, the precise definition of a Turing Machine M is a 4-tuple $M = (K, \Sigma, s_0, \delta)$, where K is a finite set (of states), Σ is a set (of symbols), s_0 is a distinct initial state, and δ is the transition function $\delta : \mathrm{K} \times \Sigma \rightarrow (\mathrm{K} \cup \{h\}) \times (\Sigma \cup \{R,L\})$, where h is the halt

Turing's formalization captures the essence of what we think of as a computer program. That it corresponds exactly to what one informally means by an algorithm is often referred to as the Church-Turing thesis. Virtually every practicing computer scientist accepts this thesis. Regardless, it is Turing's formalization of an algorithm and its extension to a general all-purpose computer (called the *Universal Turing Machine*) that is the theoretical base behind every digital computer in existence today. What can be proved in a theoretical sense about Universal Turing Machines *must* hold true for digital computers and their programs. If it can be shown there exists no algorithm for solving a problem, it follows that no computer can be programmed to solve it. Additionally, if we have any reason to think it possible that parts of human thinking are not algorithmic in the sense just described, we would likewise cast doubt on the certitude of claims made by Kurzweil and others regarding the possibility of computer intelligence.

Computer scientists have classified problems and algorithms used in their solution in various ways. To understand the importance of this, we will look briefly at two different problems. The first one involves the question as to whether an algorithm exists for sorting a list of numbers into increasing order. Indeed, there is. Below is one such example:

1. Search through the list and find the smallest number.
2. Write down this number.
3. Cross that number off the list.
4. If the list is not all crossed out, go to Step 1, otherwise go to Step 5.
5. Stop.

For the purpose of arguments to be made later, it will be useful to analyze how much time our sorting algorithm takes when it is implemented on a computer. The exact time, of course, will depend on the type of computer we have, but it will in any case be proportional to the amount of checking that the algorithm does. When we sort four numbers, Step 1 requires initially three comparisons to find the smallest number. (We take the first number and compare it with the second through fourth numbers, keeping track

state (not contained in K), and R and L are special symbols (not contained in Σ) so that if $k \in K$, $\sigma \in \Sigma$, and $\delta(k, \sigma) = (p, q)$, then the Turing Machine M, when in state k and reading symbol σ, will enter state p and write the symbol q on the tape, unless q is either R or L, in which case the Turing Machine will move its read/write head to the right or left respectively.

of the smallest number found at each stage.) The second time through we need only make two comparisons, as there are only three numbers left on the list. Finally, we make one comparison the last time through our process, giving a total of 3 + 2 + 1 = 6 comparisons. In general, it can be shown that this algorithm will require

$$\frac{n(n-1)}{2} = \frac{n^2}{2} - \frac{n}{2}$$

comparisons when sorting a list with n numbers. Computer scientists are interested in worst case scenarios of algorithms, especially when n, the input size of the problem, gets quite large. (By "worst case" we mean the most comparisons that could possibly be made by the algorithm in question when it operates on lists of a fixed size.) When this happens, the dominant part of the expression

$$\frac{n^2}{2} - \frac{n}{2}$$

is the term

$$\frac{n^2}{2},$$

and for this reason we say that the sorting algorithm given above is "of order n^2," meaning that if n is large, we must make (very roughly) n^2 comparisons in order to sort a list with n items in it. Admittedly this is a crude estimate, since not only did we at first drop the

$$\frac{n}{2}$$

term, but we are now ignoring the factor of 2 in the denominator of the expression

$$\frac{n^2}{2}.$$

The reason this is done is for the sake of making relative comparisons. If we had an algorithm that was "of order n^2" we would know that, in the worst case, if n were large and we doubled its input size, the time required to run the algorithm would increase roughly by a factor of 4. For our sorting example, this means that if it took 1 millisecond to sort 1000 numbers on a computer in the worst case, it would take about 4 milliseconds to sort 2000 numbers in the worst case, provided we used the same algorithm and computer each time.

There are better sorting algorithms than the one given above. It can be shown, however, that no algorithm for sorting can do better than to make "on the order of" $n \log n$ comparisons in the worst case. Computer scientists have a description for problems that can be solved by algorithms like the above, whose worst case running time is no longer than some polynomial function of the length of their input. (In our sorting example, the polynomial would be $p(n) = n^2$.) Such problems are said to be in the class P (for polynomial).

There is another class of problems, known as the *NP*-complete class, whose solutions probably require much more time to solve than problems in the class P.[11] They have the common property that as the length of their input increases, the time required for their solution grows faster than any polynomial function, even when we use the best known algorithm for their solution. The intriguing quality about problems in this class is that not one of them has a known algorithm that takes "polynomial time" in the worst case, but if a polynomial time algorithm were found for any one of them, they *all* would have polynomial time algorithms. Most computer scientists believe that no problem in the *NP*-complete class has a polynomial time algorithm, but this is not known for sure, and whether or not this is the case is perhaps the biggest unsolved problem in theoretical computer science today.

A classic example of something in the *NP*-complete class is the "traveling salesman" problem. A salesman has a list of several cities or towns that he must visit from his home base. To save time and money, he wants to know the ordering of the cities that will minimize the total distance he travels in going to all of them.[12] How can we determine this?

One algorithm for solving this problem would be to calculate the distance traveled for each possible itinerary. But this approach would not be very wise; for the possible ordering of the cities grows "super-exponentially" as the

11. For a more complete discussion, see M. R. Garey and D. S. Johnson, *Computers and Intractability: A Guide to the Theory of NP-Completeness* (San Francisco: W. H. Freeman, 1979).

12. Technically, the problem is defined as follows: We are given a list of cities, a cost involved in traveling between each pair of cities, and an integer N. The question we must answer is whether there is an ordering of these cities so that the total cost in traveling to all of them is less than N. The formulation requiring us to find the minimum distance, however, is practically equivalent to this.

number we have on our list increases.[13] With only 10 cities to visit, there are over 3.5 million ways to order them. For computers, this is not a very large number, and the fastest computer today could do all the required calculations in less than a second. Unlike our sorting problem, however (when doubling our list merely quadrupled the time needed to run our algorithm), if we doubled the number of cities involved to 20 and ran through the algorithm again, the time now required to solve the problem (again on the fastest computer available) would take more than 100 years. If we were to use this algorithm on a list of only 100 cities and had a computer with the fastest clock speed physically possible at our disposal, and if we had started the computer on its task when the universe was created, it would not be close to finishing yet.

Many people intuitively feel that a computer could perform almost any computation rather quickly, and that, in any case, as technology increases and computers become faster, there is virtually no problem that would take a computer very long to solve. Hopefully the above example illustrates that this is not the case. There are some problems that, as their size increases, require an unreasonable amount of computer time to solve, due to the "exponentially" increasing number of tasks the computer must do.

David Hilbert, surely one of the greatest mathematicians of the late nineteenth and early twentieth centuries, gave our second problem for consideration in one of the leading addresses at the second International Congress of Mathematicians held in Paris in 1900. At that time he actually proposed a list of 23 problems as suggested targets for mathematical research in the future.[14] It is indicative of Hilbert's genius that many of these problems are still actively being discussed. His tenth problem is the question as to whether or not there can be an algorithm for deciding if there are integer solutions to equations involving any number of variables and their integer powers. Such equations are known as Diophantine equations. The equation $x^2y^3z + 2xy^2 + 5z^2 - 2101 = 0$, for example, has integer solutions $x = 2$, $y = 4$, and $z = 7$, but if we hadn't been given these numbers to check, how would we have known ahead of time whether or not there were any integer solutions? In 1829 the French mathematician Evariste Galois proved that not all equations have formulas we can use to solve them. Notice, however, that there is a

13. That is, starting from and returning to a fixed home base, the requirement to visit n additional cities yields $n!$ possible itineraries.

14. David Hilbert, "Mathematische Probleme," *Proceedings of the International Congress of Mathematicians in Paris 1900* (Göttingen: 1900), pp. 253-97 (English translation, *Bulletin of the American Mathematical Society* 2 [1901-1902]: 437-39).

difference between having an algorithm for finding the solutions to an equation and having an algorithm that determines *whether* there are (integer) solutions to that equation. Hilbert asked about the latter situation.

In 1970 the works of Julia Robinson, Martin Davis, and Yuri Matiyasevich solved Hilbert's tenth problem. They showed that there is no algorithm for determining whether Diophantine equations have integer solutions.[15] This result is quite interesting. It means that, logically, using the rules of mathematics, it is impossible to come up with any mechanism that we could apply to *all* types of Diophantine equations to see ahead of time whether or not they had integer solutions. At the risk of sounding blasphemous, this means that not even God could come up with such an algorithm. This does not mean that God would not be able to solve the problem, only that if we *limited* him to using the rules of mathematics, he would not be able to produce an all-purpose *algorithm* for solving the problem, much like he would not be able to come up with an algorithm guaranteed to win at tic-tac-toe, since there is no such strategy.

As we shall see shortly, there is an amazing connection between Hilbert's tenth problem and part of Turing's 1936 paper that dealt with what is known as the "halting problem," which related to the debugging of computer programs. Consider the following two computer programs:

PROGRAM 1

1. Set x equal to 1.
2. Replace x with double its current value.
3. If the current value of x is less than 10, go to Step 2, otherwise go to Step 4.
4. Stop.

PROGRAM 2

1. Set x equal to 1.
2. Replace x with double its current value.
3. If the current value of x is less than 10, go to Step 1, otherwise go to Step 4.
4. Stop.

What would happen if we ran these programs on a computer? We notice that program 1 will stop after a few cycles between steps 2 and 3. Program

15. Yuri Matiyasevich, "Enumerable Sets Are Diophantine," *Doklady Akademii Nauk SSSR* 191 (1970): 279-82. A popular explanation of this result is given by M. Davis, "Hilbert's Tenth Problem Is Unsolvable," *American Mathematical Monthly* 80 (1973): 233-69. See also Yuri Matiyasevich, *Hilbert's Tenth Problem* (Cambridge, Mass.: MIT Press, 1993).

2 has a fatal flaw that resets x to 1 at every cycle, so x will always be less than 10 when we get to Step 3. Program 2 will never stop. It is an example of an "infinite loop," something that every programmer has inadvertently written! The question of the halting problem is this: Is it possible to write an all-purpose computer program that would be able to decide, with respect to *any* given computer program, whether or not it would eventually stop? Turing was able to prove that there can be no such program. This is a theoretical result of a practical problem that no computer program will ever be able to solve.

This is a strange result because, on a quick analysis, this appears to be a problem that human beings *can* solve, so that computer experts at least could read any possible computer program and tell whether or not it would stop. If pressed to explain how the experts would do this, we could respond by saying that their strategy would be to read through the computer code of any given program and analyze what it was doing at every step. After all, this is precisely what we non-experts did in deciding whether programs 1 and 2 in our most recent example would ever stop. If it is stated that this argument is not good enough, that we must (to be precise) describe in detail how the analysis would proceed, we could then claim that now a demand is being made to produce an algorithm that our experts would follow in making such determinations, and it is precisely Turing's result that no such algorithm exists. Perhaps there is some capacity humans have (call it creativity if you will) that enables them to solve problems for which there is no algorithm. To some extent, academic assignments that demand the greatest amount of student creativity also cause the greatest anxiety. Students frequently complain that they do not know exactly how to proceed, for example, in writing a paper. What they mean by this, at least in part, is that they do not have an algorithm they can follow. Yet the papers are eventually produced. Could this be the example (of some thinking process people perform for which there is no algorithm) we were looking for earlier in this chapter?

If we think so, we would be jumping to conclusions too fast, as the above analysis is loaded with pitfalls. First, computer experts cannot even really tell whether or not a program will *compile*. To say a program compiles means it has the correct syntax for the language (e.g., C++) in which it is written. Every computer language has what is called a *compiler*. This is a program (i.e., an algorithm) that can decide, with respect to any given program, whether or not it has the correct syntax in that particular language. Since there *are* algorithms that can make this determination, in some sense the question of compilation is a much simpler problem than the halting problem. No practicing computer programmer today, however, would be able to tell with any certainty that a sufficiently complex program, sent to a computer for the very first time, would compile. How do we really expect such a person to tackle the question of whether one would terminate, a question for which there can be no predetermined algorithm?

This concern, however, misses the distinction between a logical error and a logical limitation. The reason programmers mistakenly believe some programs will compile when they really won't is due to some oversight on their part. In the sense that correctly programmed computers *never* make logical errors with respect to their given tasks (barring hardware failure, of course), they outperform human beings. In principle, however, this is not a human limitation per se, only a practical observation of human carelessness. Perhaps people can, *in principle,* determine whether or not programs will compile. We only need to allow room for correcting inadvertent errors they make once these errors are pointed out. After all, it is people that write the compiler programs we rely upon to make compilation judgments. Perhaps, then, people *could* solve the halting problem, barring the commission of logical errors.

Or could they? Let's pretend for the moment we have found a computer scientist whose logic is always infallible. Do we believe that person could solve the halting program for a *lengthy* program that was billions of lines long? Of course not. In fact, when we think about it, we would agree that this "limitation of capacity" would also be telling with respect to a person's ability to decide if a program would compile. Even if given logical infallibility, it stretches credulity to believe that a human would be able to keep track of a program that utilized billions of variables. Such a limitation is not present in Turing Machines.

But this is the one area where Turing Machines are not faithful models of computers. Recall that a Turing Machine comes equipped with an *infinite* memory (its infinitely long tape). No computer could possibly have this capacity. Given any computer, it would be quite easy to create a program *its* compiler would not be able to handle. We would simply include more variables in this program than would fit into the computer's memory.[16]

16. More precisely, each variable would require a symbol pointing to a memory reference containing what we call the value of the variable. Even if variables had shared memory references, the memory needed to store the symbols referencing them would be exceeded if there were too many variables.

This forces us to complicate our argument somewhat. We are interested in determining whether a computer could possibly simulate human thought in its fullest form. In our investigation, we have found something that no computer can do, not even an idealized one with an infinite memory capacity. Let us, for the moment, imagine we had an idealized person with an infinite memory capacity. If we believe such a being would be able to solve the halting problem, by analogy we might be led to conclude that the capacity allowing our "superhuman" to do this is shared by real people as well, possibly pointing at least to some capacity for thinking that humans possess and computers do not. In taking this approach we must concede, however, that we are speaking metaphorically at best, so that our argument will be far from compelling. Perhaps, though, it will be sufficient at least to accomplish what we have set out to do — cast doubt on the certainty of the position that claims that in principle computers can duplicate any human activity.

What is it, then, that *does* trip up a computer on the halting problem? The method Turing used to prove the undecidability of the halting problem is what mathematicians refer to as diagonalization. Basically, Turing found a way to list *all* the programs for which any given algorithm could possibly make a halting decision, and then he cleverly created a program not on the list.[17] Since the list contained *all* programs for which the algorithm could make a halting decision, the program created could *not* be determined by that algorithm eventually to halt or not halt.

Clearly, the algorithm cannot make a decision because it lacks *some* capacity (again, we could think of it as creativity, even though we have not been precise about this term). A program is produced that the algorithm does not, in some sense, expect. The method of Turing's proof could arguably be described as one that, seeing what the program can do, creates something the program cannot handle within the context of its rule-governed structure. Maybe human beings have this (creative) capacity — to be able to jump out of any rule-governing system they are in and to handle the unexpected.

It could be argued, however, that the limiting factor in making "halt-

17. This is somewhat of a simplification, as we are ignoring the complicating factor of program inputs. Turing's method was very similar to the procedure Georg Cantor employed in proving the uncountability of the irrational numbers, however, and also not unlike the method Kurt Gödel used in his proof of the existence of formally undecidable propositions.

ing decisions" is really one of logic — not logical precision, but logical limitation. Consider the following program:

1. Set x equal to 2.
2. Replace x with a number 2 units larger.
3. If x can be written as a sum of two prime numbers, go to Step 2, otherwise go to Step 4.
4. Stop.

Does this program ever stop? It will go into an infinite loop if all even numbers larger than 2 can be written as the sum of two prime numbers. That this is indeed the case is known as Goldbach's conjecture,[18] but no one knows if it is true. With the aid of a Cray C90 computer, Jean-Marc Deshouillers, Yannick Saouter, and Herman te Riele verified in 1998 that the conjecture is true for all possible numbers up to 100,000,000,000,000. This might make us believe Goldbach's conjecture is true in general, but we still don't know for sure, as there are infinitely many numbers left to check! The only way to be certain is to come up with a proof. At the moment, there is none. Hence, no human being can, at the moment, possibly determine whether or not the above program will stop.

It might be argued that eventually we can count on mathematicians to solve Goldbach's conjecture. Unfortunately, even if this were to happen, we cannot get around Gödel's result that (if mathematics is consistent) there exist problems that cannot in principle be solved. It is a limitation of our logic itself that is the culprit here.

But where has this led us? Is there anything we have said that might give an indication that both creativity (or whatever we call it) and logical limitation are central to the undecidability of the halting problem? This is a difficult question, and perhaps the best we can do in answering it is by a very loose indirect argument. We have already suggested that the method of proof concerning the halting problem might point to some quality humans possess that computers lack. As another indication, recall that Hilbert's tenth problem (involving integer solutions to Diophantine equations) was solved

18. Actually, Goldbach wrote a letter to Euler in 1742 conjecturing that every integer greater than *five* is the sum of *three* primes. Euler observed in a reply that this would be the case if every even integer greater than two is the sum of two primes. Euler's observation is what we now refer to as Goldbach's conjecture.

in 1970. It was proved that there is no all-purpose algorithm we can use to determine whether Diophantine equations have integer solutions. Remarkably, it turns out that Hilbert's tenth problem is "reducible" to the halting problem. This means that a mechanism has been found that will take any given computer program (i.e., an algorithm) and convert its encoded symbols of 0's and 1's into a Diophantine equation. Furthermore, the conversion is such that the program will terminate if and only if the corresponding Diophantine equation has integer solutions. Hence, since by Turing's result there is no decision procedure for the halting problem, there can neither be one for determining whether or not Diophantine equations have integer solutions.

Continual progress is being made, however, in coming up with decision procedures for special classes of Diophantine equations. The most famous example of this was the proof in 1993-94 by Andrew Wiles of Fermat's Last Theorem. We shall return to this theorem and Wiles's method of proof towards the end of this chapter. For now we pose the following question: Given that there can be no general algorithm for making a decision as to whether Diophantine equations have integer solutions, how do we explain the progress that mathematicians are making? A possible explanation is that people are creative, in the sense that they can hop out of a rule-governing system, whereas algorithms (and therefore computers) cannot. But what arguments do we have for this? Our suggestion cannot be pressed too far, for once a mathematician has discovered some results, the theories developed could easily be turned into an algorithm for solving whatever had been investigated, and with it a computer could duplicate whatever our mathematician friend had discovered. Even though mathematicians might make progress in Diophantine equation analysis, they will never to be able to have several categories of theories that collectively make decisions regarding all types of Diophantine equations, for to do so would effectively create an algorithm, violating the fact that no such algorithm exists. There is a crucial distinction that needs to be made here, however, and it has to do with ex-post-facto algorithm design versus the creating of heretofore-unknown algorithms. Perhaps it is the *creation* of such algorithms that causes computers trouble. In their basic operations computers do not create things, they merely follow pre-stated rules.

We are now into all sorts of difficulties. It could be argued that in an ultimate sense people are not creative either, but are merely following rules that are built into their genetic code. Perhaps what we are calling creativity is really nothing more than a complex interaction of these rules. Many arguments could be used to support this viewpoint, and it is to some of these arguments that we now turn.

Indeed, many have argued that in a biological sense we really are digital computers. It has been well established, for example, that some types of neurons in the brain fire in "all-or-nothing" bursts. This behavior seems to parallel the "on-off" states that exist in the transistors of a digital computer. Additionally, our entire developmental process seems to be controlled by a DNA code. Could we not view this "code" as a hardware/software mix which, when combined with environmental factors, constitutes a complete computer/software package?

The fact that some neurons fire in such bursts might imply that the brain functions as a digital computer, but only if each burst (or combination of bursts) corresponded with *some* activity of a Turing Machine, whose basic functions involve either changing states, creating or reading a symbol, or repositioning the place where the creating or reading of symbols is occurring. If, on the other hand, each burst does not necessarily have such a correspondence, the argument weakens considerably. Recent evidence seems to indicate that the *rate* of burst transmission is, in fact, the essential component in determining meaningful activity of the nervous system. It also appears that axon diameters affect the speed at which nerve pulses are transmitted. It is difficult to see from this how the above correspondence could be derived, but even if such a correspondence could be theorized, it appears that the processing of these neuron bursts occurs in some global way, which is now only beginning to be understood. In any case, this global processing of information appears to be quite unlike digital (sequential) processing. The brain, then, may be much more of an analog device than a digital one.

It is here that some folks might step in and claim that whether the brain functions in a digital or analog manner is irrelevant. Even if the latter were the case, it is well known that analog devices can be simulated digitally. Furthermore, it can be reasonably argued that human behavior is generally orderly. If orderly, then rule-governed. But if rule-governed, then controlled by an algorithm (for this is one sense of what "rule-governed" means), and therefore, *simulatable,* at least, by a digital computer. Besides, they could argue, the empirical evidence to date supports the theory that computers *can be programmed* to be intelligent. Newell and Simon state their case as follows.

> There is a steadily widening area within which intelligent action is attainable. From the original tasks, research has extended to building systems that handle and understand natural language in a variety of ways, systems for interpreting visual scenes, systems for hand-eye coordination, systems that design, systems that write computer programs, systems for speech understanding — the list is, if not endless, at least very long.[19]

They press their argument even further.

> The empirical character of computer science is nowhere more evident than in this alliance with psychology. Not only are psychological experiments required to test the veridicality of the simulation models as explanations of the human behavior, but out of the experiments come new ideas for the design and construction of physical-symbol systems.[20]

To the already impressive looking list of computer accomplishments in the area of intelligence simulation, we might add that a computer chess program has beaten the current (human) world chess champion, something many people predicted would never happen.

How can we fairly evaluate all this? At least two claims have been brought forth: that there is an ever-increasing area in which intelligence can be programmed, and that in any given area continual progress is being made. Let's look at these claims in greater detail.

Hubert Dreyfus has suggested a helpful classification scheme for intelligent activities: the associationistic, the simple-formal, the complex-formal, and the non-formal.[21] The associationistic level is characterized by problems that are irrelevant in terms of meaning, such as word-by-word translation exercises, or random memorization games — not to be confused with final examinations week! Computers clearly outperform people at this level. The kinds of programs involved are generally of a "list search" nature, such as a program that would determine if words entered in a document were possi-

19. Newell and Simon, "Computer Science as an Empirical Inquiry," pp. 47-48.

20. Newell and Simon, "Computer Science as an Empirical Inquiry," p. 49.

21. Hubert L. Dreyfus, *What Computers Can't Do* (San Francisco: Harper and Row, 1979), p. 292.

bly misspelled as judged from their failure to match a set of words in an existing database.

In the simple-formal category belong tasks whose meaning is completely explicit and situation independent, such as the game of tic-tac-toe, or the algorithm given earlier for solving the traveling salesman problem. Again, computers outperform people in this area. The type of program used to solve problems here is generally one that involves a decision tree of some kind.

It is in the complex-formal area that computers begin to have difficulty. In principle, this area is the same as the simple-formal, but in practice there is an internal dependence on the situation under scrutiny as is the case with games such as chess. We will return to a discussion of this area shortly.

Finally, the non-formal area includes ill-defined games such as riddles and the translating of natural languages. Such activity is characterized by problems that are situation dependent for their meaning. It is here that computers perform very badly. The focus of this chapter has been to suggest that, when compared to humans, computers will never be in the same league.

This certainly must raise some eyebrows. Does not the robot in the SHRDLU program (involving blocks and pyramids) understand, at least to some extent, natural language? If by *understand* one means *respond in a way that looks reasonable in a limited context,* then the answer is certainly in the affirmative. But surely much more than this is involved in understanding. The reason the robot responds, "NO," to the question, "Do I own the box?" for example, is because the label 'own' has not been tagged to a (person, box) pair in the robot's memory. But this is not what comes to mind when *we* think of something being owned. It is not clear at all that the robot has some idea of what it *means* to own something.[22] Moreover, we can quickly see that the robot would not even respond in a way that looks reasonable if we enlarged the context slightly. Let's consider the beginning portion of the dialogue:

PERSON: Find a block that is taller than the one you are holding and put it into the box.

ROBOT: BY 'IT' I ASSUME YOU MEAN THE BLOCK WHICH IS TALLER THAN THE ONE I AM HOLDING.

22. This notion of meaning is one that we must leave without further nuance, as there is simply no room in this chapter to go into more detail at this point.

The reason the robot can identify the proper pronominal referent to *it* is because it (by this last 'it' we mean the robot, of course!) has been programmed with linguistic rules. Admittedly, the rules are complex. They can determine subordinate clauses based on sentence structure alone, and, in *this* context, correctly identify pronominal referents. But it is more than use of linguistic rules that enables people to perform this task. Consider the following standard example:

> The rock fell on the drinking glass, and it broke.
> The drinking glass fell on the rock, and it broke.

We all know that *drinking glass* is the pronominal referent of *it* in each sentence. Is it the rules of grammar that give us this information? Hardly. It is our knowledge of the world. We know or infer by experience what happens when rocks and drinking glasses meet.

It could be argued here that this example is merely an extension of complexity to the block-pyramid situation rather than something qualitatively different. A more complicated program could be written that would include checks of relative hardness of materials and rules incorporating those checks. It could also be suggested that this is exactly what goes on inside the human mind, at least unconsciously.

Again, life is not so simple. If isolated sentences complicate things so much, imagine what short dialogues do. Modifying a bit a construct from Dreyfus, consider the following:

> ANN: Thanks for taking me shopping, Joe. It's Billy's birthday next week. I have to decide between a dump truck, a basketball, and a kite.
>
> JOE: Billy's mother told me he doesn't like playing in sand boxes any more, so one of your choices is gone.
>
> ANN: I know what I'll do! Billy's father told me that Billy loves Magic Johnson.
>
> JOE: Wait! You'd better not get that either. His father just gave him one. You'd have to take it back.

Several pages would be required to describe in detail all the inferences we make when reading the above dialogue. A detailed analysis might even surprise us when we discover that the text does not mention (a) that the items Ann suggests are to be *bought,* (b) that they are candidates for *gifts,* (c) that the intention is to *give one to Billy,* (d) that initially Anne is thinking of buying a basketball, (e), that . . . These are inferences we almost automatically make as a result of our "being-in-the-world."[23]

For the moment, let us suppose that, somehow, rules of inference could be incorporated into a computer program to allow the above deductions to be made. We are still faced with a massive data problem. It is dump trucks, not basketballs and kites, which children use in sandboxes. Our program would now have to include, with each item in its database, a mechanism for associating any possible connection with other items in the database. The noun *basketball* would have to be connected with the entire N.B.A. roster, or at least a large subset of that roster, both of active and retired players, in order to be able to make the connection with Magic Johnson — not his correct first name, by the way. For other possible inferences, it would have to be connected with companies that make basketballs, stores that sell basketballs, and other people who play basketball, not to mention connections from them to other relevant entities such as their parents, their teachers, what type of shoes they wear, what they eat. . . .

Matters are even worse than this. When Joe says, "You'd have to take it back," we are able to infer that *it* refers to the *new* basketball (and not the old) because of our cultural practices, not by any deductions or sentence semantics. (We can easily imagine a culture where the *old* item is returned after a new one is received. Perhaps certain automobile sales transactions model this already.) Now our program must somehow encode habits, mores, religious beliefs, etc.

Assuming, again, that all this encoding could be physically accomplished, and we hope we have at least created some doubt that it could be, we are still faced with an analogous situation to our traveling salesman problem. It will arguably take far too long for any computer to sort through the exponentially increasing number of connections that even the most simple data item generates.

But this claim, too, has been challenged. First, on the grounds that we are, at the most, only talking about several billion connections, not the magnitude of the sort we run into when looking at possible orderings of cities. Second, even if the number of connections turns out to be prohibitive, peo-

23. This is a key idea for Dreyfus. The phrase is due to the well-known philosopher Martin Heidegger. For more on this, see Heidegger's *Being and Time.*

ple themselves do not always understand dialogue perfectly. Perhaps their minds contain *heuristic* procedures (rules that determine how to apply rules), which only trace connections through "likely" paths. Of course, sometimes the heuristics fail and misunderstanding ensues.

Indeed, there are algorithms that "practically" solve the traveling salesman problem in a short amount of time. (By *practically* solve, we mean they produce with amazing consistency a solution that is close to the optimal one.) Let's briefly look at one such algorithm, known as *simulated annealing*.

Simulated annealing was introduced in its present form in 1983,[24] although slightly different versions were used as early as 1953. It has been applied to a wide class of problems, and has been the subject of much research.[25] One reason simulated annealing succeeds is because there is a numerical measure that can determine whether progress is being made in solving a problem (e.g., whether a new ordering of cities to visit will result in a shorter total traveling distance than the shortest route found so far in examining possible itineraries for the traveling salesman problem). At the end of the day, programs that simulate decision-making by employing heuristics invariably rely on some kind of numerical criterion for making their decisions, even if that criterion is a probabilistic one or a table-based rule guide. Curiously, an argument against there being a numerical function of some type that humans employ (even if unconsciously) in day-to-day experiences is the difficulty in mimicking intuitive strategies in making decisions for situations belonging to the "complex formal" area, such as what needs to occur in computer chess programs. These strategies were tried early on, but they have virtually been abandoned in favor of "brute force" approaches. But it is the intuitive approach that every computer chess expert uses.

It is generally agreed, for example, that one capability successful chess players possess is the ability to "zero in" on a likely route to pursue. To be sure, they also have the ability to count out several moves ahead, but without being able to zero in, a hopelessly large set of possible moves to consider (at least, for a human being, not for a computer) would prevent great play. What enables these players to zero in? Partially, it is the ability to recognize similarity between a current situation and one that has been extensively analyzed.

24. S. Kirkpatrick, C. Gelatt, and M. Vecchi, "Optimisation by Simulated Annealing," *Science* 220 (May 1983): 671-80.

25. M. Lundy and A. Mees, "Convergence of an Annealing Algorithm," *Mathematical Programming* 34 (1986): 111-24.

For example, Max Euwe and Walter Meiden made the following comment on a particular stage of a Karpov-Spasky game: "As was also the case in the Peters-Larsen game, the White Bishop now presses against d5 and increases the mobility of the White Queen. True, this Bishop no longer controls c4, but that is not so important now. . . ."[26]

Whether or not mechanisms will be found enabling chess programs to recognize "similarity" of positions to the degree people are able remains to be seen. Many think this is doubtful. Contrary to the opinion of Marilyn Vos Savant, Douglas Hofstadter believed that the only way a computer program could be world chess champion is if it also possessed "general intelligence" and would, therefore, be ". . . just as temperamental as people."[27] Evidently Hofstadter was incorrect in his assessment, but it might be argued that this is only because he did not envision, in 1980, the amount of brute force computation computers would be able to produce. Although Hofstadter would not argue against the theoretical possibility of being able to program a computer to have general intelligence, no one would point to any computer program close to having such capacity today. In this chapter we are attempting to cast doubt on that possibility, contrary to the optimistic quotations we examined early on.

If creating proper heuristic procedures is difficult in a well-defined, unambiguous game such as chess, it is arguably impossible in real-world situations, where meaning is not only situation specific, but dependent on a subjective person. Furthermore, even if heuristics *could* work in a complex situation, additional rules would have to be added to interpret the heuristics and allow for exceptions. In our block/pyramid example, the robot may know the table in its current world cannot pick up blocks, but what if some of the blocks were made of iron, and the table magnetized? But then there would be a need for rules to interpret these rules (e.g., some blocks might be too heavy), and so on. We must stop somewhere, but now we are in a situation similar to what Turing found when he realized no algorithm could solve the halting problem. (Recall he produced a program that in some sense the algorithm did not expect.) No matter what collection of pre-defined rules we come up with, it appears there will always be real-world situations these

26. M. Euwe and W. Meiden, *Chess Master Vs. Chess Master* (New York: David McKay Co., 1977), p. 137.

27. Douglas R. Hofstadter, *Gödel, Escher, Bach: An Eternal Golden Braid* (New York: Basic Books, 1979), p. 678.

rules cannot interpret. Is it not reasonable to suggest that human beings *do* have this ability, the ability to "jump out" of a particular rule-governed way of operating in order to adjust to a new situation? Perhaps this is part of our creative capacity, and something not shared by computers.

As one example illustrating this, let us return briefly to Andrew Wiles's proof of Fermat's Last Theorem. This theorem had baffled the greatest mathematicians ever since Fermat first claimed its truth in 1630. Part of its intrigue is its simplicity. It is to find non-zero integer solutions to the equation $x^n + y^n = z^n$. For example, with $n = 2$, we can quickly check that $3^2 + 4^2 = 5^2$, so the integers $x = 3$, $y = 4$, and $z = 5$ are indeed non-zero integer solutions to the above equation. Fermat stated that if the exponent in the above equation were larger than 2, it would have *no* non-zero integer solutions. In other words, there are no non-zero integer solutions to the Diophantine equation $x^n + y^n = z^n$ if n is bigger than 2.

Unfortunately, not everyone can appreciate the enormous creativity that went into Wiles's proof (over 200 pages) of this theorem.[28] He actually proved a generalization of Fermat's Last Theorem, known as the Taniyama-Shimura conjecture. It basically states that every elliptic curve with rational coefficients is modular. (We need not worry about the precise meaning of these terms. They are used only to give a context for our discussion.) By itself, this would not prove Fermat's Last Theorem without a very important result proved by Ken Ribet in 1986. His result shows that *if* there were a solution to the equation $x^n + y^n = z^n$ for some integer n larger than 2, then there *would be* an elliptic curve with rational coefficients that is not modular. Because Wiles was able to show the latter cannot happen, he is the one who gets credit for the proof of Fermat's Last Theorem.

Although Wiles's initial proof was flawed, he was, with the help of Richard Taylor, able to correct it. In an interview with the producers of NOVA, Wiles described his incredible journey of coming to the final proof. Listen carefully to the language he uses, first in laying out his overall strategy:

> I tried doing calculations which explain some little piece of mathematics. I tried to fit it in with some previous broad conceptual understanding of some part of mathematics that would clarify the particular problem I was thinking about. Sometimes that would involve going and looking it up in a book to see how it's done there. Sometimes it was a

28. Andrew J. Wiles, *Annals of Mathematics* 141, no. 3 (May 1995).

> question of modifying things a bit, doing a little extra calculation. And sometimes I realized that nothing that had ever been done before was any use at all. Then I just had to find something completely new; *it's a mystery where that comes from.*

This was his general procedure. Now hear what he has to say about the breakthrough that occurred in his initial proof:

> I was casually looking at a research paper and there was one sentence that just caught my attention. It mentioned a 19th-century construction, and *I suddenly realized* that I should be able to use that to complete the proof.

This same kind of thing occurred when he and Taylor worked to fix the error in the original proof.

> I *suddenly* came to a marvelous revelation: I *saw in a flash* on September 19th, 1994, that de Shalit's theory, if generalized, could be used. I had *unexpectedly* found the missing key to my old abandoned approach.

The italicized portions of the above quotations are quite significant. They seem to indicate that Wiles himself has no idea how certain insights had originated. Perhaps they came from being creatively human.

Concluding Remarks

Of course, in some sense we're running around in circles. The fact that someone is unaware from whence insights come does not imply that there is no algorithm for explaining their occurrence. Indeed, much research is now devoted to finding mechanisms that explain creativity.[29]

Ironically, it might be that the history of artificial intelligence itself argues somewhat against this possibility. As impressive as the ELIZA program example we gave at the beginning of this chapter is, it is intriguing that from 1966 to the present time there has been no further significant progress in all-

29. D. N. Perkins, "Creativity and the Quest for Mechanism," in *The Psychology of Human Thought,* ed. R. Sternberg and E. Smith (Cambridge: Cambridge University Press, 1988).

purpose language recognition. To be sure, great progress has been made in *speech* recognition, but the most recent attempts to program machines with "general intelligence" have been phenomenal failures. Even the author of ELIZA, Joseph Weizenbaum, was amazed that some actually believed his program could understand natural language, stating he ". . . tried to say that no general solution to that problem was possible, i.e., that language is understood only in contextual frameworks. . . ."[30]

This framework includes an encounter of some sort between a subjective person and an objective world. In everyday life, we are involved in "subworlds" of this larger world, such as our business, or our family. The success in artificial intelligence, however, has been in the creation of programs that apply to "microworlds," where what is relevant is fixed, and relevant factors can be defined in terms of context-free symbols. These microworlds are not, however, the same as our subworlds, which, as Dreyfus points out, are really *modes* of our everyday general world. It seems to be a mistake to think, as do Newell and Simon, that these microworlds could somehow and in some way be connected to ". . . assemble large intelligent systems in a modular way."[31] As Dreyfus observes, "If microworlds were subworlds, one would not have to extend and combine them to reach the everyday world because the everyday world would have to be included already."[32]

This is more in line with holistic, Gestalt psychology, and is a sharp contrast with the views of some cognitive psychologists such as Winston, who argues that the key to computer thinking is the proper representation of data. Indeed, the model of the world for computers *is* data, but is this the model for people? Given what we have argued, it seems, as Dreyfus asserts, that our model of the world is the world itself. To this point it is people who appear to be creative, not computers. In fact, historians of artificial intelligence have observed that the discipline is now focusing on special-purpose programs tailored to narrow, well-defined domains.[33] This seems to be a proper role for artificial intelligence to pursue, and one that should be encouraged, as the potential benefits are enormous.

It is important to note carefully what we have argued for in this chap-

30. Weizenbaum, *Computer Power and Human Reason,* p. 7.

31. Newell and Simon, "Computer Science as an Empirical Inquiry," p. 48.

32. Dreyfus, *What Computers Can't Do,* p. 14.

33. P. McCorduck, *Machines Who Think* (San Francisco: W. H. Freeman, 1979).

ter, viz., that it is not at all clear that human thought in its fullest form can be simulated by a digital computer. We have not taken a position here as to whether the mind itself is a material object or not, nor have we taken a position as to whether some physical device other than a digital computer is capable of simulating human thought. These are important questions, of course, and how people might go about answering them will in part depend on whether their orientation is similar to that of folks like Simon and Minsky, who take a strong scientific view of the world, or to that of psychologists such as Rollo May, who take an intuitive approach towards explaining reality.

Considering the biblical portrayal of persons as being made of dust and in the image of God, there are dangers lurking in whatever approach Christians take on this issue. For those favoring an intuitive approach, it is to ignore that there is at least a 'dust' component to our minds. Thus, for example, pastors counseling people who are depressed would be irresponsible if they automatically concluded that their counselees had a spiritual problem and overlooked the possibility that a medical condition might be causing the depression. For those taking a strong scientific approach, the danger lies not in the temptation to create computers that are like people; rather, it is in the temptation to believe people are not bearers of the image of God — that people are no more than 'raw' computers, with no ultimate moral responsibility or capacity for choice. We have only scratched the surface in how one aspect of our *"dust/image of God"* personhood might play out with respect to human thought. How to deal with this interplay will be an important area for Christians to consider for years to come.

CHAPTER 10

The Possibility of Detecting Intelligent Design

Introduction

In discussing the miracle stories of the gospels, Alan Richardson remarked, "Only those who came in faith understood the meaning of [Jesus'] acts of power. That is why any discussion of the Gospel miracles must begin . . . with a consideration of the biblical theology, with the faith which illuminates their character and purpose."[1] This view — that we must start with faith before we can even begin to assess divine action — is common in our day. It is fully developed in the theology of Karl Barth.

Nonetheless, there is an alternate tradition within Christianity that regards our native intellects as adequate for acquiring certain limited knowledge of God's action in the world independently of faith or explicitly Christian presuppositions. For instance, in his *Summa Contra Gentiles* Thomas Aquinas wrote: "By his natural reason man is able at once to arrive at some knowledge of God. For seeing that natural things run their course according to a fixed order, and since there cannot be order without a cause of order, men, for the most part, perceive that there is one who orders the things that we see."[2]

Richardson leaves the detection of divine action to those who already

1. Alan Richardson, *The Miracle-Stories of the Gospels* (New York: Harper & Brothers, 1942), p. 127.

2. Thomas Aquinas, *Summa Contra Gentiles,* III.38, in *Introduction to St. Thomas Aquinas,* ed. A. C. Pegis (New York: Modern Library, 1948), p. 454.

belong to a community of faith. Aquinas, on the other hand, claims that the detection of divine action is open to all quite independently of one's membership in a community of faith. Aquinas's position is therefore the more risky and the one more difficult to maintain, especially in a scientific age that regards empirical evidence as silent about divine action.

We wish to explore the possibility of detecting design, and to show that mathematics, especially in the form of probability and complexity theory, is indispensable for establishing its empirical detectability. This approach, of course, sides with Aquinas. Even so, in arguing for the empirical detectability of divine action, we don't want to give the impression that faith has no role in our understanding of divine action. In the passage quoted two paragraphs back, Aquinas is quick to add: "But who or of what kind this cause of order may be, or whether there be but one, cannot be gathered from this general consideration."[3]

Yes, our native intellects are able to infer that an ordering cause is responsible for the world. But the precise nature of this cause is something that our native intellects cannot grasp. Those inclined to stress evidence over presuppositions will think such an inference to an ordering cause contributes to our understanding of divine action. Those inclined to stress presuppositions over evidence, on the other hand, will view such an ordering cause as falling so far short of the full God of Christianity as to be not worthy of consideration.

Neither a pure evidentialism nor a pure presuppositionalism is particularly helpful. Evidence and presuppositions work together. To see this, consider a balance scale. Suppose a balance scale has a weight of one pound on the left and an indeterminate weight on the right. If the indeterminate weight causes the right side to go down and the left side to go up, then we know that the weight on the right is strictly heavier than one pound. But how much heavier? It may be two pounds, three pounds, or even a billion pounds. We won't know the exact weight, but we'll know that it is more than one pound. We thereby obtain valuable information about the indeterminate weight even if we can't determine its weight exactly.

The balance-scale example serves as a useful analogy. Just as the indeterminate weight must be causally sufficient to account for the rising of the one-pound weight on the left side of the balance scale, so too Aquinas's indeterminate order-giver must be causally sufficient to account for the order

3. Thomas Aquinas, *Summa Contra Gentiles,* III.38, p. 455.

in the world. Causal sufficiency of the indeterminate weight in this case means that it is strictly greater than one pound. Causal sufficiency of Aquinas's indeterminate order-giver means — minimally — being an intelligent cause of remarkable talents and power.

An evidentialist approach therefore provides certain limited insight into divine action. To be sure, there is much more to divine action than it can tell us, and for that we need a presuppositionalist approach that takes full account of theological developments in the doctrine of God. Nonetheless, an evidentialist approach can tell us, in Aquinas's words, that "there is one who orders the things that we see."[4] On presuppositionalist grounds we then know that this order-giver is God and that the actions of this order-giver are instances of divine action.

Aquinas's order-giver may seem theologically barren. But in an age that regards science as the preeminent road to reliable knowledge and that views divine action as inaccessible to scientific inquiry, the ability to extract even limited knowledge about divine action from empirical evidence becomes highly significant. Granted, presuppositionalism can live with a God who is a master of stealth and who constantly eludes our best efforts to detect him empirically. But scientific naturalism cannot. Naturalism, the view that nature is the fundamental reality, and more specifically scientific naturalism, the form of naturalism that appeals to science as its guarantor, both crash as soon as divine action becomes empirically detectable.

Not every case of divine action is empirically detectable. Divine providence, for instance, is a case of divine action where God's activity is invisible except to the eyes of faith. Divine action that is also empirically detectable falls under the newly emerging field of intelligent design. Intelligent design studies the characteristic effects that intelligent causes leave behind when they act. Intelligent causes can do things that undirected natural causes cannot. Undirected natural causes can throw Scrabble tiles on a board, but cannot arrange the tiles to form meaningful words or sentences. To obtain a meaningful arrangement requires an intelligent cause.

Intelligent design formalizes and makes precise something humans do all the time. All of us are all the time engaged in a form of rational activity that, without being tendentious, can be described as "inferring design." Inferring design is a perfectly common and well-accepted human activity. People find it important to identify events caused through the purposeful, pre-

4. Thomas Aquinas, *Summa Contra Gentiles,* III.38, p. 454.

meditated action of an intelligent agent, and to distinguish such events from events due to natural causes. Intelligent design unpacks the logic of this everyday activity and applies it within the special sciences. There is no magic, no vitalism, no appeal to occult forces here. Inferring design is common, rational, and objectifiable.[5]

Intelligent design is a fully scientific theory. The world contains events, objects, and structures that exhaust the explanatory resources of undirected natural causes, and that are best explained by recourse to a designing intelligence. This is not an argument from ignorance. Precisely because of what is known about undirected natural causes and their limitations, science is now in a position to demonstrate design.[6] At the same time, intelligent design resists speculating about the nature, moral character, or purposes of this designing intelligence. In particular, intelligent design presupposes neither a creator nor miracles.

To say intelligent causes are empirically detectable is to say there exist well-defined criteria that, on the basis of observational features of the world, are capable of reliably distinguishing intelligent causes from undirected natural causes. Many special sciences have already developed such methods for drawing this distinction — notably forensic science, artificial intelligence (cf. the Turing test), cryptography, archeology, and the search for extraterrestrial intelligence (cf. the movie *Contact*).[7]

Now, it is unavoidable that any such criteria for detecting intelligent causes will have to be probabilistic. Consider, for instance, a sequence of Scrabble tiles that spell out a Shakespearean sonnet. As strictly a logical possibility, there is nothing to prevent Scrabble tiles from randomly being shaken out one-by-one so that the resulting sequence spells out a Shakespearean sonnet. Sheer possibility admits the wildest improbability so long as the probability is not identical to zero. Thus, if we are going to attribute the Shakespearean sonnet (and indeed any intelligently caused object) to something other than chance, we have to have a way of precluding chance, and this necessarily falls to probability theory and statistics.

In *The Design Inference*, William Dembski identifies and makes precise a probabilistic criterion for detecting design. He calls it the *complexity-*

5. William A. Dembski, *The Design Inference* (Cambridge: Cambridge University Press, 1998), ch. 1.

6. Michael Behe, *Darwin's Black Box* (New York: Free Press, 1996), ch. 9; Dembski, *The Design Inference*, ch. 2.

7. See Dembski, *The Design Inference*, ch. 1.

specification criterion. When intelligent agents act, they leave behind a characteristic trademark or signature — what Dembski defines as *specified complexity.* The complexity-specification criterion detects design by identifying this key trademark of designed objects. Let us now turn to this criterion.

The Complexity-Specification Criterion

A detailed explication and justification of the complexity-specification criterion is technical and can be found in *The Design Inference.*[8] Nevertheless, the basic idea is straightforward and easily illustrated. Consider how the radio astronomers in the movie *Contact* detected an extraterrestrial intelligence. This movie, based on a novel by Carl Sagan, was an enjoyable piece of propaganda for the SETI research program — the Search for Extra-Terrestrial Intelligence. To make the movie interesting, the SETI researchers in *Contact* actually did find an extraterrestrial intelligence (the *non*-fictional SETI program has yet to be so lucky).

How, then, did the SETI researchers in *Contact* convince themselves that they had found an extraterrestrial intelligence? To increase their chances of finding an extraterrestrial intelligence, SETI researchers monitor millions of radio signals from outer space. Many natural objects in space produce radio waves (e.g., pulsars). Looking for signs of design among all these naturally produced radio signals is like looking for a needle in a haystack. To sift through the haystack, SETI researchers run the signals they monitor through computers programmed with pattern-matchers. So long as a signal doesn't match one of the pre-set patterns, it will pass through the pattern-matching sieve (and that even if it has an intelligent source). If, on the other hand, it does match one of these patterns, then, depending on the pattern matched, the SETI researchers may have cause for celebration.

The SETI researchers in *Contact* did find a signal worthy of celebration:

11011101111101111111011111111111011111111111110111111111111
11111011111111111111111110111111111111111111111110111111111

8. Strictly speaking, in *The Design Inference* Dembski develops a "specification/small probability criterion." This criterion is equivalent to the complexity-specification criterion described here.

11111111111111111111011111111111111111111111111111110111111
11111111111111111111111111111110111111111111111111111111111
11111111111111011111111111111111111111111111111111111111110
11111111111111111111111111111111111111111111111011111111111
11111111111111111111111111111111111111111101111111111111111
11111111111111111111111111111111111111111111101111111111111
11111111111111111111111111111111111111111111111111111101111
11111111111111111111111111111111111111111111111111111111111
11111111011111111111111111111111111111111111111111111111111
11111111111111111111111011111111111111111111111111111111111
11111111111111111111111111111111111111111111011111111111111
11111111111111111111111111111111111111111111111111111111111
11111111110111111111111111111111111111111111111111111111111
11111111111111111111111111111111111111111011111111111111111
11111111111111111111111111111111111111111111111111111111111
11111111111111111111101111111111111111111111111111111111111
11111111111111111111111111111111111111111111111111111111111
11111

The SETI researchers in *Contact* received this signal as a sequence of 1126 beats and pauses, where 1's correspond to beats and 0's to pauses. This sequence represents the prime numbers from 2 to 101, where a given prime number is represented by the corresponding number of beats (i.e., 1's), and the individual prime numbers are separated by pauses (i.e., 0's). The SETI researchers in *Contact* took this signal as decisive confirmation of an extraterrestrial intelligence.

What is it about this signal that implicates design? Whenever we infer design, we must establish three things: *contingency, complexity,* and *specification.* Contingency ensures that the object in question is not the result of an automatic and therefore unintelligent process that had no choice in its production. Complexity ensures that the object is not so simple that it can readily be explained by chance. Finally, specification ensures that the object exhibits the type of pattern characteristic of intelligence. Let us examine these three requirements more closely.

In practice, to establish the contingency of an object, event, or structure, one must establish that it is compatible with the regularities involved in its production, but that these regularities also permit any number of alternatives to it. Typically these regularities are conceived as natural laws or algorithms. By being compatible with but not required by the regularities involved in its production, an object, event, or structure becomes irreducible to any underlying physical necessity. Michael Polanyi and Timothy Lenoir have both described this method of establishing contingency.[9] The method applies quite generally: the position of Scrabble tiles on a Scrabble board is irreducible to the natural laws governing the motion of Scrabble tiles; the configuration of ink on a sheet of paper is irreducible to the physics and chemistry of paper and ink; the sequencing of DNA bases is irreducible to the bonding affinities between the bases; and so on. In the case at hand, the sequence of 0's and 1's to form a sequence of prime numbers is irreducible to the laws of physics that govern the transmission of radio signals. We therefore regard the sequence as contingent.

To see next why complexity is crucial for inferring design, consider the following sequence of bits:

110111011111

These are the first twelve bits in the previous sequence representing the prime numbers 2, 3, and 5 respectively. Now it is a sure bet that no SETI researcher, if confronted with this twelve-bit sequence, is going to contact the science editor at *The New York Times*, hold a press conference, and announce that an extraterrestrial intelligence has been discovered. No headline is going to read, "Aliens Master First Three Prime Numbers!"

The problem is that this sequence is much too short (and thus too simple) to establish that an extraterrestrial intelligence with knowledge of prime numbers produced it. A randomly beating radio source might by chance just happen to output this sequence. A sequence of 1126 bits representing the prime numbers from 2 to 101, however, is a different story. Here the sequence is sufficiently long (and therefore sufficiently complex) that only an extraterrestrial intelligence could have produced it.

Complexity as we are describing it here is a form of probability. Later in this chapter we will require a more general conception of complexity to

9. Michael Polanyi, "Life Transcending Physics and Chemistry," *Chemical and Engineering News*, August 21, 1967, pp. 54-66; Michael Polanyi, "Life's Irreducible Structure, *Science* 113 (1968): 1308-12; Timothy Lenoir, *The Strategy of Life: Teleology and Mechanics in Nineteenth Century German Biology* (Dordrecht: Reidel, 1982), pp. 7-8. See also Hubert Yockey, *Information Theory and Molecular Biology* (Cambridge: Cambridge University Press, 1992), p. 335.

unpack the logic of design inferences. But for now complexity as a form of probability is all we need. To see the connection between complexity and probability, consider a combination lock. The more possible combinations there are of the lock, the more complex the mechanism and correspondingly the more improbable that the mechanism can be opened by chance. Complexity and probability therefore vary inversely: the greater the complexity, the smaller the probability. Thus to determine whether something is sufficiently complex to warrant a design inference is to determine whether it has sufficiently small probability.

Even so, complexity (or improbability) isn't enough to eliminate chance and establish design. If we flip a coin 1000 times, we will participate in a highly complex (i.e., highly improbable) event. Indeed, the sequence we end up flipping will be one in a trillion trillion trillion . . . , where the ellipsis needs twenty-two more "trillions." This sequence of coin tosses won't, however, trigger a design inference. Though complex, this sequence won't exhibit a suitable pattern. Contrast this with the previous sequence representing the prime numbers from 2 to 101. Not only is this sequence complex, but it also embodies a suitable pattern. The SETI researcher who in the movie *Contact* discovered this sequence put it this way: "This isn't noise, this has structure."

What is a *suitable* pattern for inferring design? Not just any pattern will do. Some patterns can legitimately be employed to infer design whereas others cannot. The intuition underlying the distinction between patterns that alternately succeed or fail to implicate design is, however, easily motivated. Consider the case of an archer. Suppose an archer stands 50 meters from a large wall with bow and arrow in hand. The wall, let us say, is sufficiently large that the archer cannot help but hit it. Now suppose each time the archer shoots an arrow at the wall, the archer paints a target around the arrow so that the arrow sits squarely in the bull's-eye. What can be concluded from this scenario? Certainly, we can conclude absolutely nothing about the archer's ability as an archer. Yes, a pattern is being matched; but it is a pattern fixed only after the arrow has been shot. The pattern is thus purely *ad hoc*.

But suppose instead the archer paints a fixed target on the wall and then shoots at it. Suppose the archer shoots a hundred arrows, and each time hits a perfect bull's-eye. What can be concluded from this second scenario? Confronted with this second scenario we are obligated to infer that here is a world-class archer, one whose shots cannot legitimately be referred to luck, but rather must be referred to the archer's skill and mastery. Skill and mastery are of course instances of design.

Figure 1. What Is a Pattern?

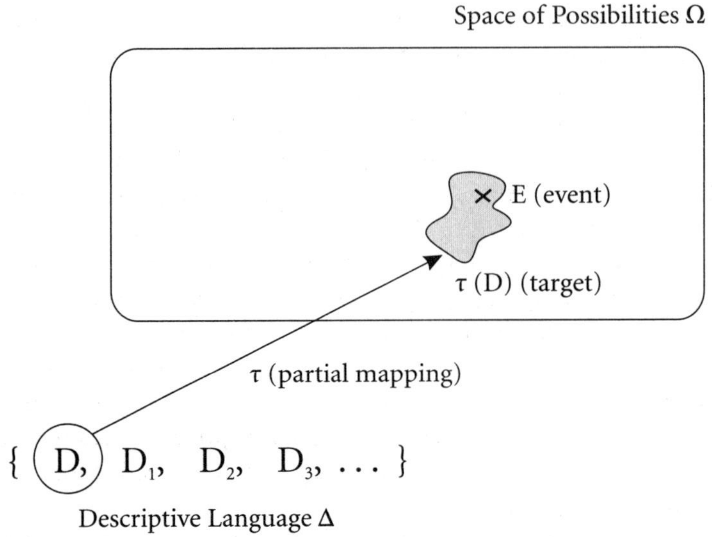

Space of Possibilities Ω

E (event)

τ (D) (target)

τ (partial mapping)

$\{ D, D_1, D_2, D_3, \ldots \}$

Descriptive Language Δ

$$P = \langle D, \tau \rangle \text{ (pattern)}$$

The archer example introduces three elements that are essential for inferring design:

(1) A reference class of possible events (here the arrow hitting the wall at some unspecified place);
(2) A pattern that restricts the reference class of possible events (here a target on the wall); and
(3) The precise event that has occurred (here the arrow hitting the wall at some precise location).

In a design inference, the reference class, the pattern, and the event are linked, with the pattern mediating between event and reference class, and helping to decide whether the event is due to chance or design.

Figure 1 above illustrates the connection between patterns, events, and reference classes. The space of possibilities Ω is the reference class. The event

E, denoted by an "x," is the outcome that actually occurred. Probabilists distinguish between outcomes or elementary events and events generally. To roll a six with a single die is an outcome or elementary event. To roll an even number with a single die is an event that includes the outcome of rolling a six, but also includes rolling a four or a two. Events include at least one elementary event or outcome, but may include more.

Accordingly, the elementary event E, denoted by an "x," belongs to the event $\tau(D)$, denoted by the squiggly shaded area. We call $\tau(D)$ the "target." The target is the event associated with the pattern in question. We indicate the pattern here by the ordered pair $\Pi = \langle D, \tau \rangle$. The first element of this ordered pair, denoted by the letter D, belongs to a "descriptive language," which we symbolize by the Greek letter Δ. Formally, Δ can be any nonempty set, though in practice Δ is either a formal or a natural language. The second element, denoted by the lowercase Greek letter τ, is a partial mapping from Δ to the subsets of Ω (these subsets are events associated with the reference class Ω). As a partial mapping, τ need not be defined on all of Δ. It is important, of course, that τ be defined at the particular D we are looking at.

It follows that patterns are not just descriptions associated arbitrarily with events, but rather descriptions conjoined with mappings that uniquely and unambiguously associate descriptions with events. Note that in determining whether an event is sufficiently improbable or complex to implicate design, the relevant improbability is not that of the elementary event E, but that of the target $\tau(D)$. Indeed, the bigger the target, the easier it is to hit it by chance and thus apart from design.

The type of pattern where an archer fixes a target first and then shoots at it is common to statistics, where it is known as setting a *rejection region* prior to an experiment. In statistics, if the outcome of an experiment falls within a rejection region, the chance hypothesis supposedly responsible for the outcome is rejected. The reason for setting a rejection region prior to an experiment is to forestall what statisticians call "data snooping" or "cherry picking." Just about any data set will contain strange and improbable patterns if we look hard enough. By forcing experimenters to set their rejection regions prior to an experiment, the statistician protects the experiment from spurious patterns that could just as well result from chance.

Now a little reflection makes clear that a pattern need not be given prior to an event to eliminate chance and implicate design. Consider the following cipher text:

nfuijolt ju jt mjlf b xfbtfm

Initially this looks like a random sequence of letters and spaces — initially you lack any pattern for rejecting chance and inferring design.

But suppose next that someone comes along and tells you to treat this sequence as a Caesar cipher, moving each letter one notch back in the alphabet. Now the sequence reads,

methinks it is like a weasel

Even though the pattern (in this case, the decrypted text) is given after the fact, it still is the right sort of pattern for eliminating chance and inferring design. In contrast to statistics, which always identifies its patterns before an experiment is performed, cryptanalysis must discover its patterns after the fact. In both instances, however, the patterns are suitable for inferring design.

Patterns thus divide into two types, those that in the presence of complexity warrant a design inference and those that despite the presence of complexity do not warrant a design inference. Call the first type of pattern a *specification,* the second a *fabrication.* Specifications are the non-*ad hoc* patterns that can legitimately be used to eliminate chance and warrant a design inference. In contrast, fabrications are the *ad hoc* patterns that cannot legitimately be used to warrant a design inference. This distinction between specifications and fabrications can be made with full statistical rigor.[10]

To sum up, the complexity-specification criterion detects design by establishing three things: contingency, complexity, and specification. When called to explain an event, object, or structure, we have a decision to make — are we going to attribute it to *necessity, chance,* or *design?* According to the complexity-specification criterion, to answer this question is to answer three simpler questions: Is it contingent? Is it complex? Is it specified? Consequently, the complexity-specification criterion can be represented as a flowchart with three decision nodes. Call this flowchart the Explanatory Filter (see figure 2, p. 289). So long as the questions raised at the three decision nodes are answerable (which, depending on our background knowledge, is not always the case), the filter reliably adjudicates between necessity, chance, and design.

10. Dembski, *The Design Inference,* ch. 5.

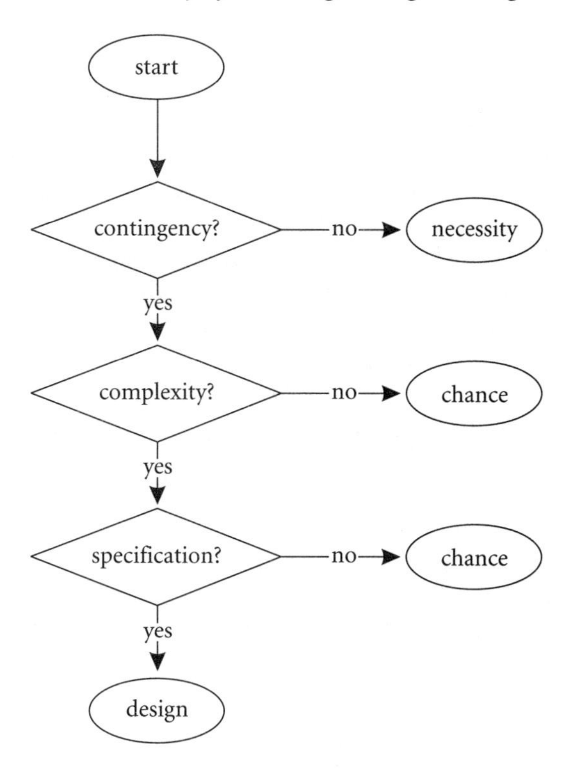

Figure 2. The Explanatory Filter

Specification[11]

Because specification is so central to inferring design, we need to say a few more words about it. For a pattern to count as a specification, the important thing is not when it was identified, but whether in a certain well-defined sense it is *independent* of the event it describes. Drawing a target around an arrow already embedded in a wall is not independent of the arrow's trajectory. Consequently, such a target qua pattern cannot be used to attribute the arrow's trajectory to design. Patterns that are specifications cannot simply be read off the events whose design is in question. Rather, to count as specifications, patterns must be suitably independent of events. We refer to this rela-

11. This section summarizes chapter 5 of *The Design Inference.*

tion of independence as *detachability,* and say that a pattern is *detachable* if it satisfies that relation.

Detachability can be understood as asking the following question: Given an event whose design is in question and a pattern describing it, would we be able to construct that pattern if we had no knowledge which event occurred? Here is the idea. An event has occurred. A pattern describing the event is given. The event is one from a range of possible events. If all we knew was the range of possible events without any specifics about which event actually occurred, could we still construct the pattern describing the event? If so, the pattern is detachable from the event.

To see what's at stake, consider the following example. It should illustrate what transforms an ordinary pattern into a pattern that serves as a specification. Consider the following event E, an event that to all appearances was obtained by flipping a fair coin 100 times:

THTTTHHTHHTTTTTHTHTTHHHTT
HTHHHTHHHTTTTTTTHTTHTTTHH
THTTTHTHTHHTTHHHHTTTHTTHH
THTHTHHHHTTHHTHHHHTHHHHTT E

Is E the product of chance or not? A standard trick of statistics professors with an introductory statistics class is to divide the class in two, having students in one half of the class each flip a coin 100 times, writing down the sequence of heads and tails on a slip of paper, and having students in the other half each generate purely with their minds a "random looking" string of coin tosses that mimics the tossing of a coin 100 times, also writing down the sequence of heads and tails on a slip of paper. When the students then hand in their slips of paper, it is the professor's job to sort the papers into two piles, those generated by flipping a fair coin, and those concocted in the students' heads. To the amazement of the students, the statistics professor is typically able to sort the papers with 100 percent accuracy.

There's no mystery here. The statistics professor simply looks for a repetition of six or seven heads or tails in a row to distinguish the truly random from the pseudo-random sequences. In a hundred coin flips, one is quite likely to see six or seven such repetitions. On the other hand, people concocting pseudo-random sequences with their minds tend to alternate between heads and tails too frequently. Whereas with a truly random sequence of coin tosses there is a 50 percent chance that one toss will differ from the next, as a matter of human psychology people expect that one toss will differ from the next around 70 percent of the time.

How, then, will our statistics professor fare when confronted with E? Will E be attributed to chance or to the musings of someone trying to mimic chance? According to the professor's crude randomness checker, E would be assigned to the pile of sequences presumed to be truly random, for E contains a repetition of seven tails in a row. Everything that at first blush would lead us to regard E as truly random checks out. There are exactly 50 alternations between heads and tails (as opposed to the 70 that would be expected from humans trying to mimic chance). What's more, the relative frequencies of heads and tails check out: there were 49 heads and 51 tails. Thus it's not as though the coin supposedly responsible for generating E was heavily biased in favor of one side versus the other.

Suppose, however, that our statistics professor suspects she is not up against a neophyte statistics student, but instead a fellow statistician who is trying to put one over on her. To help organize her problem, study it more carefully, and enter it into a computer, she will find it convenient to let strings of 0's and 1's represent the outcomes of coin flips, with 1 corresponding to heads and 0 to tails. In that case the following pattern D will correspond to the event E:

0100011011000001010011100
1011101110000000100100011
0100010101100111100010011
0101011110011011110111100 D

(Note that for convenience we've suppressed the partial mapping that maps this string of 0's and 1's to the event E.) Now, the mere fact that the event E conforms to the pattern D is no reason to think that E did not occur by chance. As things stand, the pattern D has simply been read off the event E.

But D need not have been read off of E. Indeed, D could have been constructed without recourse to E. To see this, let us rewrite D as follows:

0
1
00
01
10

11
000
001
010
011
100
101
110
111
0000
0001
0010
0011
.
.
.
1111
00 D

By viewing D this way, anyone with the least exposure to binary arithmetic immediately recognizes that D was constructed simply by writing binary numbers in ascending order, starting with the one-digit binary numbers (i.e., 0 and 1), proceeding then to the two-digit binary numbers (i.e., 00, 01, 10, and 11), and continuing on until 100 digits were recorded. It's therefore intuitively clear that D does not describe a truly random event (i.e., an event gotten by tossing a fair coin), but rather a pseudo-random event, concocted by doing a little binary arithmetic.

Although it's now intuitively clear why chance cannot properly explain E, we need to consider more closely why this mode of explanation fails here. We started with a putative chance event E, supposedly gotten by flipping a fair coin 100 times. Since heads and tails each have probability ½, and since this probability gets multiplied for each flip of the coin, it follows that the probability of E is 2^{-100}, or approximately 10^{-30}. In addition, we constructed a pattern D to which E conforms. Initially D proved insufficient to eliminate chance as the explanation of E since in its construction D was simply read off of E. Rather, to eliminate chance we had also to recognize that D could have been constructed quite easily by performing some simple arithmetic operations with binary numbers. Thus, to eliminate chance we needed to employ additional *side information,* which in this case consisted of our knowledge of binary arithmetic. This side information detached the pattern D from the event E and thereby rendered D a specification.

For side information to detach a pattern from an event, it must satisfy two conditions, a *conditional independence condition* and a *tractability condition.* According to the conditional independence condition, the side information must be conditionally independent of the event E. Conditional independence is a well-defined notion from probability theory. It means that the probability of E doesn't change once the side information is taken into account. Conditional independence is the standard probabilistic way of unpacking epistemic independence. Two things are epistemically independent if knowledge about one thing (in this case the side information) does not affect knowledge about the other (in this case the occurrence of E). This is certainly the case here since our knowledge of binary arithmetic does not affect the probabilities we assign to coin tosses.

The second condition, the tractability condition, requires that the side information enable us to construct the pattern D to which E conforms. This is evidently the case here as well since our knowledge of binary arithmetic enables us to arrange binary numbers in ascending order, and thereby construct the pattern D. But what exactly is this *ability to construct a pattern on the basis of side information?* Perhaps the most slippery words in philosophy are "can," "able," and "enable." Fortunately, just as there is a precise theory for characterizing the epistemic independence between an event and side information — namely, probability theory — so too there is a precise theory for characterizing the ability to construct a pattern on the basis of side information — namely, complexity theory.

Complexity theory, conceived now quite generally and not merely as a form of probability, assesses the difficulty of tasks given the resources available for accomplishing those tasks.[12] As a generalization of computational complexity theory, complexity theory ranks tasks according to difficulty, and then determines which tasks are sufficiently manageable to be doable or tractable. For instance, given current technology we find sending a person to the moon tractable, but sending a person to the nearest galaxy intractable. In the tractability condition, the task to be accomplished is the construction of a pattern and the resource for accomplishing that task is side information.

12. For an exact treatment of complexity theory see Dembski, *The Design Inference,* ch. 4.

Thus, for the tractability condition to be satisfied, side information must provide the resources necessary for constructing the pattern in question. All of this admits a precise complexity-theoretic formulation and makes definite what we called "the ability to construct a pattern on the basis of side information."[13]

Taken jointly, the tractability and conditional independence conditions mean that side information enables us to construct the pattern to which an event conforms, yet without recourse to the actual event. This is the crucial insight. Because the side information is conditionally and therefore epistemically independent of the event, any pattern constructed from this side information is obtained without recourse to the event. In this way any pattern that is constructed from such side information avoids the charge of being *ad hoc.* These, then, are the detachable patterns. These are the specifications.

False Negatives and False Positives

As with any criterion, we need to make sure that the judgments of the complexity-specification criterion agree with reality. Consider medical tests. Any medical test is a criterion. A perfectly reliable medical test would detect the presence of a disease whenever it is indeed present, and fail to detect the disease whenever it is absent. Unfortunately, no medical test is perfectly reliable, and so the best we can do is keep the proportion of false positives and false negatives as low as possible.

All criteria, and not just medical tests, face the problem of false positives and false negatives. A criterion attempts to classify individuals with respect to a target group (in the case of medical tests, those who have a certain disease). When the criterion places in the target group an individual who should not be there, it commits a false positive. Alternatively, when the criterion fails to place in the target group an individual who should be there, it commits a false negative.

Let us now apply these observations to the complexity-specification criterion. This criterion purports to detect design. Is it a reliable criterion? The target group for this criterion comprises all things intelligently caused. How accurate is this criterion at correctly assigning things to this target

13. See Dembski, *The Design Inference,* ch. 5.

group and correctly omitting things from it? The things we are trying to explain have causal stories. In some of those causal stories intelligent causation is indispensable, whereas in others it is dispensable. An inkblot can be explained without appealing to intelligent causation; ink arranged to form meaningful text cannot. When the complexity-specification criterion assigns something to the target group, can we be confident that it actually is intelligently caused? If not, we have a problem with false positives. On the other hand, when this criterion fails to assign something to the target group, can we be confident that no intelligent cause underlies it? If not, we have a problem with false negatives.

Consider first the problem of false negatives. When the complexity-specification criterion fails to detect design in a thing, can we be sure no intelligent cause underlies it? The answer is No. For determining that something is not designed, this criterion is not reliable. False negatives are a problem for it. This problem of false negatives, however, is endemic to detecting intelligent causes.

One difficulty is that intelligent causes can mimic necessity and chance, thereby rendering their actions indistinguishable from such unintelligent causes. A bottle of ink may fall off a cupboard and spill onto a sheet of paper. Alternatively, a human agent may deliberately take a bottle of ink and pour it over a sheet of paper. The resulting inkblot may look identical in both instances, but in the one case results by chance, in the other by design.

Another difficulty is that detecting intelligent causes requires background knowledge on our part. It takes an intelligent cause to know an intelligent cause. But if we don't know enough, we'll miss it. Consider a spy listening in on a communication channel whose messages are encrypted. Unless the spy knows how to break the cryptosystem used by the parties on whom he is eavesdropping, any messages passing the communication channel will be unintelligible, and might in fact be meaningless.

The problem of false negatives therefore arises either when an intelligent agent has acted (whether consciously or unconsciously) to conceal his or her actions, or when an intelligent agent in trying to detect design has insufficient background knowledge to determine whether design actually is present. Detectives face this problem all the time. A detective confronted with a murder needs first to determine whether a murder has indeed been committed. If the murderer was clever and made it appear that the victim died by accident, then the detective will mistake the murder for an accident. So too, if the detective is careless and misses certain obvious clues, the detective will mistake the murder for an accident. In mistaking a murder for an accident, the detective commits a false negative. Contrast this, however, with a detective facing a murderer intent on revenge, and who wants to leave no doubt that the victim was intended to die. In that case the problem of false negatives is unlikely to arise (though we can imagine an incredibly stupid detective, like Chief Inspector Clouseau, mistaking a rather obvious murder for an accident).

Intelligent causes can do things that unintelligent causes cannot, and can make their actions evident. When for whatever reason an intelligent cause fails to make its actions evident, we may miss it. But when an intelligent cause succeeds in making its actions evident, we take notice. This is why false negatives do not invalidate the complexity-specification criterion. This criterion is fully capable of detecting intelligent causes intent on making their presence evident. Masters of stealth intent on concealing their actions may successfully evade the criterion. But masters of self-promotion intent on making sure their intellectual property gets properly attributed find in the complexity-specification criterion a ready friend.

And this brings us to the problem of false positives. Even though specified complexity is not a reliable criterion for *eliminating* design, it can be argued to be a reliable criterion for *detecting* design. The complexity-specification criterion is a net. Things that are designed will occasionally slip past the net. We would prefer that the net catch more than it does, omitting nothing due to design. But given the ability of design to mimic unintelligent causes and the possibility of our own ignorance passing over things that are designed, this problem cannot be fixed. Nevertheless, we want to be very sure that whatever the net does catch includes only what we intend it to catch, to wit, things that are designed. Only things that are designed had better end up in the net. If this is the case, we can have confidence that whatever the complexity-specification criterion attributes to design is indeed designed. On the other hand, if things end up in the net that are not designed, the criterion will be worthless.

We propose, then, that specified complexity is a reliable criterion for detecting design. Alternatively, that the complexity-specification criterion successfully avoids false positives. Thus, whenever this criterion attributes design, it does so correctly. Let us now see why this is the case by looking at two arguments. The first is a straightforward inductive argument: in every instance where the complexity-specification criterion attributes design, and where the underlying causal story is known, it turns out design actually is present; therefore, design actually is present whenever the complexity-specification criterion attributes design. The conclusion of this argument is a straightforward inductive generalization. It has the same logical status as concluding that all ravens are black given that all ravens observed to date have been found to be black.

The naturalist is likely to object at this point, claiming that the only things we can know to be designed are artifacts manufactured by intelligent beings that are in turn the product of blind evolutionary processes (e.g., humans). Hence to use the complexity-specification criterion to extrapolate design beyond such artifacts is illegitimate. This argument doesn't work. It is circular reasoning to invoke naturalism to underwrite an evolutionary account of intelligence, and then in turn to employ this account of intelligence to insulate naturalism from critique. Naturalism is a metaphysical position, not a scientific theory based on evidence. Any account of intelligence it entails is therefore suspect, and needs to be subjected to independent checks. The complexity-specification criterion provides one such check.

If we dismiss the naturalist's evolutionary account of intelligence, a more serious objection remains. We are arguing inductively that the complexity-specification criterion is a reliable criterion for detecting design. The conclusion of this argument is that whenever the criterion attributes design, design actually is present. The premise of this argument is that whenever the criterion attributes design and the underlying causal story can be verified, design actually is present. Now, even though the conclusion follows as an inductive generalization from the premise, the premise itself seems false. There are a lot of coincidences out there that seem best explained without invoking design. Consider, for instance, the Shoemaker-Levy comet. The Shoemaker-Levy comet crashed into Jupiter exactly 25 years to the day after the Apollo 11 moon landing. What are we to make of this coincidence? Do we really want to explain it in terms of design? What if we submitted this coincidence to the complexity-specification criterion and out popped design? Our intuitions strongly suggest that the comet's trajectory and NASA's space program were operating independently, and that at best this coincidence should be referred to chance — certainly not design.

This objection is readily met. The fact is that the complexity-specification criterion does not yield design all that easily, especially if the complexities are kept high (or correspondingly, the probabilities are kept small). It is simply not the case that unusual and striking coincidences automatically yield design. Martin Gardner is no doubt correct when he notes, "The number of events in which you participate for a month, or even a week, is so huge that the probability of noticing a startling correlation is quite high, especially if you keep a sharp outlook."[14] The implication he means to draw, however, is incorrect, namely, that therefore startling correlations/coincidences may uniformly be relegated to chance. Yes, the fact that the Shoemaker-Levy comet crashed into Jupiter exactly 25 years to the day after the Apollo 11 moon landing is a coincidence best referred to chance. But the fact that Mary Baker Eddy's writings on Christian Science bear a remarkable resemblance to Phineas Parkhurst Quimby's writings on mental healing is a coincidence that cannot be explained by chance, and is properly explained by positing Quimby as a source for Eddy.[15]

The complexity-specification criterion is robust and easily resists counterexamples of the Shoemaker-Levy variety. Assuming, for instance, that the Apollo 11 moon landing serves as a specification for the crash of Shoemaker-Levy into Jupiter (a generous concession at that), and that the comet could have crashed at any time within a period of a year, and that the comet crashed to the very second precisely 25 years after the moon landing, a straightforward probability calculation indicates that the probability of this coincidence is no smaller than 10^{-8}. This simply isn't all that small a probability (i.e., high complexity), especially when considered in relation to all the events astronomers are observing in the solar system. Certainly this probability is nowhere near the universal probability bound of 10^{-150} that Dembski proposed in *The Design Inference.*[16] We have yet to see a convincing application of the complexity-specification criterion in which coincidences better explained by chance get attributed to design, nor have we seen a specified event of probability less than Dembski's universal probability bound for which intelligent causation can be convincingly ruled out.

Why the Criterion Works

Our second argument for showing that specified complexity reliably detects design considers the nature of intelligent agency, and specifically, what it is

14. Martin Gardner, "Arthur Koestler: Neoplatonism Rides Again," *World,* August 1, 1972, pp. 87-89.

15. Walter Martin, *The Kingdom of the Cults,* rev. ed. (Minneapolis: Bethany House, 1985), pp. 127-30.

16. Dembski, *The Design Inference,* sec. 6.5.

about intelligent agents that makes them detectable. Even though induction confirms that specified complexity is a reliable criterion for detecting design, induction does not explain why this criterion works. To see why the complexity-specification criterion is exactly the right instrument for detecting design, we need to understand what it is about intelligent agents that makes them detectable in the first place. The principal characteristic of intelligent agency is *choice.* Even the etymology of the word "intelligent" makes this clear. "Intelligent" derives from two Latin words, the preposition *inter,* meaning between, and the verb *lego,* meaning to choose or select. Thus, according to its etymology, intelligence consists in *choosing between.* For an intelligent agent to act is therefore to choose from a range of competing possibilities.

This is true not just of humans, but of animals as well as of extraterrestrial intelligences. A rat navigating a maze must choose whether to go right or left at various points in the maze. When SETI researchers attempt to discover intelligence in the extraterrestrial radio transmissions they are monitoring, they assume an extraterrestrial intelligence could have chosen any number of possible radio transmissions, and then attempt to match the transmissions they observe with certain patterns as opposed to others. Whenever a human being utters meaningful speech, a choice is made from a range of possible sound-combinations that might have been uttered. Intelligent agency always entails discrimination, choosing certain things, ruling out others.

Given this characterization of intelligent agency, the crucial question is how to recognize it. Intelligent agents act by making a choice. How, then, do we recognize that an intelligent agent has made a choice? A bottle of ink spills accidentally onto a sheet of paper; someone takes a fountain pen and writes a message on a sheet of paper. In both instances ink is applied to paper. In both instances one among an almost infinite set of possibilities is realized. In both instances a contingency is actualized and others are ruled out. Yet in one instance we ascribe agency, in the other chance.

What is the relevant difference? Not only do we need to observe that a contingency was actualized, but we ourselves need also to be able to specify that contingency. The contingency must conform to an independently given pattern, and we must be able independently to construct that pattern. A random ink blot is unspecified; a message written with ink on paper is specified. To be sure, the exact message recorded may not be specified. But orthographic, syntactic, and semantic constraints will nonetheless specify it.

Actualizing one among several competing possibilities, ruling out the rest, and specifying the one that was actualized encapsulates how we recognize intelligent agency, or equivalently, how we detect design. Experimental psychologists who study animal learning and behavior have known this all along. To learn a task an animal must acquire the ability to actualize behaviors suitable for the task as well as the ability to rule out behaviors unsuitable for the task. Moreover, for a psychologist to recognize that an animal has learned a task, it is necessary not only to observe the animal making the appropriate discrimination, but also to specify the discrimination.

Thus, to recognize whether a rat has successfully learned how to traverse a maze, a psychologist must first specify which sequence of right and left turns conducts the rat out of the maze. No doubt, a rat randomly wandering a maze also discriminates a sequence of right and left turns. But by randomly wandering the maze, the rat gives no indication that it can discriminate the appropriate sequence of right and left turns for exiting the maze. Consequently, the psychologist studying the rat will have no reason to think the rat has learned how to traverse the maze.

Only if the rat executes the sequence of right and left turns specified by the psychologist will the psychologist recognize that the rat has learned how to traverse the maze. Now it is precisely the learned behaviors we regard as intelligent in animals. Hence it is no surprise that the same scheme for recognizing animal learning recurs for recognizing intelligent agency generally, to wit: actualizing one among several competing possibilities, ruling out the others, and specifying the one actualized.

Note that complexity is implicit here as well. To see this, consider again a rat traversing a maze, but now take a very simple maze in which two right turns conduct the rat out of the maze. How will a psychologist studying the rat determine whether it has learned to exit the maze? Just putting the rat in the maze will not be enough. Because the maze is so simple, the rat could by chance just happen to take two right turns, and thereby exit the maze. The psychologist will therefore be uncertain whether the rat actually learned to exit this maze, or whether the rat just got lucky.

But contrast this with a complicated maze in which a rat must take just the right sequence of left and right turns to exit the maze. Suppose the rat must take one hundred appropriate right and left turns, and that any mistake will prevent the rat from exiting the maze. A psychologist who sees the rat take no erroneous turns and in short order exit the maze will be convinced that the rat has indeed learned how to exit the maze, and that this was not dumb luck.

This general scheme for recognizing intelligent agency is but a thinly disguised form of the complexity-specification criterion. In general, to recognize intelligent agency we must observe an actualization of one among several competing possibilities, note which possibilities were ruled out, and then be able to specify the possibility that was actualized. What's more, the competing possibilities that were ruled out must have been live possibilities, and sufficiently numerous so that specifying the possibility that was actualized cannot be attributed to chance. In terms of complexity, this is just another way of saying that the range of possibilities is complex. In terms of probability, this is just another way of saying that the possibility that was actualized has small probability.

All the elements in this general scheme for recognizing intelligent agency (i.e., actualizing, ruling out, and specifying) find their counterpart in the complexity-specification criterion. It follows that this criterion formalizes what we have been doing right along when we recognize intelligent agency. The complexity-specification criterion pinpoints how we detect design.

Irreducible Complexity

The million-dollar question now is this: Does the complexity-specification apply to biology and in particular does it detect design in biological systems? To date the most compelling evidence for design in biology comes from biochemistry. In a February 1998 issue of *Cell*, Bruce Alberts, president of the National Academy of Sciences, remarked,

> The entire cell can be viewed as a factory that contains an elaborate network of interlocking assembly lines, each of which is composed of large protein machines. . . . Why do we call the large protein assemblies that underlie cell function *machines?* Precisely because, like the machines invented by humans to deal efficiently with the macroscopic world, these protein assemblies contain highly coordinated moving parts.[17]

Even so, Alberts sides with the majority of biologists in regarding the cell's marvelous complexity as only apparently designed. The Lehigh Univer-

17. Bruce Alberts, "The Cell as a Collection of Protein Machines: Preparing the Next Generation of Molecular Biologists," *Cell* 92, February 8, 1998, p. 291.

sity biochemist Michael Behe disagrees. In *Darwin's Black Box* Behe presents a powerful argument for actual design in the cell. Central to his argument is his notion of *irreducible complexity.* A system is irreducibly complex if it consists of several interrelated parts such that removing even one part completely destroys the system's function. As an example of irreducible complexity Behe offers the mousetrap. A mousetrap consists of a platform, a hammer, a spring, a catch, and a holding bar. Remove any one of these five components, and it is impossible to construct a functional mousetrap.[18]

Irreducible complexity is properly contrasted with *cumulative complexity.* A system is cumulatively complex if the components of the system can be arranged sequentially so that the successive removal of components never leads to the complete loss of function. An example of a cumulatively complex system is a city. It is possible successively to remove people and services from a city until one is down to a tiny village — all without losing the sense of community, which in this case constitutes function.

From this characterization of cumulative complexity, it is clear that the Darwinian mechanism of selection and mutation can readily account for cumulative complexity. Indeed, the gradual accrual of complexity via selection and mutation mirrors the retention of function as components are successively removed from a cumulatively complex system.

But what about irreducible complexity? Can the Darwinian mechanism account for irreducible complexity? Certainly, if selection acts with reference to a goal, it can produce irreducible complexity. Take Behe's mousetrap. Given the goal of constructing a mousetrap, one can specify a goal-directed selection process that in turn selects a platform, a hammer, a spring, a catch, and a holding bar, and at the end puts all these components together to form a functional mousetrap. Given a pre-specified goal, selection has no difficulty producing irreducibly complex systems.

But the selection operating in biology is Darwinian natural selection. And this form of selection operates without goals, has neither plan nor purpose, and is wholly undirected. The great appeal of Darwin's selection mechanism was, after all, that it would eliminate teleology from biology. Yet by making selection an undirected process, Darwin drastically abridged the type of complexity biological systems could manifest. Henceforth biological systems could manifest only cumulative complexity, not irreducible complexity.

18. Behe, *Darwin's Black Box,* pp. 39-45.

Why is this? As Behe explains in *Darwin's Black Box*,

> An irreducibly complex system cannot be produced . . . by slight, successive modifications of a precursor system, because any precursor to an irreducibly complex system that is missing a part is by definition nonfunctional. . . . Since natural selection can only choose systems that are already working, then if a biological system cannot be produced gradually it would have to arise as an integrated unit, in one fell swoop, for natural selection to have anything to act on.[19]

For an irreducibly complex system, function is attained only when all components of the system are in place simultaneously. It follows that natural selection, if it is going to produce an irreducibly complex system, has to produce it all at once or not at all. This would not be a problem if the systems in question were simple. But they're not. The irreducibly complex biochemical systems Behe considers are protein machines consisting of numerous distinct proteins, each indispensable for function, and together beyond what natural selection can muster in a single generation.

One such irreducibly complex biochemical system that Behe considers is the bacterial flagellum. The flagellum is a whip-like rotary motor that enables a bacterium to navigate through its environment. The flagellum includes an acid-powered rotary engine, a stator, O-rings, bushings, and a drive shaft. The intricate machinery of this molecular motor requires approximately fifty proteins. Yet the absence of any one of these proteins results in the complete loss of motor function.[20]

The irreducible complexity of such biochemical systems counts powerfully against the Darwinian mechanism, and indeed against any naturalistic evolutionary mechanism proposed to date. Moreover, because irreducible complexity occurs at the biochemical level, there is no more fundamental level of biological analysis to which the irreducible complexity of biochemical systems can be referred, and at which a Darwinian analysis in terms of selection and mutation can still hope for success. Undergirding biochemistry is ordinary chemistry and physics, neither of which can account for biological information. Also, whether a biochemical system is irreducibly complex is a fully empirical question: Individually knock out each protein constituting a

19. Behe, *Darwin's Black Box*, p. 39.
20. Behe, *Darwin's Black Box*, pp. 69-72.

biochemical system to determine whether function is lost. If so, we are dealing with an irreducibly complex system. Protein knock-out experiments of this sort are routine in biology.[21]

But are there no counterexamples to irreducibly complex systems arising by gradual means? The most common objection to Behe's claim that irreducibly complex biochemical systems lie beyond the remit of mutation and selection is the "scaffolding" objection.[22] Accordingly, in generating an irreducibly complex system, first some cumulatively complex system must arise by incrementally adding components via mutation and selection. Then somewhere along the way a sub-configuration arises that is able to function autonomously (i.e., without the rest of the configuration). Since it can function autonomously, the other components are now vestigial and drop away. When all have dropped away, we have a system that is irreducibly complex. In short, what appears to be a qualitative difference is really only the result of a lot of small quantitative changes.

There are two problems with this objection, one theoretical, the other practical. The theoretical problem is this: Selection has to work on function. We know that we have function with an irreducibly complex system like the bacterial flagellum. Let us concede that we have function with the irreducibly complex system plus its scaffold (which eventually drops off). In building to the irreducibly complex system plus scaffold, when did function begin? With a bacterial flagellum plus scaffold, for instance, when did we get outboard rotary motion to propel the bacterium through a solution? Indeed, even with the use of a scaffold, there is no reason to think that function was attained *until* all the pieces of the final irreducibly complex system were in place. Given an irreducibly complex system, the challenge for the selectionist

21. See, for example, Nicholas Gaiano, Adam Amsterdam, Koichi Kawakami, Miguel Allende, Thomas Becker, and Nancy Hopkins, "Insertional Mutagenesis and Rapid Cloning of Essential Genes in Zebrafish," *Nature* 383 (1996): 829-32; Carolyn K. Suzuki, Kitaru Suda, Nan Wang, and Gottfried Schatz, "Requirement for the Yeast Gene *LON* in Intramitochondrial Proteolysis and Maintenance of Respiration," *Science* 264 (1994): 273-76; Qun-Yong Zhou, Carol J. Qualfe, and Richard D. Palmiter, "Targeted Disruption of the Tyrosine Hydroxylase Gene Reveals that Catecholamines Are Required for Mouse Fetal Development," *Nature* 374 (1995): 640-43.

22. The noted Darwinist Michael Ruse, for instance, raises this objection in a November 1997 PBS *TechnoPolitics* program on science and materialism (for details look up the website of Seattle's Discovery Institute, which handles the videotape of the program — www.discovery.org/crsc).

is to show that there is a sequence of gradual *functional* intermediaries that leads to an irreducibly complex system plus scaffold. Granted, the scaffold can help build the irreducibly complex system. But the scaffold itself is non-functional, and the only evidence of the function in question is from the irreducibly complex system itself.

The practical problem with the scaffolding objection is equally serious. Behe states the problem as follows:

> There is no possibility of arguing against some fuzzy transmutation [that those who raise the "scaffolding" objection] have half-imagined in their minds — as if we could demonstrate with complete certainty that Dr. Jekyll really couldn't transform into Mr. Hyde by some unspecified process. If they don't put forward a model that is detailed enough to be seriously criticized, then we can't show them how their model would be inadequate. That's why I think the point about the dearth of detailed models in the literature is so critical. If models involving "scaffolding" were the answer to the problem, then one would expect to see them in the relevant literature, or to run across them in evolution experiments. But we don't. That in itself is a good reason to think they aren't the answer.[23]

What, then, is the connection between Behe's notion of irreducible complexity and our complexity-specification criterion? The irreducibly complex systems Behe considers require numerous components specifically adapted to each other and each necessary for function. On any formal complexity-theoretic analysis, they are complex in the sense required by the complexity-specification criterion. Moreover, in virtue of their function, these systems embody patterns independent of the actual living systems. Hence these systems are also specified in the sense required by the complexity-specification criterion.

Biological specification always denotes function. An organism is a functional system comprising many functional subsystems. The functionality of organisms can be described in any number of ways. Arno Wouters describes functionality globally in terms of the *viability* of whole organisms.[24] Michael Behe does so in terms of the *minimal function* of biochemical sys-

23. Personal communication with William Dembski, June 3, 1999.

24. Arno Wouters, "Viability Explanation," *Biology and Philosophy* 10 (1995): 435-57.

tems.[25] Even the staunch Darwinist Richard Dawkins will admit that life is specified functionally, describing functionality in terms of the *reproduction* of genes. Thus, in *The Blind Watchmaker* Dawkins writes, "Complicated things have some quality, specifiable in advance, that is highly unlikely to have been acquired by random chance alone. In the case of living things, the quality that is specified in advance is . . . the ability to propagate genes in reproduction."[26]

Conclusion

There exists a reliable criterion for detecting design. This criterion detects design strictly from observational features of the world. Moreover, it belongs to probability and complexity theory, not to metaphysics and theology. And although it cannot achieve logical demonstration, it does achieve statistical justification sufficiently compelling as to command assent. This criterion is relevant to biology. When applied to the complex, information-rich structures of biology, it detects design. In particular, the complexity-specification criterion shows that Michael Behe's irreducibly complex biochemical systems are designed.

What are we to make of these developments? Specified complexity, that key trademark of design, is, as it turns out, a form of information, though one considerably richer than Shannon's purely statistical form of it.[27] Shannon's purely statistical theory of information is giving way to a richer theory of specified complexity whose possibilities are only now coming to light. Although called by different names and developed to different degrees of rigor, specified complexity is starting to impact the special sciences.

For instance, specified complexity is what for Manfred Eigen constitutes the great mystery of life's origin;[28] what Michael Behe has uncovered with his criterion of irreducible complexity for biochemical systems;[29] what

25. Behe, *Darwin's Black Box.*

26. Richard Dawkins, *The Blind Watchmaker* (New York: W. W. Norton, 1996), p. 9.

27. Cf. Dembski, *The Design Inference,* ch. 7, and Dembski, *Intelligent Design: The Bridge between Science and Theology* (Downers Grove, Ill.: InterVarsity, 1999), ch. 6.

28. Manfred Eigen, *Steps Towards Life: A Perspective on Evolution,* trans. P. Woolley (Oxford: Oxford University Press, 1992), p. 12.

29. Behe, *Darwin's Black Box.*

David Chalmers hopes will ground a comprehensive theory of human consciousness;[30] what for some cosmologists underlies the fine-tuning of the universe;[31] what enables Maxwell's demon to outsmart a thermodynamic system tending toward thermal equilibrium;[32] what David Bohm's quantum potentials are extracting when they scour the microworld for what Bohm called "active information";[33] what for Roy Frieden, when formulated as Fisher information, promises to unify the whole of physics;[34] and what within the Kolmogorov-Chaitin theory of algorithmic information identifies the highly compressible, non-random strings of digits.[35] How specified complexity gets from an organism's environment into an organism's genome is one of the long-standing questions addressed by the Santa Fe Institute.

We suggest, then, that the science that needs to ground all other sciences, is not, as is widely supposed, an atomistic, reductionist, and mechanistic science of particles or other mindless entities, which then need to be built up to ever increasing orders of complexity by equally mindless principles of association, known typically as natural laws. If this is an information-rich universe, then a problem with mechanistic science is that it has no resources for recognizing and understanding information, in particular the information that God speaks to create the world, the information that continually proceeds from God in sustaining the world and acting in it, and the information that passes between God's creatures.

In this chapter we have argued that types of information, whether divinely inputted or transmitted between creatures, are empirically detectable

30. David J. Chalmers, *The Conscious Mind: In Search of a Fundamental Theory* (New York: Oxford University Press, 1996), ch. 8.

31. John Barrow and Frank Tipler, *The Anthropic Cosmological Principle* (Oxford: Oxford University Press, 1986).

32. Rolf Landauer, "Information Is Physical," *Physics Today,* May 1991, p. 26.

33. David Bohm, *The Undivided Universe: An Ontological Interpretation of Quantum Theory* (London: Routledge, 1993), pp. 35-38.

34. Roy Frieden, *Physics from Fisher Information: A Unification* (Cambridge: Cambridge University Press, 1998).

35. The non-random strings form a very small (i.e., highly improbable and therefore highly complex) set within the space of all possible strings, most of which are random in the sense of being non-compressible. The non-random strings are also specified (compressibility provides the specification). See Andrei N. Kolmogorov, "Three Approaches to the Quantitative Definition of Information," *Problemy Peredachi Informatsii* (in translation) 1, no. 1 (1965): 3-11; Gregory J. Chaitin, "On the Length of Programs for Computing Finite Binary Sequences," *Journal of the Association for Computing Machinery* 13 (1966): 547-69.

via the complexity-specification criterion. For example, the fine-tuning of the universe and irreducibly complex biochemical systems are instances of specified complexity, and signal that information was inputted into the universe by God at creation. If the naturalistic science of the last two centuries has failed to grasp the centrality and pervasiveness of information within the natural world, the science of the new millennium will not be able to avoid it. Indeed, we already live in an information age.[36]

36. Paul Davies writes, "Anyone who invested in information technology stocks will have seen their shares rocket recently. After years of hype, the information revolution is finally here. As the futurist George Gilder points out, telecommunications networks carry more valuable goods than all the world's supertankers. . . . If information is indeed poised to replace matter as the primary 'stuff' of the world, then an even bigger prize may lie in store. One of the oldest problems of existence is the duality between mind and matter. In modern parlance, brains (matter) create thoughts (mental information). Nobody knows how. But if matter turns out to be a form of organised information, then consciousness may not be so mysterious after all." From Paul Davies, "Bit Before It?" *New Scientist,* January 30, 1999, p. 3.

CHAPTER 11

A Psychological Perspective on Mathematical Learning and Thinking

Introduction

Understanding mathematics in this postmodern age requires that we understand the human side of mathematics. We focus on that aspect in this chapter by providing a psychological analysis of the nature of mathematical learning and thinking — in other words, how people do and understand mathematics.

The first part of this chapter examines human thinking in general, and the extent to which people's thinking in one context can be used in another context. The second describes the process of mathematical thinking more specifically, with a psychological analysis of how people solve mathematical problems. The third explores the social aspects of mathematical thinking, and how culture and schooling affect mathematical thinking. The fourth describes research on students' beliefs about mathematics learning in relation to learning in other academic disciplines. Finally, we map out a Christian perspective on mathematical learning and thinking and the issues Christian mathematicians might consider when trying to understand the psychology of learning mathematics.

Domain-General versus Domain-Specific Thinking

To understand mathematical thinking, it's helpful to have some grasp of human thinking in general. Perhaps the most asked question in cognitive psychology about the nature of human thought is this: Are people limited to using their cognitive skills only in the contexts and contents in which they learned these skills? Or, can people use their cognitive skills in a variety of situations? In other words, if a person learns something, can she use that skill in areas beyond the narrow context in which she learned it? If people possess domain-independent problem-solving skills, it would suggest that mathematics knowledge could facilitate learning in other domains. In fact, belief in the domain-generality of cognitive skill was the essence of the argument for "formal discipline" — the idea that training students in fundamental areas like Latin, Greek, and mathematics would make them intellectually sophisticated.[1] In contrast, if skills are bound to the context in which they are learned, then it is not likely that people will transfer their skills from one context to the next. From the view of the teacher, this distinction can be framed as teaching specific content knowledge versus teaching transferable skills or, as Perkins and Salomon have said, "educating memories versus educating minds."[2]

Empirical attempts to answer the question of whether cognitive skill is general or specific appeared early in psychology. For example, Edward Thorndike found that improvement in one skill domain as a result of training in another skill domain occurred only if the two domains shared similar features (that is, "identical elements"). In other words, the closer one domain is to another, the more likely it is that learning in one domain will facilitate learning in another. Thorndike's research contradicted the idea of formal discipline.[3]

Thorndike's work precedes the birth of cognitive psychology, which is the current dominant paradigm for understanding mental activity in psychology. Early work in problem solving in cognitive psychology and artificial intelligence posited a domain-general view of cognitive skill acquisition. Early research in the area of problem solving, led by Newell and Simon, found that people could solve a wide range of problems with effective general strategies and heuristics.[4] In their approach, people solve a problem by

1. See R. E. Nisbett, G. T. Fong, D. R. Lehman, and P. W. Chang, "Teaching Reasoning," *Science* 238 (1987): 625-31.

2. D. N. Perkins and G. Salomon, "Are Cognitive Skills Context Bound?" *Educational Researcher* 18 (1989): 23.

3. See E. L. Thorndike and R. S. Woodworth, "The Influence of Improvement in One Mental Function upon the Efficiency of Other Functions," *Psychological Review* 8 (1901): 247-61, 384-95, 553-64.

4. A. Newell and H. Simon, *Human Problem Solving* (Englewood Cliffs, N.J.: Prentice-Hall, 1972).

moving from an original state (givens), in which the solution to a problem is not apparent, to a solution (goal) state. General strategies and heuristics serve as the mechanisms by which people move from givens to goal. In this paradigm, people's content knowledge of the particular domain is less important than their ability to apply general strategies effectively. At the very least, Newell and Simon leave open the possibility that thinking and problem solving are domain-general. What is needed, then, are demonstrations of people who learn a cognitive skill and an examination of whether they use that skill beyond the context in which they learned it. We look specifically at two different types of cognitive skills that can be considered at least partly domain-general.

Conditional Reasoning as a Domain-General Skill

One approach to studying the issue of domain-general vs. domain-specific thinking is to examine people's ability to solve conditional reasoning problems of the form *if p then q*. Consider a classic four-card problem in which a learner is presented with four cards, as shown in Figure 1. The cards have a number printed on one side and a letter printed on the other.

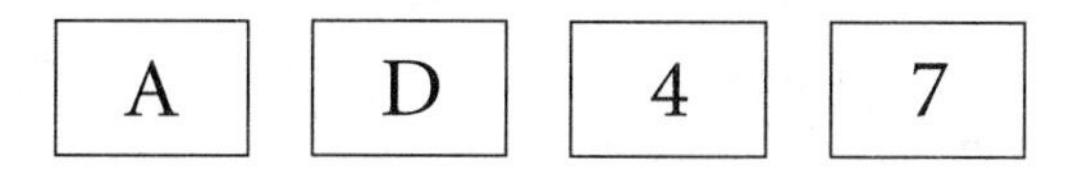

Figure 1. Example of a four-card selection problem

People are told there is a rule such as "if there is a vowel on one side, then there is an even number on the other" to describe these cards. Students are asked to indicate which cards need to be turned over in order to determine if the rule is being violated. Research shows that most students correctly select the conditional argument *modus ponens* (p therefore q), and select the 'A'. A non-even number on the other side would indicate rule violation. However, up to a third of the time, college students fail to recognize that they must also select the *modus tollens* argument (not q, therefore not p), and therefore fail to select the '7'. Many times, students also make a second error of *affirming the consequent* (q therefore p), by incorrectly selecting the '4'.

Psychologists believe that people fail on such conditional problems for several reasons. First, people may match the terms of the hypothesis (if there is a vowel on one side there must be an even number on the other) with the evidence present. In other words, if the words "vowel" and "even number" appear in the hypothesis, people are drawn to those components of the evidence, and therefore select the vowel and even number.

Second, when people solve problems they rely heavily on content and experience. As a simple example, suppose students were shown the cards in Figure 2.

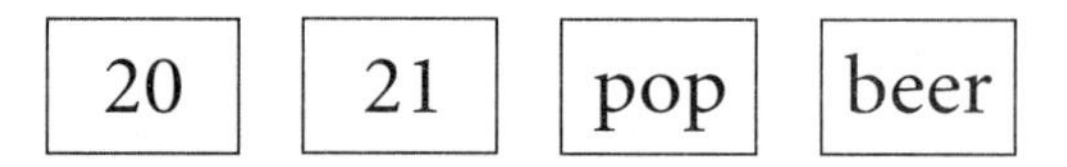

Figure 2. Example of a four-card selection problem using familiar content

If students were told that the rule was, "if a person is under 21, they must drink non-alcoholic beverages," most college students could correctly identify that the 20-year-old (p) and the beer drinker (not q) would need to be checked for further information.[5] They could also articulate why one need not check the pop drinker (affirming the consequent) or the 21-year-old (denying the antecedent). However, with such a familiar problem, students are not likely relying on any conditional reasoning rules they have learned, but rather simply on memory of specific instances.[6]

Johnson-Laird, Legrenzi, and Legrenzi specifically tested the role of experience in conditional reasoning.[7] They gave people the following rule: If a letter is sealed, then it has a 5d stamp on it (d stands for an English penny). They found that

5. R. A. Griggs and J. R. Cox, "The Elusive Thematic-Materials Effect in Wason's Selection Task," *British Journal of Psychology* 73 (1982): 407-20. See also R. A. Griggs and J. R. Cox, "Permission Schemas and the Selection Task," *The Quarterly Journal of Experimental Psychology* 46 (1993): 637-51.

6. B. H. Ross, "Distinguishing Types of Superficial Similarities: Different Effects on the Access and Use of Earlier Problems," *Journal of Experimental Psychology: Learning, Memory, and Cognition* 15 (1989): 456-68.

7. P. N. Johnson-Laird, P. Legrenzi, and M. S. Legrenzi, "Reasoning and a Sense of Reality," *British Journal of Psychology* 63 (1972): 395-400.

- English adults did much better on this problem than abstract problems of the form A-D-4-7,
- US adults were not any better on this problem than an A-D-4-7 problem type, and
- English adults who were over 45 years old did better than younger adults.

Older adults did better, the researchers argue, because this rule was an English postal requirement at a time when those over 45 would have had experience with it. This suggests that people do rely on the content of a problem and their experience with information available in the problem.

The effect of experience demonstrates that it is probably not the case that the skill of conditional reasoning is completely a domain-general activity. In fact, if we think of content as referring to the specific details of a problem, this skill looks quite tied to content. In order to understand the role of content in learning this type of problem, it must be asked exactly *how* content affects reasoning. One way it affects reasoning is that our experience usually facilitates our ability to solve a problem because there is a positive correlation between empirical proof and logical validity. That is, the rules of logic correspond to the patterns found in such familiar experiences as cause and effect and having permission. If people don't understand modus ponens and modus tollens, they at least use observable, empirical evidence to back up their claims. Since most of the time our empirical observations correspond to the laws of logic, we become comfortable making logical assertions even though we might not understand the laws of conditional logic. As an example, if a person believes that if her car engine has no mechanical problems, then her car will start, and then she starts her car, she assumes her car has no mechanical problems. The success of car-starting provides empirical proof of her belief, even though starting the car is not a logical check of the rule (it is affirming the consequent). The empirical observation reinforces the faulty reasoning, which may be harmless, provided she does not end up along the roadside with an overheated engine halfway through a long trip.

So it seems clear that content and context are important for conditional reasoning. Does that mean that people never understand abstract rules like *if p then q?* Patricia Cheng, Keith Holyoak, and colleagues propose that people do not reason using the abstract syntactic rules, nor is reasoning completely context bound.[8] Rather, people reason with what they called "pragmatic reasoning schemas." Pragmatic reasoning schemas are knowledge structures that are, in terms of cognitive representation, a "middle ground" between completely abstract and completely domain-dependent. Cheng and colleagues argue that these schemas are arranged around certain goals such as permission, obligation, and causality. In other words, it is not the specific content knowledge of the problem that facilitates performance on conditional reasoning problems, nor is it the abstract rule. Rather, people do better if the problem highlights the actions of permission, obligation, or causality. For example, what makes the alcohol problem easier than the A-D-4-7 problem? Cheng and Holyoak argue that it is not so much the familiar content, but rather the fact that a "permission schema" has been invoked. In other words, people would do just as well with the rule "if someone is over 100 years old, then they must drink Grape Nehi." This is not a rule with which anyone is familiar, so good performance cannot be explained simply by context and experience. But it is an example of a permission schema, and so it would be easier for people to solve.

How do people learn these pragmatic reasoning schemas? Cheng and Holyoak contend that as people experience everyday phenomena, such as events involving causation, their pragmatic reasoning schemas become more sophisticated. For example, with respect to understanding causation, people's ability to understand what is necessary to infer causation becomes more complete and refined as they experience situations that involve causation. Cheng and colleagues have found that purely abstract training in deductive reasoning did little to improve deductive reasoning performance, unless examples were also provided. However, training based on pragmatic reasoning schemas such as permission did improve subjects' deductive reasoning performance. Rules so learned could be applied in different content domains.

All of this evidence leads one to conclude that in a purely abstract sense, people do not possess conditional reasoning as a domain-general skill that they learn abstractly and apply to specific domains. However, if these rules of logic are framed in a way that is consistent with people's experience (e.g., permission), people can learn the rules at a level of generality that will

8. P. W. Cheng and K. J. Holyoak, "Pragmatic Reasoning Schemes," *Cognitive Psychology* 17 (1985): 391-416; P. W. Heng, K. J. Holyoak, R. E. Nisbett, and L. M. Oliver, "Pragmatic Versus Syntactic Approaches to Training Deductive Reasoning," *Cognitive Psychology* 18 (1986): 293-328.

allow them to understand even strange or unconventional rules (e.g., the Grape-Nehi-drinking centenarians). They can further apply these reasoning rules such as permission to other content domains that they didn't specifically learn. So, in fact, certain types of conditional reasoning — pragmatic-reasoning schemas — appear to be domain-general cognitive skills.

There is evidence that suggests that mathematics training improves the students' ability to do conditional reasoning, although it is correlational. Lehman and Nisbett found that students who majored in natural science disciplines had greater improvement in conditional reasoning than social science students from their first year to their fourth year in college.[9] It is likely that natural science majors take more mathematics classes than social science majors. Even assuming this, it is still indirect evidence, because learning of conditional reasoning could occur in other parts of the natural science curriculum. Lehman and Nisbett did find, however, a statistically significant, although in absolute terms not large correlation (zero-order Pearson correlation of 0.20) between the number of mathematics courses taken and scores on conditional reasoning at the end of college. More direct evidence came from Jackson and Griggs who found that those students trained in mathematics performed better on logic of the conditional. Jackson and Griggs argue that mathematics students' familiarity with concepts such as proof by contradiction foster this reasoning skill.[10]

Statistical Thinking as a Domain-General Skill

Richard Nisbett and his colleagues argue that statistical and probabilistic reasoning is a domain-general cognitive skill. They argue that people possess an inferential rule system that is similar to the statistical notion of the law of large numbers (as a sample becomes larger, it more accurately estimates the population parameter) and regression to the mean (sample values way above or way below the population mean will show subsequent values that regress toward the population mean).[11] Nisbett and others have argued

9. D. R. Lehman and R. E. Nisbett, "A Longitudinal Study of the Effects of Undergraduate Training on Reasoning," *Developmental Psychology* 26 (1990): 952-60.

10. S. L. Jackson and R. A. Griggs, "Education and the Selection Task," *Bulletin of the Psychonomic Society* 26 (1998): 327-30.

11. R. E. Nisbett, D. H. Krantz, C. Jespson, and Z. Kunda, "The Use of Statistical Heuristics in Everyday Inductive Reasoning," *Psychological Review* 90 (1983): 339-63.

quite convincingly that this rule system is domain-independent, and that it can be applied to a variety of content domains.

Fong, Krantz, and Nisbett, for example, found that training in the law of large numbers improved subjects' ability to reason using that statistical principle. Subjects were given examples of the law of large numbers such as the following:

> Two sports fans are arguing over which sport — baseball or football — has the best (most accurate) playoff system. Charlie says that the Super Bowl is the best way of determining the world champion because, according to him, the seven games of the World Series are all played in the home cities of the two teams, whereas the Super Bowl is usually played in a neutral city. Since you want all factors not related to the game to be equal for a championship, then the Super Bowl is the better way to determine the world championship. Which procedure do you think is a better way to determine the world championship — World Series or Super Bowl? Why?[12]

Subjects were able to abstract the general principle underlying the concept and apply the statistical principle to several other content domains. Subjects were indeed able to apply these principles to other domains, and this experimental training has been shown to persist over time delays of up to two weeks.[13]

Undergraduate and graduate training has also been found to have an effect on statistical reasoning. In one study, students enrolled in an undergraduate statistics course were surveyed but were told that the survey was independent of any course. These students used statistical principles to reason about everyday-life events more often than students who had not taken a statistics course. Lehman, Lempert, and Nisbett studied the effects of graduate-school training on statistical and methodological reasoning.[14] To study methodological reasoning they examined students' ability to reason about

12. G. T. Fong, D. H. Krantz, and R. E. Nisbett, "The Effects of Statistical Training on Thinking about Everyday Problems," *Cognitive Psychology* 18 (1986): 253-92.

13. G. T. Fong and R. E. Nisbett, "Immediate and Delayed Transfer of Training Effects in Statistical Reasoning," *Journal of Experimental Psychology: General* 120 (1991): 34-45.

14. D. R. Lehman, R. O. Lempert, and R. E. Nisbett, "The Effects of Graduate Training on Reasoning: Formal Discipline and the Thinking about Everyday-Life Events," *American Psychologist* 43 (1988): 431-42.

evidence from real-life scenarios in light of methodological concerns. For example, Lehman and colleagues studied graduate students' ability to understand concepts such as spurious correlations and the lack of an appropriate control group as they occur in real-life situations. They studied first-year and third-year graduate students in law, medicine, chemistry, and psychology. They employed both longitudinal (study the same people over time) and cross-sectional (different groups of people of different ages) research designs, and tested subjects' statistical and methodological reasoning as well as formal logical reasoning (e.g., if p then q). They found that students in medicine and psychology showed significant improvement in both statistical and methodological reasoning from the first to the third year of graduate school. Law students improved significantly in logical reasoning. Training in chemistry did not improve performance in any of the three domains. Lehman concluded that different graduate disciplines foster the use of certain inferential rule systems — for example the law of large numbers and regression to the mean — and that training in other disciplines does not require the use of these schemas and thus does not lead to improvement in these types of reasoning.

Lehman and Nisbett studied the effects of undergraduate education on students' statistical and methodological reasoning. They gave students real-life problems and measured the extent to which they used statistical or methodological reasoning in their answers. Subjects answered these questions at the beginning of their first year and the end of their fourth year of college. They found that social science majors showed large improvement in statistical and methodological reasoning from first year to fourth year. Humanities and natural science majors showed smaller but still significant improvement. In addition, natural science and humanities majors, unlike social science majors, showed large improvement in formal logic (e.g., conditional reasoning).

In sum, the research by Fong, Nisbett, and their colleagues has shown that

- people possess an inferential rule system in their cognitive apparatus that employs knowledge of statistical principles,
- people can reason about everyday-life problems using this knowledge,
- this ability is domain-general and can be applied to a range of contents, and
- this ability can be improved by training.

Conclusions on Domain-General Thinking

The research on conditional reasoning — specifically pragmatic reasoning schemas — and statistical reasoning provides some evidence that there are skills which are represented in the human mind at a high enough level of generality that they can be applied across a variety of contexts. However, other work has been less sanguine about the generality of cognitive skills. In particular, many educational researchers are now putting forth the notion of *situated cognition,* which posits that knowledge acquisition is a function of the "activity, context, and culture in which it is developed and used."[15] This approach places greater emphasis on tasks and contexts as important factors in understanding human learning. This is in contrast to traditional cognitive theories that tend to emphasize the role of people's mental representations, and search for universal laws of human information processing. Brown and colleagues forcefully contend that the idea that teaching and learning can take place independent of contextual factors is misleading and that ". . . educational practice is the victim of an inadequate epistemology. A new epistemology might hold the key to a dramatic improvement in learning and a completely new perspective on education."

This new epistemology has taken hold in some areas of the study of the mind, most notably in the area of educational research. Most educational researchers now understand the importance of considering context when studying student learning. Social-cultural researchers such as Lave[16] place a heavy emphasis on social participation, guided practice, and communication in the development of cognitive expertise. Other researchers such as Greeno[17] have continued in this direction. What all of these researchers have in common is the notion that cognition does not take place in isolation, and in order to understand and predict student learning one must take account of the various social factors at work in the life of the learner.[18]

15. J. S. Brown, A. Collins, and P. Duguid, "Situated Cognition and the Culture of Learning," *Educational Researcher* 18 (1989): 32.

16. J. Lave, *Cognition in Practice: Mind, Mathematics, and Culture in Everyday Life* (Cambridge: Cambridge University Press, 1988).

17. J. G. Greeno and S. V. Goldman, eds., *Thinking Practices in Mathematics and Science Learning* (Mahwah, N.J.: Lawrence Erlbaum Associates, 1998).

18. For an interesting application of this approach to mathematics learning and instructional design, see P. Cobb and J. Bowers, "Cognitive and Situated Learning Perspectives in Theory and Practice," *Educational Researcher* 28 (1999): 4-15.

Thus, the research on the generality of cognitive skills is mixed. The work by Nisbett and colleagues argues for generality in thinking.[19] This research perhaps represents the best evidence that people can transfer knowledge to a variety of contexts. Cheng and Holyoak also show that pragmatic reasoning schemas can be applied to a variety of content domains. These two lines of research suggest that people can use rules or knowledge in a variety of contexts. On the other hand, Brown et al. and others have pointed out the failure of cognitive theories to account for contextual factors in knowledge acquisition. They argue that the generality of cognitive skills is more limited than what has been previously believed.

What should be concluded from the work on skill generality? VanderStoep and Seifert argue that either extreme position is insufficient to explain human learning.[20] A domain-general approach to learning and transfer is inadequate because of research that highlights failures to transfer knowledge from one content domain to the next. On the other hand, a completely domain-specific approach makes generalizations about learning difficult to make. From a domain-specific perspective, people interested in understanding why and how and students learn can only do so with reference to specific content and contexts in which the learning took place.

More specifically, what can be concluded about the generality of mathematical thinking? Is it a domain-general cognitive phenomenon or is it a cognitive activity bound to the context in which it was learned? There is abundant research evidence that suggests that people have trouble transferring their knowledge from one domain to the next.[21] In one classical approach to education, the mind is viewed as a muscle that can be strengthened by any rigorous course. Such a metaphor underlies some arguments in favor of general education courses. The evidence suggests that such a metaphor is inappropriate. However, it is not the case that learning is so domain-specific that all teaching must be done in very specific ap-

19. E.g., Nisbett et al., "Teaching Reasoning."

20. S. W. VanderStoep and C. M. Seifert, "Problem Solving, Transfer, and Thinking," in *Student Motivation, Cognition, and Learning,* ed. R. R. Pintrich, D. R. Brown, and C. E. Weinstein (Hillsdale, N.J.: Lawrence Erlbaum Associates, 1994), pp. 27-49.

21. See M. L. Gick and K. J. Holyoak, "Analogical Problem Solving," *Cognitive Psychology* 12 (1980): 306-55; M. L. Gick and K. J. Holyoak, "Schema Induction and Analogical Transfer," *Cognitive Psychology* 15 (1983): 1-38; M. L. Gick and K. J. Holyoak, "The Cognitive Basis of Knowledge Transfer," in *Transfer of Learning: Contemporary Research and Applications* (San Diego: Academic Press, 1987).

plied settings either. Transferability between domains does exist, although it is limited.

The Cognitive Study of Mathematics

Some cognitive psychologists have studied mathematics because of an interest in improving students' mathematics learning. For example, Schoenfeld and his colleagues have been interested in how mathematics learning actually takes place in the classroom.[22] From their findings, they also have made interesting and important contributions to research in science education and also made broader claims about human cognition.

Other cognitive psychologists have studied mathematics not so much because of their interest in mathematics but rather because of what mathematics can reveal about human cognition in general. Mathematics problems are useful stimuli for understanding important mechanisms of the human mind. For example, Laura Novick and her colleagues see mathematical problem solving as a process of analogy.[23] Through their work on mathematics problem solving, they have discovered important aspects of human problem solving. (More on Novick's research will be mentioned later in this section.) Although these two branches of the cognitive study of mathematics have different starting points, both provide important information about mathematics learning in particular as well as human thinking in general. Thus, we will discuss these areas of research under the general rubric of understanding mathematics from a cognitive psychological perspective.

Newell and Simon provided the classic framework for studying high-level cognition in their book *Human Problem Solving.* They proposed that when solving a problem, a person is faced with: (a) a current state or problem state — a learner's understanding of the problem's givens, (b) a goal state — understanding of the desired outcome, and (c) problem operators — means by which the solver moves from problem to solution. This is a useful framework because it explains problem-solving behavior independent of

22. A. Arcavi and A. H. Schoenfeld, "Mathematics Tutoring Through a Constructivist Lens: The Challenges of Sense-Making," *Journal of Mathematical Behavior* 11 (1992): 321-35. See also A. H. Schoenfeld, "Explorations of Students' Mathematical Beliefs and Behavior," *Journal for Research in Mathematics Education* 20 (1989): 328-355.

23. L. R. Novick and K. J. Holyoak, "Mathematical Problem Solving by Analogy," *Journal of Experimental Psychology: Learning, Memory, and Cognition* 17 (1991): 398-415.

the ability/experience of the solver or the difficulty of the problem. It is also useful in describing problems for which multiple steps are required for solution. That is, the use of operators reduces the distance between the current problem state and the goal state. To write a computer program, for example, a person must clearly identify the desired inputs and outputs and the steps necessary to transform those inputs into outputs. The completion of each of these activities brings the solver closer to the solution (goal state).

Another approach to understanding mathematical problem solving is the study of *analogy.* Analogy is the use of previous solution procedures to solve a current problem. Solving problems by analogy involves recognizing the higher-order relationships that exist between two domains, even though the two domains share very few similar surface features or characteristics.[24]

The work by Gick and Holyoak introduced the modern paradigm of analogical transfer. The procedure involved giving learners an initial story or problem that contained a solution. Learners were then given a problem that required the same solution as the initial problem. Psychologists studying analogical transfer have made a distinction between two components of analogical thinking — access and use. *Access* is recalling the proper solution procedure from memory, and *use* is the correct implementation of that solution procedure. Researchers have quite consistently found that the inability of learners to transfer a solution procedure from an old problem to a new problem is a result of failure to access. In other words, people fail to recall the correct solution procedure from memory. Once the learner is told which solution procedure to use, solution rates increase significantly.

This general access-use framework for understanding analogical problem solving was applied specifically to mathematical problems in research by Laura Novick and Keith Holyoak. They taught college students the least common multiple (LCM) procedure with an example problem:

> Members of the West High School Band were hard at work practicing for the annual Homecoming Parade. First they tried marching in rows of twelve, but Andrew was left by himself to bring up the rear. The band director was annoyed because it didn't look good to have one row with

24. D. Gentner, "Structure-Mapping: A Theoretical Framework for Analogy," *Cognitive Science* 7 (1983): 155-70; D. Gentner, "The Mechanisms of Analogical Learning," in *Similarity and Analogical Reasoning,* ed. S. Vosniadov and A. Ortony (Cambridge: Cambridge University Press, 1989), pp. 197-241.

> only a single person in it, and of course Andrew wasn't very pleased either. To get rid of this problem, the director told the band members to march in rows of eight. But Andrew was still left to march alone. Even when the band marched in rows of three, Andrew was left out. Finally, in exasperation, Andrew told the band director that they should march in rows of five in order to have all the rows filled. He was right. This time all the rows were filled and Andrew was not alone any more. Given that there were at least 45 musicians on the field but fewer than 200 musicians, how many students were there in the West High School Band?

They were then asked to solve a similar problem, such as finding the number of plants in a vegetable garden given that 6 plant types fill the rows perfectly and 4, 5, and 10 plant types all leave two extra at the end of one of the rows.

Included in Novick and Holyoak's 1991 results are two findings particularly relevant to mathematics learning. First, they found that students who did the best on solving the problems were those who abstracted the structural features of the problems — what cognitive psychologists call "schema induction." These who induced the appropriate schema developed a better conceptual understanding of the class of problems represented in the experiment. (Schema induction was measured not by their performance on the problems but by evaluating students' answers to questions regarding the similarities between the learning example and the test problem. This may be a more direct measure of conceptual understanding, since mathematics teachers probably understand more than anyone that sometimes students solve problems correctly while being only dimly aware of how they did so.)

Second, they found a positive correlation between transferring the learned solution (LCM) from the example problem to the test problem and students' scores on the Mathematics portion of the SAT (MSAT). Although it's not surprising that those with high mathematical expertise do better on mathematics problems (in fact, it's even tautological), it is still of interest to cognitive psychologists to consider why this is the case. Novick and Holyoak suggest at least two reasons why experts do better. First, experts (defined in this case as those with high MSAT scores) are more likely to represent problems (that is, develop a mental image) in ways that will facilitate recall of relevant information as well as facilitate the appropriate mapping of values to variables. Second, although knowledge of the LCM procedure may not have been an automatic or well-learned process for any of the students, experts probably had a better conceptual understanding of the problem's operations (e.g., recognizing the three divisors, identifying the dividend by which those three numbers will be divided, adding the constant remainder that is stated in the problem). Experts would be better able to hold this information in working memory, and therefore will have cognitive capacity freed up to apply the LCM procedure to the problem at hand.

Bassok and Holyoak also explored the extent to which students can transfer their mathematics learning to other domains.[25] They studied students' ability to learn and transfer knowledge from the content domains of algebra and physics. In a controlled laboratory setting, they taught college students arithmetic-progression problems from algebra and constant-acceleration problems from physics. An example of a progression problem they used was:

> A boy was given an allowance of 50 cents a week starting on his sixth birthday. On each birthday following this, the weekly allowance was increased 25 cents. What is the weekly allowance for the year beginning on his 15th birthday?

An example of an acceleration problem they used was:

> An express train traveling at 30 meters per second (30 m/s) at the beginning of the 3rd second of its travel, uniformly accelerates increasing in speed 5 m/s each successive second. What is its final speed at the end of the 9th second?

Figure 3 (on p. 324) shows Bassok and Holyoak's task analysis, which identifies the isomorphic sub-goals in solving these problems.

Bassok and Holyoak's findings are likely to be encouraging to mathematicians. Specifically, they found that students who learned the arithmetic-progression procedure were very likely to recognize that the physics problems could be solved with the same solution. This occurred even when the algebra word problems were taught using a content domain that had noth-

25. M. Bassok and K. J. Holyoak, "Interdomain Transfer Between Isomorphic Topics in Algebra and Physics," *Journal of Experimental Psychology: Learning, Memory, and Cognition* 15 (1989): 153-66. See also M. Bassok, "Transfer of Domain Specific Problem-Solving Procedures," *Journal of Experimental Psychology: Learning, Memory, and Cognition* 16 (1990): 522-33.

Step	Arithmetic Progression	Constant Acceleration
1	How many terms are there?	How many seconds did first body travel?
2	What is the value of the first term?	What was velocity at start of initial second?
3	What is the common difference?	What is the constant acceleration?
4	What is the value of the final term?	What was velocity at the end of final second?

Figure 3. Similarities between arithmetic-progression and constant-acceleration problems.

ing to do with physics (e.g., the allowance problem shown above). However, students who first learned the physics problems almost never recognized the applicability of the learned procedure to the arithmetic-progression problems. Bassok and Holyoak suggest that learners viewed the physics problems as instances of the class of arithmetic-progression problems. That the students who learned constant-acceleration problems did not recognize the applicability of this knowledge to the algebra problems suggests that students viewed the physical concepts as a necessary component of the solution procedure, or at least served to limit the range of applicability of the solution procedure. This did not happen when students learned the more general algebra solution procedure, even when the example problems were tied to the content of physics. In other words, students probably saw the solution procedures in algebra as generally applicable to a wide range of content domains, and the solution procedures in physics as applying to simply the narrow class known as physics problems.

VanderStoep and Seifert extended these findings to explore an unaddressed issue in the study of problem solving.[26] They attempted to answer a question particularly relevant to the learning and teaching of mathematical problem solving. They proposed that the memory access component (recall

26. S. W. VanderStoep and C. M. Seifert, "Learning 'How' Versus Learning 'When': Improving Transfer Problem-Solving Principles," *Journal of Learning Sciences* 3 (1993): 93-111.

the distinction between access and use) of the problem-solving process actually involves two sub-components: (a) solution identification and (b) memory. Solution-identification involves examining the problem at hand and determining which of a variety of solution procedures are available to the solver (those solution procedures with which the solver is familiar). The memory process is recalling the specific components of a solution procedure.

Their original research question was: How do people recognize the solution to a novel problem when more than one principle may be applicable? They hypothesized that, if people are learning principles that appear very *similar* to each other (like combinations and permutations), they will need instructional information about the conditions under which each principle applies, and why a principle should be applied in those situations (referred to as *applicability instructions*). In other words, learners need to know which principle applies to different types of problems. However, if people are learning principles that are very distinct from each other (like permutations and conditional probability), then information about the conditions under which each principle applies will not be needed. In this case, it is not necessary to provide learners with information about when and why each principle applies to different situations. Learners will be able to determine which principle to use based simply on obvious characteristics of the problem they are trying to solve, and additional instructional information will not be helpful. They tested the extent to which instructional differences facilitated problem solving in a series of studies.

Their general approach was to have college students study principles of probability theory, specifically combinations and permutations. All students received instructions on these principles at the beginning of the experiment. This included a statement of the principle, the formula, an example problem, and a detailed explanation of its solution. After the students studied the formulas, half of them in each of the conditions received a review of how to solve the problems (referred to as Procedural Review). The other half of the students received information about when and why each principle should be applied (referred to as Applicability Instructions). Students were then given test problems to solve.

The results showed that when students were asked actually to solve a problem (which involves both solution identification and memory), the students given the Applicability Instructions and Procedural Review treatments did not differ. However, when students were asked only to indicate *how* they would solve the problem (the solution-identification process), the students given the Applicability Instruction performed better than the students given the Procedural Review conditions.

This research suggests that what is critical to successful mathematical problem solving is recognizing what we called "applicability conditions." In other words, being able to solve a mathematics problem is contingent on the solver being able to recognize which solution is appropriate to the current problem. Presumably, a solver learns multiple solution procedures (say, throughout a semester-long course), and what distinguishes the best problem solvers are those who can identify when each of those solution procedures is appropriate to apply.

Consider some ways in which the skill of identifying applicability conditions is important in a variety of cognitive domains. First, consider how elementary probability theory might normally be taught. Permutations might be taught in one section and then combinations taught in the next section. Students could become proficient at solving the two different types of problems; however, without explicit instruction on when to apply each principle, learners may have little knowledge of when each should be applied.

As another example, consider the different methods of integration taught in calculus. Textbook sections devote attention to different ways functions can be integrated. Students likely will solve problems using the integration-by-parts method, integration by substitution, and certain rules for integrating specific trigonometric functions. However, without adequate knowledge of when to apply each of these different procedures, students will have difficulty applying the different methods to new problems.

As a final example, consider a pre-service student teacher taking educational psychology. She receives instruction on, among other things, how students learn most effectively, how and why students are motivated, how individual differences play a role in classroom behavior, how to manage her classroom, and how to discipline problem behavior. What is potentially lacking from this student's repertoire of knowledge, is information about when and where these different pieces of knowledge are relevant. The student has been given lots of solution procedures, but they will not help her be a better teacher unless she receives training about when, where, and how to apply these rules. In general, instruction on the applicability conditions of the knowledge, rules, strategies, and skills students learn during school should be an essential part of the instruction they receive.

Conceptualizing problem solving as learning "applicability conditions," as VanderStoep and Seifert did in this research, suggests that mathematical thinking can be understood as two processes — understanding principles and executing procedures.[27] The research just described indicates that cognitive psychologists can contribute to the discussion of how to promote the understanding of mathematical principles.

We now turn to understanding mathematical thinking from a more macro-psychological perspective.

Mathematical Thinking from a Social-Cultural Perspective

Harold Stevenson, James Stigler, and their colleagues have studied extensively the achievement differences between American and Asian schoolchildren in mathematics.[28] Their work constitutes the most comprehensive cognitive and social psychological analysis of cross-cultural mathematics learning to date. They studied several hundred first- and fifth-grade students from the U.S. (Minneapolis and Chicago), Japan (Sendai), and China (Beijing and Taipei, Taiwan) on measures of mathematics learning. They went beyond standard computational fluency (which other research has shown does not necessarily indicate sophisticated mathematical learning) and tested a broad battery of mathematical competencies — word problems, number concepts/equations, estimation, operations, geometry, graphing, mental folding (determining what a geometric figure will look like after imagining a series of folds) and mental calculation. Their results are straightforward and compelling: Japanese students outperform U.S. students on all aspects of mathematics achievement that were tested (except mental calculation at both first and fifth grades). The achievement difference is even greater in the fifth grade than in the first grade. One of their studies showed that kindergarten students in Japan also performed better than U.S. kindergartners. This suggests that early (preschool) family socialization practices, in addition to schooling, play a role in fostering mathemat-

27. R. E. Mayer and M. Hegarty, "The Process of Understanding Mathematical Problems," in *The Nature of Mathematical Thinking*, ed. R. J. Sternberg and T. Ben-Zeev (Hillsdale, N.J.: Lawrence Erlbaum Associates, 1996), pp. 29-54.

28. J. W. Stigler, S. Lee, and H. W. Stevenson, *Mathematical Knowledge of Japanese, Chinese, and American Elementary School Children* (Reston, Va.: National Council of Teachers of Mathematics, 1990). See also H. W. Stevenson and J. W. Stigler, *The Learning Gap* (New York: Summit Books, 1992).

ics achievement. Chinese students perform about as well as U.S. students in the first grade, but excel beyond U.S. students by the fifth grade.

Perry, VanderStoep, and Yu attempted to understand possible sources of these achievement differences.[29] They examined classroom transcripts of first-grade mathematics classes from Stevenson and Stigler's database. The hypothesis was that the kinds of questions teachers ask students will correlate with achievement. Specifically, teachers who ask high-level questions requiring conceptual understanding will have students who show high achievement. Thus, these types of questions will be more likely to be found in Asian classrooms. To test this, they constructed an a priori classification of six types of questions. These problem types, along with examples, are shown below. The first two question types were considered low level and the last four question types were considered high level.

Question Type	Example
Computation/Rote Recall	"What is 124 - 121?"
Rule Recall	"What is the rule for two-digit addition?"
Make up a Problem	(Students generate their own problem that they solve.)
Computing in Context	(These are usually story problems.)
Problem-Solving Strategies	"Explain how you solved 14 + 32."
Conceptual Knowledge	"Why is this story problem an addition problem?"

The authors found no cross-cultural differences in Computation/Rote Recall, Rule Recall, or Make Up a Problem. They did find significant differences on Computing in Context, Problem-Solving Strategies, and Conceptual Knowledge, which are all higher-level cognitive operations. In all three cases, the U.S. classrooms had the lowest proportion of these question types.[30]

These data suggest that classroom factors do influence mathematical thought. Other work has shown that family and other social influences affect mathematical thought as well. Consider some of the differences Stevenson

29. M. Perry, S. W. VanderStoep, and S. L. Yu, "Asking Questions in First-Grade Mathematics Classes: Precursors of Mathematical Thought," *Journal of Educational Psychology* 85 (1993): 31-40.

30. For Computing in Context the Chinese classrooms had the highest proportion; for Problem-Solving Strategies and Conceptual Knowledge, the Japanese classrooms had the highest proportions.

and Stigler (1992) identify between Asian and American children's academic lives. Asian children, for example:

- more frequently have space in which they can complete their homework (e.g., a desk of their own),
- spend more days in school per year,
- spend more time doing homework,
- spend more time reading for pleasure, and
- receive more help on their homework from their parents (perhaps because Asian students are assigned more homework).

All of these differences in the academic and social lives of Asian and American children may give clues as to why these cross-national differences in mathematics achievement exist. They suggest that activities such as asking high-level questions in class and encouraging parents to become actively involved in their children's learning *may* be educationally beneficial.

Beliefs about Mathematics and Other Disciplines

Psychologists have also sought to understand mathematics learning and thinking describing the beliefs that people have about mathematics. This area of research is known as epistemological beliefs. Most of the research in this area has been done with a method known as "self-report," a fairly straightforward approach to assessing people's cognitions. The approach is to construct statements, for which people will indicate agreement or disagreement (usually on a multi-point continuum). Several items that are highly statistically related can be combined to measure a particular dimension of belief.

Hofer, VanderStoep, and Pintrich have examined students' beliefs about learning in different academic disciplines.[31] One approach posits that when students learn something (or claim that they know something), two different cognitive claims are being made. First, students make some assessment about the "provability" of the concept they learned. For example, one

31. B. K. Hofer, S. W. VanderStoep, and P. R. Pintrich, *Disciplinary Differences in Epistemological Beliefs,* poster session presented at the 104th Meeting of the American Psychological Association, Toronto, Canada, August 1996.

would predict that students would believe that concepts in physical or natural sciences are high on provability (e.g., cell structures in biology). Conversely, one would predict that students would believe that concepts in the social sciences and humanities are low on provability (e.g., superego). Second, students make some assessment about the "realism" of the concept they learned. In other words, does a concept have an ontological reality in the world, or is it a concept that exists only in the mind of the knower? For example, one would predict that students would see concepts in the physical and natural sciences as high on realism (e.g., human anatomy), whereas students would see concepts in the social sciences as low on realism (e.g., human personality). The lists below show the items designed to measure provability and realism.

Provability Items

1. Concepts that are studied in this field can be proven to exist.
2. We can verify that what we are studying in this field is true.
3. Investigations in this field can prove certain things about the world.
4. Being able to prove something is an important part of investigations in this field.
5. If the investigative methods in this field can show something to exist, then we are confident in such a claim.
6. Proving that something exists is not an element of this field.
7. It is not possible to demonstrate whether or not a concept is true in this field.
8. We have no way of proving for sure that what we are studying in this field actually exists.
9. Being able to prove something is not essential to studying and understanding concepts in this field.
10. The things that are studied in this field are more a matter of opinion than fact.

Realism Items

1. Things that are studied in this field actually exist in the world.
2. This field attempts to understand an objective reality out in the world.
3. The aim of this field is to understand laws about the universe.
4. There is not a lot of debate about the fundamental facts of this discipline.

5. Learning in this field involves accumulating objective facts about the world.
6. The concepts in this field consist of ideas generated by those who study in the field.
7. In this field, the only truth is in the eye of the beholder.
8. The aim of this field is to create ideas and conceptualizations.
9. There is a lot of debate and disagreement among people about what is truth in this field.
10. Learning in this field involves coming up with your own ideas about what is correct.

One of their studies asked 189 college students to respond to these twenty items on a five-point scale. Students completed the questionnaire four times for four different academic disciplines — biology, mathematics, psychology, and religion. The means were computed for the provability and realism scales. Figure 4 shows the differences between academic disciplines on these two scales. Results indicate that mathematics and biology are seen as higher on both realism and provability than psychology and religion. Students are more likely to believe in the reality of mathematical objects and that truth claims in mathematics can be demonstrated (proven) to be correct.

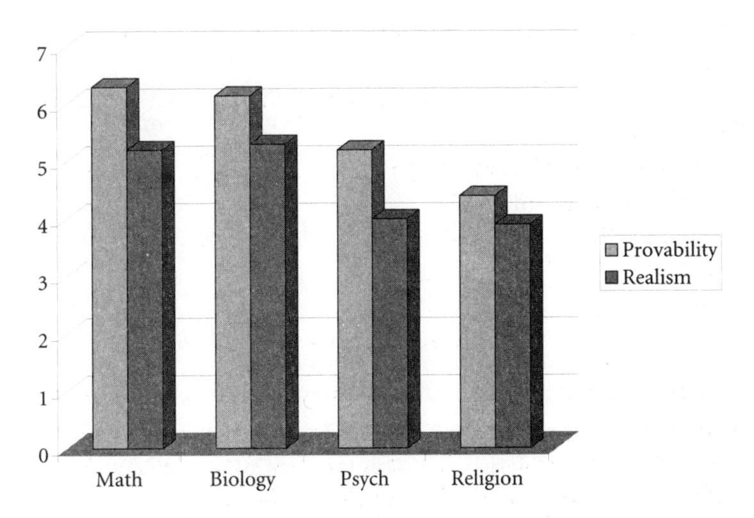

Figure 4. Mean scores for provability and realism

At least two aspects of these findings are interesting for mathematicians. First, that students view mathematics as being high on provability and high on realism says nothing about the actual nature of mathematics. This research is silent on the actual nature of mathematics; it rather simply reflects how undergraduate students view mathematics. Second, this research has not yet addressed what professional mathematicians believe about their field. Assuming the questionnaire items would be appropriate for Ph.D. mathematicians, a comparison of students' beliefs (both mathematically naïve and mathematics majors) to professors' beliefs would be interesting. Scholars in the research area do believe that students' beliefs about a discipline will affect how a student learns and thinks in that discipline, although the research is still tentative on this issue. Intuitively, it is plausible to conceive that students' beliefs about learning will affect their learning. As a simple example, a student who believes that there are absolute, provable truths to be discovered in a discipline would approach learning differently than a student who believes that the claims of a discipline are socially constructed, and dependent on context and perspective.

A Christian Critique

In the final section, we attempt to identify challenges a Christian must face when it comes to understanding what it means to engage in mathematical learning and thinking. Two issues are important in this regard. The first is to understand what assumptions psychological methods make about human thinking. Although this is not a book about psychology, to understand a psychological perspective on mathematics requires some critique of the psychological method for knowing. The second and related issue is to understand that psychological data about mathematical thinking are influenced by various perspectival concerns.

Psychology as a Limited Window on the World

Psychology studies human beings, so for Christians who believe that people are image-bearers of the Creator, it is important to know what psychology assumes about human beings. A psychological perspective on mathematics, like a psychological perspective on anything else, brings with it a limited understanding. In other words, a discipline's methodology focuses and limits how and how much we can know about something. This affects a psychological study of mathematics learning in (at least) the following three ways.

First, psychology's methodological emphasis on the gathering and analysis of social science data can yield interesting and important information. However, like any method, this approach comes with certain assumptions and values that limit understanding. That is, psychology is an empirical science and, like any science, this means it restricts its domain of study to observable, replicable aspects of reality. This does not exclude the postulation of entities that are not directly observable, but it does mean that observable evidence for their existence and nature must be presented. The restriction to replicable aspects of reality means that unique events are outside its domain of study. Thus, to whatever extent human beings are free moral agents who make choices that are not fully explainable by laws of cause and effect, to that extent, humans cannot be studied by the methods of empirical psychology. One can easily imagine how such choices could affect a person's study of mathematics, so as we consider a psychological perspective on mathematical learning, we need to keep this limitation in mind.

It is also important to understand, especially with respect to mathematical thinking, what psychology believes about the human mind. Many psychologists, especially those who operate from a cognitive perspective, assume what is known as a "computational" model of the mind. For psychologists, computation does not refer just to computational operations (necessarily), but rather more generally to systematic and interpretable mappings of inputs to outputs. Many psychologists believe this is how the mind works. With respect to mathematical thinking, then, cognitive psychologists are interested in how thinking about a mathematical problem will produce certain behaviors, which result in the production of a solution to the problem. This approach also seems shortsighted from a Christian perspective. The Christian understanding of humans is much fuller than that proposed by the computational understanding. The richness of the human experience includes the powerful intellectual capacities given to us, but also the powerful affective, emotional capacities such as care and interest. It seems certainly plausible, in fact likely, that such human emotional experiences will add complexity to people's cognitive apparatus.

A third consideration is that the computational approach to thinking means that there is some material substance involved in the thinking process. For humans it is the brain. But computationalists don't need biological substance, so the electronic circuitry of a computer is both necessary and sufficient (provided there is adequate hardware, software, and programming) for thinking to take place. There is nothing particularly interesting to strong computationalists about the people who do mathematics. In contrast, cognitive psychologists who study mathematics examine the components of the mathematical tasks and the extent to which they are done well/poorly, and the factors that influence the competence and performance of the thinker. These are interesting and important aspects of the human cognitive experience to study. In spite of this difference between strong computationalists and cognitive psychologists, however, there is nothing in cognitive psychological *theory* that necessitates that the mathematical thinker even be a human. It studies mathematical operations, not the people performing the mathematical operations. It is true that thinking machines can do mathematics, but for people who see academic inquiry as a way to understand God's creatures, it is neither the elements of the task nor the sheer computation that is of interest. What is interesting is understanding how *we* do mathematics, not simply how mathematics is done. Psychological models of mathematical learning do not adequately address the issue from this broader perspective.

One advantage of the computational approach is that human thinking can be modeled by computing machines. For example, much of artificial intelligence is interested in how intelligent programs are similar to human thought processes. However, the computational approach also brings with it certain challenges. First, understanding human thought from this perspective means that human thought could be understood objectively. This model does not place value on either the inputs or the outputs of the thinker. The kinds of things that are thought about or the values or attitudes the thinker has while processing the thoughts are not important to understanding how thinking gets done. To the extent that they are important, values and attitudes are simply another set of computations going on in the mind of the knower. To ignore aspects such as values and attitudes, or only to treat them as "more cognitions," seems shortsighted. Teachers more than anyone recognize the role of goals, attitudes, and motives in predicting successful thinking. Cognitive psychology has underestimated the role of these features of human thinking. Another way to understand this distinction between cognition and affect is that cognitive psychology has developed models of competence, that is, models that describe how people *could* behave under certain controlled (probably optimal) situations. It falls short of incorporating psychological constructs such as interests, attitudes, and goals. So cognitive psychology can only explain part of mathematical learning.

It is worth noting, too, that not just Christians are critical of this strong computational approach. Much of educational research has found it lacking. Most educational researchers interested in student thinking, clearly put the role of social context at the forefront of their theorizing. This is true even though many of these researchers would still describe themselves as having a cognitive orientation. DeCorte, Greer, and Verschaffel note that the computational perspective is part of the first wave of cognitive studies. The second wave (found largely in educational research but not in basic cognitive psychology) does take seriously situational and affective factors in understanding student learning.[32]

The Role of Perspective in Mathematical Thinking

The second issue to consider is that research psychologists usually hold their methods and results in high esteem. It is easy to forget that psychological data are not veridical. They are influenced by a variety of factors, including assumptions of the theory, worldview of the researchers, and specific characteristics of the people being studied. In other words, perspective matters when considering findings in psychology.

Granting that perspective matters when developing, describing, and understanding psychological theory allows room for Christian worldviews to enter psychological theory, research, and data. One of the goals of a Christian psychology is to understand how a Christian worldview will affect how one does psychology. As a hypothetical example, suppose a psychologist is interested in empirical studies of the family. Specifically, she is interested in predicting children's well-being as a function of various psychological variables of the parents. In this case, one's Christian faith will shape aspects of this research program in several ways. For example, what aspects of a child's psychological make-up would constitute a healthy outcome (e.g., independent, communal, obedient)? Also, of all values that the parents transmit to the child, which ones would this psychologist choose to measure (e.g., im-

32. E. DeCorte, B. Greer, and L. Verschaffel, "Mathematics Learning and Teaching," in *Handbook of Educational Psychology,* ed. D. Berliner and R. Calfee (New York: Macmillan, 1996), pp. 491-549.

portance of respect for authority, self-reliance, managing friendships, discipline)? Even a Christian understanding of a "family" may affect sampling and selection procedures.

This example makes clear that perspective matters in psychology, even at the basic levels of research methodology and theorizing. Does this example have anything to say about how scholars do mathematics? To allow for a Christian perspective on mathematics, one must first allow that perspective, in general, will affect mathematics. We have tried to show that perspective does matter. Section II of this book gave numerous historical examples of how perspective has shaped the way mathematics has influenced Western culture. It also discussed how perspective affects the way that mathematicians perceive their discipline and it discussed two case studies. Section III has examined how perspective affects the way mathematicians and philosophers understand the nature of mathematics.

Superficially, one would expect that the role of perspective would be quite different in psychology and mathematics. After all, psychologists study human beings and mathematicians study abstract concepts. However, when one examines the situation more closely, the role of perspective in the two disciplines is remarkably similar. That is, a comparison between psychology and mathematics is helpful in clarifying how perspective enters mathematics. In psychology, the methodology limits the domain of study to aspects of reality that are observable and replicable. The methodology of mathematics also limits its domain, namely to abstract concepts that can be precisely defined. In psychology, one's perspective affects what aspects of situations are perceived and regarded as important. The same phenomenon occurs in mathematics. (We will pursue this observation in more detail in Chapter 12.) Furthermore, if one considers a particular mathematical result (for example, the Pythagorean Theorem), its truth seems independent of perspective. However, in psychology, once one has precisely defined concepts and data to analyze, the situation is very similar.

In summary, perspective enters at a higher level than data analysis or theorem proving. If one includes such questions as "What concepts should we be investigating?" "Why?" "What relationship do our concepts have to realities other than themselves?" perspective matters as much in mathematics as in psychology.

While questions such as the ones mentioned above have been seen as central to psychology, they have largely been excluded from mathematics as practiced in the twentieth century (and often earlier). This is a significant point. For a Christian, mathematics does not exist in isolation from other aspects of reality, and such questions should not be excluded from the content of the discipline. A Christian approach to mathematics should insist that perspective matters.

CHAPTER 12

Teaching and Learning Mathematics: The Influence of Constructivism

Introduction

As Christians who have chosen careers that involve the teaching and learning of mathematics, we are often asked if our Christian view of the world makes any difference in how we approach mathematics and mathematics education. The typical assumption of the questioner is that it does not make any difference. We disagree. We respond that in several areas critical to the teaching and learning of mathematics, one's Christian perspective affects both the reasons behind deliberate choices and some of the actual choices that are made.

This chapter takes a closer look at a theory of learning and teaching called constructivism. We begin with a careful delineation of several important versions of constructivism. In the views of some constructivists, the adoption of a constructivist viewpoint necessitates significant changes in how one approaches teaching, learning, mathematics itself, and even research methodologies within mathematics education. Next, we document the wide-ranging impact that these related theories have had on the teaching and learning of mathematics in the decade following the release of the first of several documents known as the *Standards.* Finally, we analyze both the nature and effects of constructivism from a Christian perspective.

Establishing a Common Understanding of the Terminology

Before undertaking an analysis of the effects of constructivism on the teaching and learning of mathematics, as well as a possible Christian response, it is imperative that some agreement is reached on the relevant terminology. This task is more difficult than it may initially appear, since "There are almost as many varieties of constructivism as there are researchers."[1] Further complicating matters, the name constructivism (also referred to as intuitionism) has been used to describe a philosophy of mathematics, one in which constructive methods alone, and not indirect methods such as argument by contradiction, must be used to establish the existence of mathematical objects and mathematical truths. The interested reader can learn more about this and other philosophies of mathematics in earlier chapters of this book. However, the present chapter will explore a distinctly different interpretation of the term 'constructivism.' This constructivism is housed in the discipline of mathematics education rather than mathematics and is primarily a theory of learning mathematics and related implications for teaching rather than an early twentieth-century philosophy of mathematics.

Rather than attempt to catalogue all possible forms of constructivism, we will focus primarily on the following: radical constructivism, social construct*ionism,* social construct*ivism,* and naïve constructivism. In each case, we will attempt to summarize the perspective of some of the major researchers within mathematics education who advocate that particular variation. Following this series of overviews, some attention will be given to the common threads of the various positions as well as their distinctiveness.

Radical Constructivism

Among leading researchers in mathematics education, perhaps no one is more representative of the radical constructivist position than Ernst von Glasersfeld. He offers the following essential characteristics of radical constructivism.

1. Paul Ernest, "The One and the Many," in *Constructivism in Education,* ed. Leslie Steffe and Jerry Gale (Hillsdale, N.J.: Lawrence Erlbaum Associates, 1995), pp. 459-86.

1. Knowledge is not passively received either through the senses or by way of communication. Knowledge is actively built up by the cognizing subject.
2a. The function of cognition is adaptive, in the biological sense of the term, tending towards fit or viability.
2b. Cognition serves the subject's organization of the experiential world, not the discovery of an objective ontological reality.[2]

When applied to the learning of mathematics, the radical constructivist claims that all mathematical knowledge, even basic facts and algorithms of arithmetic, can only be learned through the process of construction by the individual student.[3] So, in contrast to theories of learning that involve the transfer of knowledge from teacher to student, radical constructivists insist that the teacher's principal task is to structure a learning environment in which students are able to actively construct mathematical concepts. The work of Jean Piaget, particularly his notions of the assimilation and accommodation of concepts into one's own cognitive structures, is considered foundational for most radical constructivists within the field of mathematics education.

Notice that the radical constructivist speaks of viability of ideas rather than of their truth. Extending this reasoning, Glasersfeld observes that knowledge of any mathematical concept is relative to the individual knower and not in any sense universal or absolute. In contrast with a theory of knowing that accepts the possibility of truth in an absolute sense, a fallibilist epistemology denies the possibility of such an absolute truth. A somewhat more moderate interpretation of fallibilism would be that no one is able to know anything with absolute certainty. Glasersfeld's observation identifies radical constructivism with a fallibilist epistemology. At the same time, radical constructivists would limit the process of knowledge construction to the individual's interactions with others and with the external world.[4] So, each student must decide which mathematical concepts,

2. Ernst von Glasersfeld, "An Exposition of Constructivism: Why Some Like It Radical," in *Constructivist Views on the Teaching and Learning of Mathematics,* ed. Robert Davis, Carolyn Maher, and Nel Noddings (Reston, Va.: National Council of Teachers of Mathematics, 1990), pp. 19-23.

3. Nel Noddings, "Constructivism in Mathematics Education," in *Constructivist Views on the Teaching and Learning of Mathematics,* p. 8.

4. See, for example, Jere Confrey, "How Compatible Are Radical Constructivism,

as constructed, are viable within the context of dialogue with other students and the teacher and by testing these concepts against the surrounding environment.

Social Constructionism

Unlike radical constructivism, where the focus of knowledge construction lies with the individual learner, social constructionists shift the focus to groups of students working as a learning community. Within this perspective, some researchers would require all meaning to be socially negotiated and understood while others would allow for individual learning, provided that it occurs within a social context.[5] According to Gergen, meaning, for the social constructionist, is located not in the constructions of the individual but rather in the language that is communicated among members of a socially interdependent learning community.[6] This communication constitutes a dialogue through which meaning is both negotiated and verbalized.

Social Constructivism

In one sense, social constructivism can be viewed as a compromise between radical constructivism and social constructionism. Here, the construction of knowledge requires important contributions from students operating both individually and as part of a larger learning community. Ernest notes that, "social constructivism regards individual subjects and the realm of the social as interconnected. Human subjects are formed through the interactions with each other as well as by their individual processes."[7] In other writings, Ernest describes the learning environment as consisting of both "the public representation of collective, socially accepted mathematical knowledge

---

Sociocultural Approaches, and Social Constructivism?" in *Constructivism in Education,* pp. 185-226.

5. See, for example, John Richards, "Construct[ion/iv]ism: Pick One of the Above," in *Constructivism in Education,* pp. 57-64.

6. Kenneth Gergen, "Social Construction and the Educational Process," in *Constructivism in Education,* pp. 24-26.

7. Paul Ernest, "The One and the Many," pp. 479-80.

within a teaching-learning conversation" and "sustained two-way participation in such conversation."[8] Presumably, the individual is then free to make his or her individual constructions of mathematical concepts but must be able to participate in a dialogue with other students and the teacher in order to make this construction fully viable. In this perspective, the teacher functions as an expert role model, demonstrating the characteristics of a reflective individual while behaving appropriately within an interdependent learning culture.[9]

Naïve Constructivism

This perspective chooses to adopt the first characteristic of radical constructivism cited above, namely that learners construct their own knowledge. However, naïve constructivists make no claim concerning the nature of mathematical concepts and are therefore not committed to the fallibilist epistemology discussed earlier. Some researchers, according to Thompson (1995), are actually "utilitarian constructivists."[10] By this he means that such researchers act as naïve constructivists with regard to the nature of mathematical concepts even when their research is imbedded in a radical constructivist framework. Goldin offers the term "moderate constructivist viewpoint" as a label for his variation of naïve constructivism. He stresses that "one does not need to accept radical constructivist epistemology in order to adopt a model of learning as a constructive process, or to advocate increased classroom emphasis on guided discovery in mathematics."[11] Whatever the name, naïve constructivism seeks to capitalize on the potential of the construction of knowledge by individual learners without necessarily accepting the notion that viability and not truth is the ultimate goal of learning mathematical concepts. Naïve constructivists also

8. Paul Ernest, *Social Constructivism as a Philosophy of Mathematics* (Albany, N.Y.: State University of New York Press, 1998), p. 221.

9. See Heinrich Bauersfeld, "The Structuring of Structures: Development and Function of Mathematizing as a Social Practice," in *Constructivism in Education*, pp. 137-58.

10. Patrick Thompson, "Constructivism, Cybernetics, and Information Processing: Implications for Technologies of Research on Learning," in *Constructivism in Education*, pp. 123-24.

11. Gerald Goldin, "Epistemology, Constructivism, and Discovery Learning Mathematics," in *Constructivist Views on the Teaching and Learning of Mathematics*, p. 40.

allow for the possibility that these mathematical concepts may already exist and can be discovered during the learning process.

Similarities and Differences of These Perspectives

Several common threads run through all or most of these varieties of constructivism. First, with the exception of naïve constructivism, each perspective operates with a fallibilist epistemology. Second, all of the constructivist perspectives adopt the principle that knowledge is actively constructed rather than passively received. Third, each perspective advocates a significant departure from traditional views of both mathematics teaching and the assessment of mathematical learning. Both of these notions will be more fully explored in the next section. Finally, as Glasersfeld observes, each of these perspectives views learning not as a stimulus-response model but rather as "the building of conceptual structures through reflection and abstraction."[12]

At the same time, important differences exist which serve to maintain the distinctiveness of each position. The most obvious differences involve the roles of the individual and the society in the learning process. Radical constructivists stress the action of the individual learner, social constructionists shift the emphasis to the interdependence of members of a learning community, and social constructivists adopt an approach that balances the roles of the individual and the social interaction of the group. Naïve constructivists do not appear to have a fixed position on this issue and, unlike the other three perspectives, naïve constructivism does not require the adoption of a fallibilist epistemology.[13]

12. Ernst von Glasersfeld, "A Constructivist Approach to Teaching," in *Constructivism in Education,* p. 14.

13. See Paul Ernest, "The One and the Many," pp. 482-83, for an excellent comparison of these and other constructivist perspectives. Much of the material for this section of the current chapter is based on a series of papers delivered at the Alternative Epistemologies in Education Conference held in 1992. Leslie Steffe and Jerry Gale collected and edited these papers into the book *Constructivism in Education.*

Constructivism's Influence in the Era of the *Standards*

Some constructivists trace the origins of their perspective to the first half of the twentieth century (e.g., Jean Piaget's "reflective abstraction" or to the "zone of proximate development" theory advanced by Lev Vygotsky) or even trace their roots to various educational philosophers of the nineteenth century, but the impact of constructivism on the teaching and learning of mathematics is a more recent phenomenon. The following quotation from Davis, Maher, and Noddings appears in the introductory chapter of a collection of writings on constructivism.

> In 1985, when the present cooperative venture began to take shape, one hardly even heard the word "constructivism." That has somehow changed. For whatever subtle reasons, the *Standards — Everybody Counts* position has, for some researchers at least, coalesced into a very active concern to spell out, and analyze, the foundations of constructivism.[14]

The First Wave (*Standards,* 1989, 1991, 1995)

In this section, we use the publication of three related documents by the National Council of Teachers of Mathematics (NCTM) as the first signs of the impact of constructivism in its different variations. These three documents, each titled *Standards,* address recommendations by NCTM in the following areas: the mathematics content that should be taught in the schools at all levels from kindergarten through grade 12, the teaching of mathematics at these same grade levels, and the means of assessing mathematical learning.[15] A brief survey of the reference section of these documents gives initial evidence of the influence of constructivism.[16]

14. Robert Davis, Carolyn Maher, and Nel Noddings, "Constructivist Views on the Teaching and Learning of Mathematics," in *Constructivist Views on the Teaching and Learning of Mathematics,* p. 2.

15. *Curriculum and Evaluation Standards for School Mathematics* (1989), *Professional Standards for Teaching Mathematics* (1991), and *Assessment Standards for School Mathematics* (1995), all of which are published by the National Council of Teachers of Mathematics in Reston, Virginia.

16. For example, in *Standards* (1991).

What Students Need to Learn

In the area of curriculum, the *Standards* identify many topics that have traditionally been associated with school mathematics such as number concepts, algebra, and geometry. In addition, topics such as probability, statistics, and discrete mathematics are offered as areas where past mathematics curricula have proved inadequate. There do not appear to be any direct links to constructivist ideas in this list of recommended topics. This is not surprising, however, since radical constructivists hold that "students should be the main determiners of learning activities" rather than a teacher, a textbook, or a collection of content recommendations.[17]

Beyond these recommendations for individual content topics, however, the focus of the curriculum *Standards* (1989) is on four important overarching traits that each student must develop.[18] These traits are problem solving, reasoning, communication, and connections. By contrast, skills such as the rote memorization of basic facts and the ability to execute algorithms successfully are viewed as less important than these four critical traits. This orientation to the necessary traits for mathematics students to develop, is consistent with the writing of Glasersfeld, who argues that understanding "cannot be demonstrated by presentation of results that may have been acquired by rote learning."[19] Hiebert uses the term "procedural knowledge" to distinguish basic arithmetic skills and the execution of algorithms from traits such as problem solving and reasoning which would fit under the term "conceptual understanding."[20] In any case, the apparent shift in focus in the ways in which students need to understand mathematical concepts is consistent with any of the constructivist perspectives outlined in the previous section.

17. See Harro Van Brummelen, "Curriculum Development Is Dead — Or Is It?" *Pro Rege* 26, no. 1 (1997): 19.

18. National Council of Teachers of Mathematics, Commission on Standards for School Mathematics, *Curriculum and Evaluation Standards for School Mathematics* (Reston, Va., National Council of Teachers of Mathematics, 1989).

19. Glasersfeld, "An Exposition of Constructivism," p. 26.

20. James Hiebert, ed., *Conceptual and Procedural Knowledge: The Case of Mathematics* (Hillsdale, N.J.: Lawrence Erlbaum Associates, 1986).

How Students Must Learn

It is in the area of student learning that the influence of constructivism can be most clearly seen. Building on the Piagetian notion of assimilation, the introduction to the *Standards* notes that "educational research findings from cognitive psychology and mathematics education indicate that learning occurs as students actively assimilate new information and experiences and construct their own meanings." Later in the same discussion, a quotation from another national publication, *Everybody Counts,* is included to reemphasize the point: "Educational research offers compelling evidence that students learn mathematics well only when they *construct* their own mathematical understanding."[21] Both of these quotations highlight the focus on the construction of mathematical knowledge by the individual learner, a basic tenet of radical constructivism that was discussed earlier.

It should be noted that other forms of mathematical understanding, such as rote memorization of basic arithmetic facts, are possible but must be de-emphasized in favor of more active forms of learning. At the kindergarten–grade 4, grades 5-8, and grades 9-12 levels, the *Standards* (1989) offers an overall summary of areas for increased and decreased emphasis. In each case, rote memorization and algorithmic skills are de-emphasized in favor of more constructivist types of knowledge. One might view the acceptance of rote learning, even if de-emphasized, as evidence of a departure from constructivist principles. However, Noddings avoids this problem by redefining rote memorization as a "weak" act of constructivism as opposed to "strong" acts of constructivism that involve more active intellectual participation on the part of the individual learner.

Along with the focus on individual constructions of mathematical concepts, the *Standards* (1989) stresses the need to create a learning community. The key traits of communication and reasoning make reference to the need for student-to-student and student-to-teacher as well as the traditional teacher-to-student interaction in the classroom. It is in an environment where students share their insights and strategies that important mathematical concepts may emerge and individuals are able to "clarify, refine, and consolidate their thinking." This perspective echoes the social constructivist views of Noddings, who writes that "the role of the community — other

21. *Standards* (1991), p. 2. This is a quotation taken from *Everybody Counts,* National Research Council (Washington, D.C.: National Academy Press, 1991), pp. 58-59.

learners and teacher — is to provide the setting, pose the challenges, and offer the support that would encourage mathematical construction."[22]

How Teachers Must Teach

As profound as the changes proposed by the *Standards* in mathematical content and learning may seem, the proposed changes in teaching mathematics may be even more revolutionary. In traditional mathematics classrooms, the teacher serves as the sole authority while delivering a lecture to the students. Along with the textbook, the teacher is considered a source of both correct methodology and correct answers to the mathematics problems that students must solve. The *Standards* call for a very different classroom environment. Once again, the proposed changes can be seen as an outgrowth of constructivist approaches to the teaching and learning of mathematics. Consider the following quotation, which in turn is based upon research by several constructivists:

> Teachers need to employ alternative forms of instruction that permit students to build upon their repertoire of mathematical knowledge and their abilities for posing, constructing, exploring, solving, and justifying mathematical problems and concepts. Promising models for such instruction are all highly interactive. In such models, teachers both model and elicit mathematical discourse by asking questions, following leads, and conjecturing, rather than by presenting faultless products.[23]

Among many recommendations for teaching in the *Standards* (1989), perhaps the most significant is the shift away from direct instruction or "telling" to a more student-centered pedagogical approach. This shift leads to a change in the role of the teacher from a "sage on the stage" to a "guide on the side" who "facilitates" learning by the students. Confrey cites numerous fellow constructivists all sharing the notion that constructivism "seems to be a powerful source for an alternative to direct instruction." She later ar-

22. Noddings, "Constructivism in Mathematics Education," p. 3.

23. *Standards* (1991), p. 152 — based upon work by N. Noddings and D. Ball, *Halves, Pieces, and Twoths: Constructing and Using Representational Contexts in Teaching Fractions* (East Lansing, Mich.: National Center for Research on Teacher Education, 1990).

gues "when one applies constructivism to the issue of teaching, one must reject the assumption that one can simply pass on information to a set of learners and expect that learning will result."[24]

The shift from direct instruction to more student-centered classroom environments can also be seen in recent mathematics textbooks. One textbook series, published in the years following the release of the *Standards* and purporting to follow the goals of these documents, includes a section that lists features of "effective classrooms." This list includes "less lecture, recitation, and individual seatwork" and "more discussion (small group or whole class)."[25] Shortly after the publication of the *Standards,* NCTM published a series of materials designed to supplement course textbooks. A number of learning activities judged to be consistent with these new guidelines are provided, and a preface is included that reiterates major ideas from the *Standards.* The following excerpt is rather typical.

> There is noise in these classrooms — the sounds of students actively participating in the class and constructing their own knowledge through experiences that will give them confidence in their own abilities and make them mathematically powerful.[26]

Once again, the reader is urged to move toward student-centered teaching strategies and to encourage active knowledge construction on the part of the individual learner. These last two references also point out the focus on the teacher's role in creating cooperative learning environments. This recommendation, also found in the *Standards,* echoes the notion of a learning community and the social context of knowledge construction emphasized by social constructivists and social constructionists.

As a final aspect of recommendations related to the teaching of mathematics, we turn our attention to tools and materials that can be used to facilitate learning. In particular, *Standards* (1989) strongly encourages the use of manipulatives at the elementary and middle school levels. In addition, calculators and computers are viewed as important components of instruction

24. Jere Confrey, "What Constructivism Implies for Teaching," in *Constructivist Views on the Teaching and Learning of Mathematics,* pp. 107-22.

25. *Transition Mathematics* (Glenview, Ill.: University of Chicago School Mathematics Project, 1995).

26. *Curriculum and Evaluation Standards for School Mathematics Addenda Series, Grades K-6* (Reston, Va.: National Council of Teachers of Mathematics, 1993), p. v.

at all grade levels. This seems to be a radical departure from the traditional focus on the memorization of basic arithmetic facts in the primary grades. For at least some constructivists, this is entirely appropriate because rote memorization can be supplanted by more conceptual understanding of mathematical concepts. These manipulatives and technological tools can be used to explore multiple representations of concepts, can be used to reduce reliance on basic facts, and may assist in the formation of conjectures in the process of solving problems.

How Teachers Must Assess Learning

If substantive changes are made in both teaching practices and the ways in which learning is viewed, then traditional modes of assessment will also need to undergo significant changes. *Standards* (1995) provides several standards that are to guide the assessment process, including the need to "reflect the mathematics that all students need to know and be able to do." Thus, whatever changes have been made to mathematics content — particularly the emphasis on learning traits such as problem solving, communication, reasoning, and connections — must be incorporated into the assessment program.

In addition to changing what is assessed, changes must be made in the means of assessing mathematical knowledge. First, the heavy reliance on paper-and-pencil tests must be changed in favor of a wider variety of assessment tools such as group and individual projects, oral presentations, and teacher observations of the verbalizations of constructions by individual learners. Beyond alternative forms of assessment by the teacher, Confrey argues that students must evaluate their own success by determining the adequacy of their own constructions. Rather than restricting assessment to the determination of right or wrong answers, teachers must encourage students to explain what strategies were attempted in an effort to gain a more complete understanding of the student's construction of a particular mathematical concept. As with the changes in mathematical content, in approaches to the learning of mathematics, and in ways of teaching mathematics, the proposed changes in the assessment of mathematical learning can be linked to ideas found in the writings of constructivists.

The Backlash

Considering the extent of changes in the teaching and learning of mathematics that have been proposed by NCTM through the publication of the *Standards,* it should not be surprising to discover some resistance to these changes. In fact, resistance to change, even a backlash of sentiment against it, is not new to mathematics education. In many respects, the "back to the basics" movement within mathematics education during the 1970s was a reaction against the "new math" of the 1960s.

It is interesting to observe that the "new math" arose in a setting where the modern worldview, discussed at length in the first four chapters of this book, seemed to influence those mathematicians and others who promoted these changes in mathematics education. The impetus for change in mathematics education was the series of successes by the Soviet Union in launching rockets and satellites into space. Responding to this challenge, academicians and government officials diagnosed the poor preparation of American students in the areas of math and science as a major impediment for success in U.S. research programs such as space exploration. This renewed focus on math and science as the technological tools necessary to win the space race, and thereby gain political and psychological advantage over the Soviets, is consistent with the modern worldview described earlier in this book. That is, technology in general and mathematics in particular enables a society to become a master of the external reality of space.

In a similar manner, the postmodern worldview, discussed in Chapter 1, seemed to influence many constructivist researchers and thereby affected the development of the *Standards.* For constructivists, particularly those in the radical camp, a plurality of viewpoints is sought and the nature of knowledge is relative to the individual knower. These more pluralistic and relativistic perspectives are consistent with a postmodern view of the world. Thus, in each case, the underlying worldview may constitute a reason for resistance to the proposed changes.

In this section, we examine some of the elements of the backlash against the ideas contained in the *Standards.* First, many parents and teachers alike have expressed concern about the shift away from rote memorization and basic skills. In the mainstream media, distinctions such as a de-emphasis but not elimination of these aspects of mathematics teaching and learning seem to be lost in favor of dramatizing the revolutionary nature of the proposed changes. Heather MacDonald, in an editorial for *The New York Times,* detected a "pervasive disdain for memorization and book learning" among many proponents of the *Standards*.[27] The writer linked changes in mathematics education to the recent shift toward a whole language approach to the teaching and learning of reading. Later, the writer's attacks broadened to include student-centered learning. Based upon her observations of a class designed to prepare teachers for the proposed changes in pedagogy, she concluded that off-task conversation and behavior, along with a lack of structure and discipline, are likely by-products of student-centered teaching strategies.

Some concerned parents, who viewed any decreased attention to basic skills and the memorization of mathematical facts and formulas as a sign of a lower quality of mathematics instruction, reacted negatively to the changes and took actions to reverse them. Perhaps the most extreme example of this reaction occurred in the state of California. With the help of a few educators and administrators who shared their outrage, parents and the political appointees to their state's Board of Education decided to enact new curriculum guidelines. Rather than stressing the need for problem solving, reasoning, communication, and connections in a variety of content strands, the new guidelines establish the demonstration of proficiency in basic skills to be the driving force in the mathematics curriculum. Foreseeing this response, Goldin wrote that the "rejection of the radical form of constructivism must *not* be taken as support for a return to behaviorism or rule-governed learning in mathematics."[28]

Another example of a type of backlash is the response of the companies that produce and market textbooks. Some publishers responded to the outcry of some parents and educators for a return to demonstrated skill proficiencies by marketing textbooks that set this outcome as a principal goal. One of the leaders in this area is Saxon Publishers. In the company's advertisement in a magazine published by NCTM, key phrases were targeted at this audience. Typical phrases were "a Saxon student never forgets," "they are amazed that they can do algebra," and a testimonial from a satisfied teacher who boasted "although I have not assigned 30 problems each night, several students continue to do them all."[29] An examination of this

27. Heather MacDonald, "The Flaw in Student-Centered Learning," editorial in *The New York Times*, July 20, 1998.

28. Gerald Goldin, "Epistemology, Constructivism, and Discovery Learning Mathematics," in *Constructivist Views on the Teaching and Learning of Mathematics*, p. 46.

29. Excerpts from an advertisement for Saxon Publishers in *The Mathematics Teacher* 85, no. 7 (1992): 571.

textbook series shows an emphasis on continual review, extensive practice of both new and previously covered skills, and a general focus on procedural knowledge. It should be noted that Saxon textbooks existed long before the first appearance of the *Standards,* but these publications provided this company with a marketing strategy to provide a clear alternative to "*Standards*-based" curricula.

Even publishers who claim to integrate the principles of the *Standards* into their textbooks are aware of the backlash from some parents and educators. The University of Chicago School Mathematics Project developed an entire kindergarten through grade 12 mathematics textbook series originally published by Scott Foresman Company and later taken over by Prentice-Hall. In a regular newsletter designed to reach teachers and administrators who have adopted or are considering the adoption of their textbooks, UCSMP warns current and potential users of the negative reaction by some who refuse to accept a shift away from basic skills as the driving force in mathematics instruction. Later, references to "reflex conservatism" and the "math wars" caused at least in part by a reaction to the *Standards* are included to alert the reader to potentially harsh criticism of their textbook series.[30]

Standardized testing is a final area where the backlash can be observed. Recent national and international assessments of the mathematics knowledge of students in the United States have caused some to question the changes advocated by the *Standards.* The Fourth National Assessment of Education Progress (NAEP) found that few students performed at advanced levels and significant percentages of 4th, 8th, and 12th graders failed to perform at even a basic level of understanding of the content areas tested. The media chose to further dramatize the performance of U.S. students when compared to their counterparts in countries around the world who participated in the Third International Math and Science Study (TIMSS). Here, although American 4th graders achieved results at or above most other countries, American 8th graders were in the middle of the pack, and American 12th graders were near the bottom of their comparison group. Although numerous reasons have been suggested for the dismal performance of the 12th graders, including the number and type of mathematics courses completed by students in the sample selected to take the test, some have responded by questioning reliance on calculators and (apparently) neglecting algebraic and arithmetic skills.

30. UCSMP Newsletter, no. 24 (Winter 1998/99), pp. 1-2.

Some teachers and researchers who have come to the defense of the *Standards* argue that the relatively better performance of 4th and 8th graders is evidence that the reforms are working and that future international assessments will show similar improvement at the 12th grade level. Perhaps more fundamentally, they argue that means of assessing mathematical knowledge must change to match reforms in the teaching and learning of mathematics. These debates are far from over, and like reactions of parents, school boards, and textbook publishers, provide evidence that the acceptance of the changes advocated by the *Standards* has been less than universal.

Standards 2000: Moving the Pendulum Back Toward the Center

Beginning in 1997, NCTM sponsored working groups to develop the next generation of the *Standards.* In 1998, NCTM released a discussion draft of a new version of standards entitled *Principles and Standards for School Mathematics.* The final version of these *Standards* appeared in May 2000. In this newest version of the *Standards,* one can see evidence of an attempt to answer the backlash by readdressing such things as the role of rote memorization, basic skills, and direct instruction. This section highlights a few important ways NCTM appears to be moving the pendulum away from some ideas consistent with constructivism, especially in its most radical form, and towards a more balanced view of the teaching and learning of mathematics.

In the area of pedagogy, direct instruction appears to be making a comeback of sorts in the newest version of the *Standards.* While not advocating a return to teaching that relies solely on direct instruction and "telling" students how to do mathematics, there is a call to include such practices in a wide array of teaching strategies. Specifically, the authors state that "there is no one 'right way' to teach."[31] Although the expectations for classroom teachers are later clarified in the section, the prominence of this phrase, repeated in color in the margin of the text, could lead many readers to conclude that direct instruction, if done effectively, might be consistent with the *Standards.*

31. National Council of Teachers of Mathematics, Commission on Standards for School Mathematics, *Principles and Standards for School Mathematics* (Reston, Va.: National Council of Teachers of Mathematics, 2000), p. 18.

Similarly, procedural knowledge, including basic skills, receives a much more prominent role in the newest version of the *Standards.* In contrast to earlier versions where procedural knowledge such as algorithmic skills were marked for "de-emphasis," the newest version argues that "The alliance of factual knowledge, procedural proficiency, and conceptual understanding makes all three components usable in powerful ways."[32] Notice the apparently equal footing given to both procedural and conceptual knowledge. Given the tendency to misinterpret the columns of items to be de-emphasized as lists of items for complete elimination, columns listing those aspects to be increased or decreased in emphasis are nowhere to be seen in this newest version of the *Standards.*

The role of technology also caused strong reactions by readers of the earlier versions of the *Standards.* In response, the newest version stresses that technological tools such as calculators and computer software packages must not be viewed as the panacea for all problems in mathematics education. After noting that calculators and other tools can be used well or poorly in mathematics instruction, the *Standards* (2000) goes one step further, noting that traditional arithmetic and algebraic skills and rote memorization of basic facts are not completely obsolete in the era of technological tools.

> Technology should not be used as a replacement for basic understandings and intuitions; rather, it can and should be used to foster those understandings and intuitions. In mathematics-instruction programs, technology should be used widely and responsibly, with the goal of enriching students' learning of mathematics.[33]

The subject of technology also ties into the role of the teachers. Critically, it is the teacher who must decide, at least in the initial phases of learning, "if, when, and how technology will be used."[34] So, unlike the earlier version of the *Standards,* which called for technology to be omnipresent in the mathematics classroom, the newest version seeks to emphasize the teacher's role in setting appropriate limits on the appropriate use of technology.

32. *Principles and Standards for School Mathematics,* p. 20.
33. *Principles and Standards for School Mathematics,* p. 25.
34. *Principles and Standards for School Mathematics,* p. 26.

A Christian Response

Having explored the main ideas of constructivism and its implications for and recent effects on the teaching and learning of mathematics, how should Christians respond? Is constructivism compatible with a Christian world and life view? While we don't want to suggest that there is a single "correct" Christian worldview, we do affirm that Christian faith has implications for all facets of everyday experience, and therefore will affect a person's view on the teaching and learning of mathematics.

This section frames a Christian response to constructivism and its effects by identifying a series of important issues. Rather than attempt an exhaustive list, we discuss a total of seven such issues. At the outset, it should be noted that there are aspects of constructivism and its implications for the teaching and learning of mathematics that Christians should endorse and attempt to implement in the classroom. Wolters argues that when we examine parts of the world around us, we need to keep in mind that God's good created order establishes the structure of the elements of creation and that these elements remain good even after the Fall.[35] To this end, we will first identify several aspects of constructivism that are consistent with a Christian worldview. Watanabe also provides an argument that constructivism allows for interpretations acceptable to many Christians.[36] On the other hand, sin has allowed this good structure to be distorted by those who would direct some aspect of creation toward purposes that do not seek to honor and glorify God. In this respect, we later consider those aspects of constructivism that contain good structural elements in need of redirection.

First, Christians should embrace the notion of centering the learning environment on students. This notion fits well within a view of our students as God's image-bearers. As Christian teachers, we have the responsibility not only for the intellectual development of our students, but also their emotional and spiritual needs. This perspective builds on the need, expressed by Davis, Maher, and Noddings, to care adequately for students. For Christians, however, this level of care runs deeper than for those who would leave any underlying world and life view outside the classroom door. Jadrich notes

35. Albert Wolters, *Creation Regained* (Grand Rapids: Eerdmans, 1985).

36. Tad Watanabe, "Constructivism, Mathematics Education, and Christianity," in *Proceedings of the Tenth ACMS Conference on Mathematics from a Christian Perspective*, ed. Robert Brabenec (Wheaton, Ill.: ACMS, 1995).

that Christian teachers who adopt a servant attitude will find the shift in focus to students a very natural one.[37]

Second, the four important traits the *Standards* advocate as critical for all students to develop, and the additions of multiple representations as a fifth trait in the newest version of the *Standards*, ring true with Christian teachings. Problem solving, reasoning, representations, communication, and the building of connections both within mathematics and across a variety of disciplines are necessary for a person to understand the regularity and order within God's creation as well as the interrelatedness of the seemingly diverse aspects of the world around us. To constructivists such as Confrey, these traits help students to "see the world through a set of quantitative lenses," a metaphor not that different from John Calvin's view of the Bible as a pair of spectacles.[38] At the same time, radical constructivists would characterize the external world as a humanly constructed reality. Van Brummelen remarks, "contrary to the basic tenets of [radical] constructivism, the Bible tells us that we live in a well-ordered reality [that] does not result from human construction."[39] So, Christians can adopt goals such as seeking connections, but for different reasons than constructivists.

Third, constructivists, with the exception of naïve constructivists, adopt a fallibilist epistemology. That is, they view all constructions of knowledge as fallible and open to alternatives. Ernest takes this notion one step further, claiming that mathematical knowledge is not absolute truth and cannot be absolutely validated but instead is constantly open to revision. As Christians, we can certainly agree that any human construction is subject to limitations and potential error. This feature of human reasoning distinguishes the fallen and limited nature of humankind from the infallible and boundless nature of God. As humans we may not be able to understand fully that which God understands perfectly.[40] At the same time, Christians acknowledge the existence of absolute or universal truths even if our own humanly constructed truths may indeed be relative to the limitations of our humanity. As Jadrich expresses the idea, "our imperfections [and imperfect constructions] do not impose a limitation on God or on his creation. It is

37. Jim Jadrich, "Constructivism in Science and Math: Is It Christian?" *Christian Educators Journal* 38, no. 1 (1998): 18-19.

38. See John Calvin, *Institutes of the Christian Religion,* ed. John McNeill (Philadelphia: Westminster Press, 1960), Book I, Chapter 6, Section 1.

39. Van Brummelen, "Curriculum Development Is Dead — Or Is It?" pp. 19-20.

40. See, for example, Psalm 145:3 or Psalm 147:5.

our fallenness that holds us back from fully comprehending truth. Truth itself is not suspect."[41]

Fourth, constructivists assert that only active construction produces knowledge. It would seem unnecessarily limiting on God's creative power to conclude that knowledge can only be gained though the process of construction. More importantly, requiring all knowledge to be actively constructed by the individual learner, as do the radical constructivists, does not make sense. As Van Brummelen writes, "students do not always have to form their own 'constructions' to be active learners. That would be too complicated and too time-consuming." In fact, Christians have proposed other models for instruction that in turn reflect alternative means of acquiring knowledge. Oppewal identifies an interactive theory of knowing and distinguishes it from a more constructivist approach. In his model, "objective reality (the revelation) and the human responder are two foci of a single process, with the responder discovering or uncovering what is by what is done to it."[42] Christians may adopt the interactive model or a number of different models; however, requiring all knowledge to be constructed by the individual knower seems unnecessarily restrictive.

Fifth, radical constructivists are reluctant to provide correction by identifying incorrect constructions. In fact, some of the most radical constructivists argue that students do not make errors or have misconceptions. As Van Brummelen puts it, "right answers are not possible in a constructivist textbook. It goes against the philosophy." Rather, Wood notes that such errors constitute their current understanding,[43] and Steffe argues that such errors are "only discrepancies, either between points of view or between a person's activity and some unexpected effects of this activity."[44] So, rather than provide corrective feedback, Confrey urges teachers to "assist the student in restructuring these views to be more adequate from the student's and from the teacher's perspective." Christians, on the other hand, understand that teachers bear a responsibility for providing correction to their

41. Jim Jadrich, "Constructivism in Science and Math: Is It Christian?" p. 18.

42. Donald Oppewal, "Biblical Knowing and Teaching," in *Voices from the Past: Reformed Educators,* ed. Donald Oppewal (Lanham, Md.: University Press of America, 1997), p. 321.

43. Terry Wood, "From Alternative Epistemologies to Practice in Education: Rethinking What It Means to Teach and Learn," in *Constructivism in Education,* pp. 332-40.

44. Leslie Steffe, "Alternative Epistemologies: An Educator's Perspective," in *Constructivism in Education,* p. 515.

students. This correction may take the form of providing counterexamples to challenge a faulty construction, as constructivists would advocate, but can also involve the identification of an error or misconception on the part of the student. While constructivists might see such behavior as too authoritative, Christians recognize the teacher-student relationship as one of many relationships that involve the appropriate respect for and use of authority. Other such relationships include that of God to humans and that of parents to their children. More importantly, Christians must not fall into the trap of allowing all human constructions to operate on an equal footing. Failing to recognize this fact could very easily lead one to a relativistic view not only of mathematics but also of the moral realm.

Sixth, if one reads the work of radical constructivists, one senses that an overriding goal of the teaching and learning of mathematics is to produce autonomous learners. Confrey states that "personal autonomy is the backbone of the process of constructivism." While Christians would hardly argue with a desire to enable students to think on their own feet, some constructivists are not willing to place this personal autonomy in the context of a dependent relationship with God. Rather, adopting a secular and humanist perspective, radical constructivists see human autonomy as the ultimate objective. In this scenario, any notion of God becomes a stumblingblock and is to be avoided at all costs. Glasersfeld argues that "from our point of view, to assume that something is God-given or innate should be the last resort — to be accepted only when all attempts at analysis have broken down." In response, Christians must state clearly that ultimately we are not autonomous but instead depend wholly and completely on a creating, redeeming, and sustaining God. Any perceived autonomy on the part of the individual learner is instead God's precious gift of wisdom and understanding, not to be distorted into a situation where, as Saint Augustine extends Romans 1:25, humans "worship and serve the creature rather than the Creator."[45]

Seventh, a recent editorial on the discussion draft of the *Standards* (1998) highlights one final area in which Christians must take a stand. Rogers suggests that the current list of ten content and process standards is incomplete. Specifically, he argues that "Mathematics as Art" should be added as an eleventh standard that cuts across all grade levels. He notes that the current list of ten standards fails to capture the beauty and elegance of certain mathematical proofs and the joy experienced by students as they suc-

45. St. Augustine, *The Confessions of St. Augustine,* Book V, Chapter 3.

cessfully explore and use mathematics.[46] While Christians can concur with such an additional standard, even its inclusion in the newest version of the *Standards* would not solve the incompleteness problem. As Christians, we acknowledge God's role in all of creation, and we view mathematics as one means to recognize and describe this role. Thus, a standard such as "Mathematics as a means to recognize and describe the beauty, order, and regularity of God's creation" deserves a place on any list of standards for mathematics. Although Christians have no right to expect such a standard to be proposed by a publicly funded and supported organization such as NCTM, Christian teachers must nonetheless take the responsibility to complete any list that is eventually produced for use within the Christian community.

Conclusion

Throughout this book we have investigated the influence Christian perspectives can have on various aspects of the mathematical enterprise. In the areas of teaching and learning of mathematics such perspectives may result in policies that are similar to those espoused by people with differing views, but for very different reasons. Sometimes, of course, differing conclusions will emerge. In particular, we argued that constructivism, in its many forms, provides an instructive example of how careful reflection will lead Christian thinkers to make and justify their choices. By acknowledging God's sovereignty and the limitations of human autonomy, a Christian view draws important distinctions from the more radical constructivists, and places even those areas of agreement in a broader context that allows faith to play an active role in the teaching and learning of mathematics.

46. James Rogers, "Mathematics as Art: The Missing Standard," *The Mathematics Teacher* 92, no. 4: 284-85.

Conclusion

Recently one of us told a friend that we were editing a book on the relationship between mathematics and Christian thought. Her reply was, "That's going to be one short book!" However, now that we are approaching the end, we have seen much that can be said about this relationship:

- We have examined the modern and postmodern worldviews in relation to mathematics.
- We compared how mathematics evolved in different cultures, and discussed how it came to be so deeply embedded in Western culture.
- We investigated some philosophical and ethical issues in mathematics and the role of values in the undertaking of mathematical projects.
- We examined how faith perspectives might shape our thinking about various endeavors such as the possibility of artificial intelligence, whether a rational basis can be constructed for determining intelligent design, how psychological considerations help inform us vis-à-vis our learning and thinking about mathematics, and how we might respond to some issues in mathematics education from a Christian perspective.

Our underlying concern throughout has been to ask, "How might a Christian perspective shape our thinking on each of these matters?" But there is more that can be said.

The subject matter of mathematics is abstract structures. A Christian mathematician investigating, say, Lie groups, will proceed by the same mathematical principles as a non-Christian. If a controversy arises in connection with a theorem she proves, it will typically be about whether the theorem has really been proven, not about the effect of her Christian beliefs on her mathematical work. Thus, in many ways, the Christian mathematician's experience of mathematics will be the same as that of an unbeliever. There are, however, significant differences, but they enter at a different level than one typically sees in mathematical research papers or textbooks. That is, they involve broad perspectives on the nature of the discipline of mathematics, the objects it studies, and the place of mathematics in human thought. In other words, even though a Christian may experience mathematics in the same way as an unbeliever, the Christian may "read" that experience differently.

In the remainder of this chapter, we consider five topics on which a different "reading" might arise:

- Perhaps the approach most frequently used to develop a Christian critique of a discipline is "presuppositional analysis"; that is, asking what presuppositions are commonly made in a discipline and how these influence its typical methods and conclusions. But mathematics often seems presuppositionless. Is it?
- If one starts from Christian presuppositions, one affirms that God is personal and purposeful. Does God have purposes for mathematics? If so, what are their implications?
- The predominant approach mathematicians have used since the beginning of the twentieth century is "formalism." To what extent is this approach consistent or inconsistent with Christian belief?
- Many have marveled that making a few marks on a sheet of paper and manipulating them by the rules of mathematics has enabled human beings to understand incredibly complex and subtle features of nature and to predict previously unsuspected phenomena. Does a Christian perspective have anything to contribute to an understanding of this "unreasonable effectiveness of mathematics"?
- While mathematics has proved to be extraordinarily powerful, it also has limitations. Can a Christian perspective help us to understand these?

Mathematics and Beliefs

When the history of mathematics is finally written, David Hilbert will certainly be listed as one of the greatest mathematicians of the nineteenth and twentieth centuries. He was born in East Prussia in 1862 and made his fame while at the University of Königsberg and also at the University of Göttingen, where he began serving in 1895. As we mentioned in Chapter 9, in 1900 he was invited to give one of the leading addresses at the second International Congress of Mathematicians held in Paris. Hilbert found this to be a perfect occasion to discuss where mathematics had been in the nineteenth century, and where it should go in the twentieth. His lecture delineated 23 problems that he felt should occupy the attention of mathematicians during the next century. It is a tribute to Hilbert's genius that these problems have been much discussed since 1900 and that many of them remain unsolved. The first two of these problems are:

(1) To prove Cantor's "continuum hypothesis."
(2) To investigate the consistency of the arithmetic axioms.

What do these statements mean? Hilbert, mathematicians know, was a formalist. That is, he believed mathematics to be nothing more than a delightfully infinite game where results are derived logically from a small set of initial axioms or assumptions. For Hilbert, asking whether a mathematical proposition was in some *realistic* sense true was an irrelevant question. The more important questions for him were whether the mathematical systems he worked with were *consistent* (that is, whether they contained no logical contradictions), and whether the systems were *complete* (that is, whether all statements about a given system could be proved or disproved using only the results of the system). Hilbert wanted to verify that mathematics is consistent. But this is not all he wanted. He hoped to be able to create a procedure that one could use to prove or disprove any conceivable mathematical statement. To see more clearly what this means, let's take a look at a result that every high school algebra student is expected to learn by heart — the quadratic formula: If $ax^2 + bx + c = 0$ then

$$x = \frac{-b \pm \sqrt{b^2 - 4ac}}{2a}.$$

This formula provides a foolproof method for finding solutions to quadratic equations. No matter what the equation, all one has to do is plug the numbers a, b, and c into this formula, and the values of x that solve the equation will automatically be generated (provided, of course, $a \neq 0$). As a formalist, Hilbert envisioned the possibility of coming up with a rule or formula like the quadratic equation that one could apply to mathematical propositions. If found, all one would have to do to see if a proposed mathematical proposition were valid would be to put an encoded version of the proposition into this formula, whereupon the formula would crank out the answer. In technical terms, such a formula and mechanism for encoding mathematical statements is called a *metalanguage.*

Such a language, then, would provide for the certainty of mathematical results. Here is what Hilbert said about this project in 1927:

> For this formula game is carried out according to certain definite rules, in which the technique of our thinking is expressed. These rules form a closed system that can be discovered and definitely stated. The fundamental idea of my proof theory is none other than to describe the activity of our understanding, to make a protocol of the rules according to which our thinking actually proceeds. . . . Already at this time I would like to assert what the final outcome will be: mathematics is a presuppositionless science. To found it, I do not need God or the assumption of a special faculty of our understanding . . . or the primal intuition of Brouwer . . . or finally, as do Russell and Whitehead, axioms of infinity, reducibility, or completeness.

Just four years later, however, Kurt Gödel, a Moravian-born mathematician, dealt a severe blow to Hilbert's program. He proved a staggering result known as Gödel's incompleteness theorem. In essence, Gödel proved that any complete mathematical system (one capable of proving all the true theorems implied by it) that includes the natural numbers can never be proven consistent. In other words, if arithmetic systems are consistent (which everyone likes to believe is the case), there are certain conjectures one can make that cannot be proved to be either true or false. We mathematicians, therefore, cannot have our cake and eat it too. Ideally, we would like our systems to be both consistent and complete, but Gödel showed that we cannot have both.

This failure of Hilbert's project is well known and has been widely discussed. His assertion that mathematics can ultimately be shown to be a presuppositionless science has received less attention. However, this is a critical question! We will focus on it for the rest of this section. Presuppositions are beliefs, and one insight that Christian scholarship brings to any discipline is the recognition that the beliefs that its practitioners hold are of great importance in understanding that discipline. So, is Hilbert right? Is mathematics a presuppositionless science, or, if it is not now, could it ever become one? We will argue in the next several pages that the answer is "no" and will then consider some implications of that negative answer.

It is certainly the case that throughout its history, people have held many beliefs about mathematics. We have seen several in this book:

- The Pythagoreans believed that the natural numbers were the underlying basis of reality.
- Plato and many others who followed him believed that mathematics provides a basis for absolutely certain knowledge. They also believed that mathematical objects possess an objective existence that transcends human beings.
- Aristotle believed that human beings have an intuition that enables them to infer fundamental properties of the material world.
- Galileo believed that mathematics was the alphabet of the universe.
- Kant believed in *a priori* synthetic reasoning — that is, that people could know true facts about the world prior to experience and that geometry was an example of this.
- Frege and Russell believed that all of mathematics was reducible to logic.
- Some mathematicians and philosophers of mathematics have rejected Platonic beliefs about mathematical objects on the grounds of a belief in materialism — that is, that the only legitimate entities are ones that are spatially-temporally located.
- L. E. J. Brouwer founded his intuitionist school of mathematics on the belief that human beings have a fundamental intuition for the natural numbers. Thus, these numbers are the only legitimate basis on which mathematical certainty can be built.

If we define a belief as something held to be true (as philosophers often do), mathematics clearly involves beliefs. Not every belief is a presupposition, however. Mathematicians would like their beliefs to be warranted, and the usual standard in mathematics for warranting belief is proof. The problem many mathematicians have with some or all of the above beliefs is that they cannot be proven.[1] So our question can be reformulated as follows:

1. Note that Hilbert's assertion that mathematics is presuppositionless was itself a presupposition! Of course, he hoped its status would be changed to that of a theorem. We are challenging this presupposition here.

Can mathematics be freed of all beliefs that cannot be warranted by proof? We are also concerned about a closely related question: Is there any legitimate warrant for mathematical belief other than proof? In the next several paragraphs, we will show that mathematics includes many beliefs that cannot be warranted by proof but that mathematicians nevertheless regard them as sound.

To begin, we take a brief excursion from our topic to introduce the work of the Christian philosopher, Nicholas Wolterstorff. We will then return and apply his ideas to mathematics. Wolterstorff critiques the traditional form of foundationalism. This is the approach to knowledge that we encountered throughout Chapters 5 and 6. He describes such an approach as follows.

> Simply put, the goal of scientific endeavor, according to the foundationalist, is to form a body of theories from which all prejudice, bias, and unjustified conjecture have been eliminated. To attain this, we must begin with a firm foundation of certitude and build the house of theory on it by methods of whose reliability we are equally certain.[2]

For the Greeks, the firm foundation was intuitively apprehended principles analogous to the axioms of plane geometry, and the reliable method was deductive reasoning. Later, empiricists extended this to include empirically obtained principles.

Wolterstorff points out that the foundationalist project has gradually collapsed in the past century. It first evolved into a weaker form, probabilism, by relaxing the rigid requirement of certitude. It then evolved into falsificationism, the idea that theories can never be known with certainty to be true, but some theories can with certainty be shown to be false. Today, most scholars regard foundationalism as having completely collapsed. The principal reason for the collapse was the recognition that every theorist approaches the world with an extensive, interdependent web of theoretical and non-theoretical beliefs. That is, the very process of weighing a theory requires beliefs about the entities within its scope, and these beliefs themselves are unable to meet the requirement of certitude that even falsificationism demands.

Wolterstorff distinguishes three types of beliefs that scientists bring to

2. N. Wolterstorff, *Reason Within the Bounds of Religion,* 2nd ed. (Grand Rapids: Eerdmans, 1984), p. 28.

their work: data beliefs, data background beliefs, and control beliefs. By data beliefs, he means the minimal beliefs one must have about the object under study even to begin investigating it. For instance, suppose we are studying a certain chemical reaction. Examples of data beliefs are "I exist," "the reaction actually occurred," and "the observations I have made are directly related to the reaction that is actually occurring." Data background beliefs depict those conditions needed for accepting as data what one observes. For instance, "no one here was trying to deceive me," and "the light that shone on my test tube did not distort the color changes I observed." Control beliefs are a person's cognitive framework as to what constitutes an acceptable theory. According to Wolterstorff, they

> . . . include beliefs about the requisite logical or aesthetic structure of a theory, beliefs about the entities to whose existence a theory may correctly commit us, and the like. Control beliefs function in two ways. Because we hold them we are led to reject certain sorts of theories — some because they are inconsistent with those beliefs; others because, though consistent with our control beliefs, they do not comport well with those beliefs. On the other hand, control beliefs also lead us to devise theories. We want theories that are consistent with our control beliefs. Or, to put it more stringently, we want theories that comport as well as possible with those beliefs.[3]

Thus the insistence on carefully crafted, rigorous arguments based on axioms and precise definitions, confidence in the reliability of logic, and the notion that mathematics, properly written, should not include the story of the dead ends the mathematician pursued and may omit the original motivation for investigating a question are examples of control beliefs about how mathematics ought to be done that are widely held by contemporary mathematicians. Furthermore, mathematicians have beliefs about the nature of mathematical entities. Jean Dieudonne, one of the founders of a group of highly influential mathematicians (who published collaboratively under the name Nicolas Bourbaki), expresses this notion:

> . . . We believe in the reality of mathematics, but of course when philosophers attack us with their paradoxes we rush to hide behind formalism and say "Mathematics is just a combination of meaningless symbols." . . .

3. N. Wolterstorff, *Reason Within the Bounds of Religion*, p. 67.

> Finally we are left in peace to go back to our mathematics and do it as we have always done, with the feeling each mathematician has that he is working with something real. This sensation is probably an illusion, but it is very convenient.[4]

Dieudonne refers here to two distinct beliefs about mathematical entities — that they have a reality beyond the symbols that represent them, and that mathematical entities are nothing more than those symbols. One study suggests that, around 1970, mathematicians held one of three classes of beliefs about mathematical entities — 65 percent of mathematicians were Platonists, 30 percent formalists, and 5 percent constructivists.[5]

Note that formalism may seem rather unlike a belief — it says more about what mathematical objects are not, than what they are. However, formalism is also a *normative theory* — it explicates some very clear rules about how mathematics "ought" to be done — precise formulation of axioms and definitions, careful deductive reasoning, and no reference to anything outside of mathematics. It thus includes a very strong set of control beliefs.

Control beliefs, then, are one way that beliefs enter mathematics. A second way that beliefs enter is via axioms. Reasoning cannot take place in a vacuum, although mathematicians try very hard to reduce the number of assumptions they make to a minimum and make those assumptions very explicit via the medium of axioms. However, the axioms themselves must have an origin. Classical mathematics viewed axioms as fundamental truths about reality inferred by our intuitions. The discovery of non-Euclidean geometry clearly exposed how unwarranted such a belief was.

Formalism does try to avoid the entry of beliefs via axioms, but its attempt is not successful. Henri Poincaré was, along with Hilbert, one of the leading mathematicians of the late nineteenth century. Poincaré was an admirer of Hilbert's work, but he pointed out the following:

> The logical point of view alone appears to interest professor Hilbert. . . . Being given a sequence of propositions, he finds that all follow logically from the first. Without the foundation of this first proposition, with its

4. Philip J. Davis and Reuben Hersh, *The Mathematical Experience* (Boston: Birkhäuser, 1981), p. 321.

5. J. D. Monk, "On the Foundations of Set Theory," *American Mathematical Monthly* 77 (1970): 703-11. As quoted in Davis and Hersh, *The Mathematical Experience*, p. 322.

> psychological origin, he does not concern himself. . . . The axioms are postulated; we do not know from whence they come; it is then as easy to postulate A as C. . . . His work is thus incomplete, but this is not a criticism I make against him. Incomplete one must indeed resign oneself to be. It is enough that he has made the philosophy of mathematics take a long step forward. . . .[6]

The "psychological origin" of axioms is a question that cannot be addressed within mathematics. Axioms cannot be warranted by proof (although proof can show them consistent and fruitful). To simply label axioms as "arbitrary" is reductionistic — "solving" a problem simply by defining it as not a problem.

In summary, our main point has been that the methodology of mathematics requires significant beliefs that are not (and cannot) be warranted by proof. Furthermore, mathematicians hold beliefs about their objects of study that are similarly incapable of being warranted by proof. And mathematicians use axioms that function in very much the same way as presuppositions do in other fields; calling them "arbitrary" simply avoids the important question of their origin. Thus, we believe Hilbert was incorrect — mathematics is not a presuppositionless science. It has presuppositions that can be analyzed and critiqued in the same way as can those of other disciplines like history and sociology. We will return to this notion after the next section for a discussion of formalism.

The Purpose of Mathematics

Many thinkers in modernity have been proponents of naturalism. In science, naturalism entails avoiding all supernatural explanations of phenomena, and instead, looking for explanations in terms of material laws of cause and effect. This perspective is quite different from the classical point of view. For example, in studying causation, Aristotle introduced both the notion of "efficient" cause — that which works some change — and the notion of "final" cause — the use or purpose for which a change is produced. Approaches that focus on final causes are called *teleological.* Thus, psychological explanations that focus on human intentionality and seek to interpret human moti-

6. In Constance Reid, *Hilbert* (New York: Springer-Verlag, 1970), p. 63.

vations are teleological. Similarly, a Marxist interpretation of economics is teleological in that it sees history as moving toward a goal — the formation of a classless society. In addition, some theological explanations of human existence are teleological in that they start from an understanding of God's purposes. However, in the naturalist framework, scientific explanations have not typically involved teleological elements — their domain has been replicable phenomena that are deterministic or probabilistic — and such explanations are presented as laws possessing predictive power. Thus physicists, biologists, experimental psychologists, and other empirical scientists have largely rejected teleological explanations.

Teleology has fared no better in mathematics. It is almost impossible to find a discussion in the mathematics literature of the purpose of mathematics, whether that purpose is defined theologically or not. However, as this book has attempted to present a Christian perspective on mathematics, some discussion of why God may have given us the capacity to do mathematics is essential.

Any claim of knowledge about God's purposes cannot be justified deductively (as in mathematics) or inductively (as in science). Christians, of course, may turn to scripture for help, but the Bible makes no direct statements as to God's purposes for mathematics. Thus, if we are to appeal to scripture, we must do so indirectly, and base our speculations regarding God's purposes for mathematics from what God has revealed about his broader purposes. Nicholas Wolterstorff has given a particularly clear statement of God's overarching purposes for humanity using the Hebraic concept of *shalom,* usually translated into English as "peace."

> The goal of human existence is that man should dwell at peace in all his relationships: with God, with himself, with his fellows, with nature, a peace which is not merely the absence of hostility, though certainly it is that, but a peace which at its highest is enjoyment. To dwell in shalom is to enjoy living before God, to enjoy living in nature, to enjoy living with one's fellows, to enjoy life with oneself. . . . In shalom there is delight.

He also speaks of the duties humans have whose fulfillment will yield this state of shalom:

> Deep in the Christian tradition is the conviction that each of us is not to be the center of his own concerns but is rather to love and serve God

> with all his life, and, in similar fashion, to love his neighbor as himself. One might add to these the conviction that each is also to be a responsible steward of the creation within which God has placed us. To love and serve God in all our ways, to love our neighbors as ourselves, and to be responsible stewards of nature — those are clearly proclaimed in the authoritative Scriptures of the Christian community as the fundamental obligations of mankind.[7]

How does mathematics fit into this picture? In two ways. First, those to whom God has given the capacity and interest to do mathematics have the privilege and obligation of using those gifts to understand mathematics itself. Second, as we saw in Chapters 5 through 7, mathematics has had an enormous impact on human culture and on our understanding of nature. Meeting the requirements of love for neighbor and care for nature involves using mathematics. As we discussed in Chapter 8, this means choosing mathematical projects not exclusively out of personal interest or convenience, but with a view toward the broader purposes of serving God and neighbor and caring for creation.

Note, however, that this line of thought does not imply that applied mathematics is preferable to pure mathematics. Understanding all of God's creation is of value and basic research is as valuable in mathematics as it is in the natural sciences. Furthermore, while mathematics provides a powerful tool for exploring the enormous variety of formal structures that arise in nature and society, it also provides a means to explore alternative realities that do not exist. For instance, consider an engineer designing a bridge. On paper or on a computer screen, she can explore many alternative designs that will never be built. These can be examined for safety, functionality, and aesthetics without the cost or danger of actually building them. All explorations of abstract mathematical structures are similar to this example — they involve asking, "What if we adopt this set of axioms?" The freedom that mathematics provides to engage in such exploration is an important dimension of human creativity and thus of fulfilling our stewardship.

Briefly, then, the capacity to do mathematics is one of the tools that God has given us to build shalom. To the extent that mathematics itself is part of God's creation, this involves not only understanding mathematics, but also using mathematics in our stewardship of nature, and in administer-

7. Wolterstorff, *Reason Within the Bounds of Religion*, pp. 114, 112.

ing the many different dimensions of human culture capable of being described precisely and abstractly. Thus, mathematics should be engaged in with joy, gratitude, and confidence that we are fulfilling God's purposes for us in doing it. However, we also need to keep in mind that we are accountable to God for how we use the capacity he has given us to do mathematics.

We cannot leave this discussion of the purposes of mathematics without considering two pitfalls that often hinder the fulfillment of those purposes; one involves beliefs, the other values. Concerning beliefs, both twentieth-century formalism and classical geometry lead to the notion that mathematics can provide absolutely certain knowledge apart from God. Formalism aimed to separate mathematics from all other forms of knowledge. Earlier, we cited Roy Clouser's argument that only God is autonomous and thus that anything other than God accorded autonomous status is an idol. For example, recall Hilbert's comment that the foundations of mathematics could be erected without God. The perspective that mathematics properly done is separate from all other knowledge or application is still predominant today. Thus, Christian mathematicians face the continual tension of working with colleagues, many of whose basic beliefs about mathematics affirm an autonomy of mathematics to which Christians cannot subscribe. Thus, rather than separating mathematics from the rest of knowledge, a Christian approach should seek to embed it in a larger context and establish connections to other disciplines.

Concerning values, mathematicians tend to hold abstract knowledge in high regard; such a value is another characteristic of modernity. As we have seen, this perspective has roots in ancient Greece. However, more recently, it has been particularly expressed in the work of Immanuel Kant. According to Wolterstorff,

> On Kant's view, the superiority of certain forms of knowledge is to be located in the formal characteristics of that knowledge. It is completeness of explanation and systematic unity that are the great desiderata in knowledge. Characteristic of human nature is an impulse toward the pursuit of ever greater completeness of explanation, and ever greater systematic unity, in the body of our collective knowledge. And this impulse, Kant obviously believes, is beneficent; the more complete and unified the body of knowledge, so far forth the better.[8]

8. Wolterstorff, *Reason Within the Bounds of Religion*, p. 121.

Therefore, according to Kant's criteria, mathematics is considerably more valuable than much of what is done in other disciplines. The mathematician who follows Kant, then, necessarily believes that her work is more valuable than other work. It is an easy step from here to believe that she is superior to scholars in other fields.

An understanding of God's purposes, however, provides little basis for Kant's perspective or the hubris that can easily flow from it. Does such abstraction necessarily lead to greater joy, peace, or love for God or neighbor? Can it not equally well lead to arrogance and ingratitude? Indeed, particulars often yield more understanding and delight than abstractions; generalization and abstraction can often confuse more than they clarify. Thus, a second tension Christian mathematicians face arises from working in a professional community whose values they do not fully share. By clearly affirming mathematics' greater purposes of service, love, and stewardship, Christian mathematicians can maintain their own identity apart from the surrounding mathematical subculture. That is, they can be "in the world, but not of it."

Beliefs of Formalism

Insofar as formalism means carefully clarifying one's axioms and rigorously deducing conclusions from them, it is non-problematic. However, formalism has other aspects that are problematic from a Christian perspective. As we suggested in the previous section, God has purposes for giving us the capacity to do mathematics and we are accountable for how we use it. Formalism aims to sever the necessary connection between mathematics and the rest of reality. To some extent this is done with good intentions, namely avoiding errors. But this surgery has had some unhappy consequences — the mathematics community has often been unconcerned with this connection[9] and many lay people have come to see mathematics as irrelevant. As we saw in the previous section, ultimately mathematics is intended to enable us to serve God and other human beings. Formalism creates a climate in which questions about God and service are regarded as inappropriate — such questions cannot be addressed by the methods of formalism. But these questions are critically important.

9. For instance, between 1900 and 1950, the annual addresses of the president of the American Mathematical Society rarely even mentioned applications of mathematics.

Second, formalism recognizes only one warrant for truth claims — formal proof. But for a Christian, logical deduction is not the only legitimate warrant for truth claims. Christians can be comfortable living with mysteries surrounding the nature of mathematical objects, how we come to understand them, and their connection with non-mathematical realities. This does not preclude a desire to understand such mysteries more deeply, but it does free the Christian from the type of search for certainty apart from God that motivated foundationalism both within and outside mathematics.

Third, formalism has affected mathematics curricula in unhappy ways. The search for foundations for mathematics and the advance of formalism have led to a narrowing of the boundaries of the discipline of mathematics, as mathematicians typically perceive them. As we saw earlier, Aristotle separated mathematics and science. While not every scholar agreed, his views dominated until roughly 1600. During the seventeenth and eighteenth centuries, mathematics and applications were largely indistinguishable. Then in the nineteenth century, mathematics began to separate gradually from applications again. By the twentieth century, the predominant mode of operation in pure mathematics was to ignore applications completely. Furthermore, graduate programs in mathematics typically included no training in the history, philosophy, or sociology of mathematics or any consideration of its role in science or culture more broadly. That is, the boundaries of the discipline were drawn so as to include only questions that can be addressed by the methods of formalism. One implication of a Christian perspective is that these boundaries need to be drawn more widely so that questions about history, philosophy, and the like are regarded as central to the discipline's concerns and not marginalized as they typically are today. We do see something of a current shift away from a narrowly formalist approach. We applaud this change but would like to see it go much further. Most obviously, mathematics curricula at all levels need to incorporate wider issues. Textbooks should emphasize such connections. Pure mathematics journal editors should encourage writers to include material on the sources of their questions as well as pointers to possible applications. Moreover, for Christians, mathematics needs to be pursued with the goal that one's work will ultimately honor God, care for his creation, and serve others.

In conclusion, formalism as a methodology has considerable value. However, as a set of control beliefs, it has borne many undesirable fruits. The purposes we articulated in the previous section provide a basis for a set of control beliefs for mathematics that are based on principles such as stewardship, service, and the joy of discovery rather than on isolation and rationalism.

The Applicability of Mathematics

In 1960, the physicist Eugene Wigner published "The Unreasonable Effectiveness of Mathematics in the Natural Sciences."[10] In 1980, the computer scientist Richard Hamming published "The Unreasonable Effectiveness of Mathematics."[11] The articles have been widely read. Both are concerned with the question "Why should the manipulation of symbols according to certain logical rules be so amazingly effective in helping us understand the natural world?" Wigner's article is primarily a collection of examples from his experience as a physicist that demonstrates how amazingly effective mathematics has been. Hamming also cites examples, but spends relatively more time addressing this question, beginning with a caricature of religious thinking.

> Man, so far as we know, has always wondered about himself, the world around him, and what life is all about. We have many myths from the past that tell how and why God, or the gods, made man and the universe. These I call theological explanations. They have one principal characteristic in common — there is little point in asking why things are as they are, since we are given mainly a description of the creation as the gods chose to do it.[12]

Hamming then turns to an extended historical discussion including many examples of mathematics' surprising effectiveness. Complex numbers, for example, originated as an abstract extension to the real numbers, intended to enable mathematicians to solve polynomial equations. They have turned out to be incredibly useful, however, in the study of many areas including fluid flow, steady state temperatures, and electrostatics.

After a number of such examples, Hamming presents an overall assessment.

> Mathematics has been made by man and therefore is apt to be altered rather continuously by him. Perhaps the original sources of mathematics were forced on us, but . . . we see that in the development of so simple a concept as number we have made choices . . . that were only partially controlled by necessity and often it seems to me, more by aesthetics. We

10. *Communications in Pure and Applied Mathematics* 13 (February 1960).
11. *American Mathematical Monthly* 87, no. 2 (February 1980).
12. *American Mathematical Monthly* 87, no. 2 (February 1980): 81.

> have tried to make mathematics a consistent, beautiful thing, and by so doing we have had an amazing number of successful applications to the real world.

He then presents four explanations that are more detailed:

1. *We see what we look for.* I.e., "... we approach the situations with an intellectual apparatus so that we can only find what we do in many cases. It is both that simple and that awful."
2. *We select the kind of mathematics to fit the situation.*
3. *Science in fact answers comparatively few problems.* "When you consider how much science has not answered then you see that our successes are not so impressive as they might otherwise appear."
4. *The evolution of man provided the model.* While Hamming mentions the evolutionary explanation — that natural selection fitted people with tools that enabled them to function effectively in this world, he does not find this argument helpful. He concludes, "it does not seem to me that evolution can explain more than a small part of the unreasonable effectiveness of mathematics."

Hamming finally concludes, "From all of this I am forced to conclude both that mathematics is unreasonably effective and that all of the explanations I have given when added together simply are not enough to explain what I set out to account for."

Hamming is certainly correct in arguing that our perceptions of nature are often shaped by the mathematics we bring to the act of observing. Indeed, much of our understanding of atomic physics was the result of analogy with much larger objects. For instance, Bohr's model of the atom is based on the solar system. As Hamming himself states, however, his explanations seem unable to account for the fact that manipulation of mathematical symbols has often predicted previously *unobserved* physical phenomena. When these phenomena were looked for, they were found. A famous example of this is Einstein's theory of general relativity, which predicts that light will bend when it passes near an object. This was subsequently observed during an eclipse of the sun.

Mark Steiner has taken a different approach. He too thinks it odd that mathematics, with its formal rules and structure, should apply so well to reality. He argues "that ours is (or rather: appears to be) an intellectually 'user friendly' universe, a universe which allows our species to discover things about it. . . ."[13] Similarly, Stanislas Dehaene summarizes an enormous amount of research in human cognition in an effort to support his conclusion that the human brain appears to be "hard-wired" to do arithmetic.[14]

For Christians these observations are reassurances or enrichments of our belief in God's goodness and the role he has given us as stewards of creation. That is, in making the world and uniquely equipping us to understand it, he has provided the tools we need to carry out our stewardship. These observations remind us of God's general purposes for us and clarify more precisely what those purposes are. A Christian, then, reads the amazing effectiveness of mathematics differently — not as unreasonable, but as a marvelous gift of God and a reminder of the stewardship that he has entrusted to us. Such a reading of mathematics does not shut off further investigation as Hamming suggested. Rather, it gives such investigations a far richer meaning than they would otherwise have.

The Limits of Mathematics

When one mentions "the limits of mathematics," Gödel's theorem will quickly spring into the minds of most mathematicians. This is an extremely important limitation, one that had implications for our thinking about computer intelligence in Chapter 9. As noted earlier, Gödel's theorem tells us that in any consistent mathematical system strong enough to include the natural numbers, there must necessarily exist propositions that cannot be proven to be true or false. That is, our mathematics is (and always will be) incomplete. Mathematicians may also think of mathematical modeling — that mathematical representations of natural phenomena are always simplifications and idealizations. That is, even though they are often very useful, they necessarily misrepresent the complexities of the natural world.

These are serious and important limitations. However, there are others that have not been as widely discussed. This section focuses on these less discussed limitations.

13. Mark Steiner, *The Applicability of Mathematics as a Philosophical Problem* (Cambridge, Mass.: Harvard University Press, 1998).

14. Stanislas Dehaene, *The Number Sense: How the Mind Creates Mathematics* (Oxford: Oxford University Press, 1997).

The collapse of foundationalism discussed earlier, for example, exposed one critical limitation of mathematics — its inability to provide an absolutely certain foundation for human knowledge. Probably few contemporary mathematicians are surprised by this particular limitation; however, for most of the past 2500 years, large numbers of scholars have looked to mathematics to provide precisely this certainty.

Another limitation we discussed earlier is the inability of formal axiomatics to address the psychological origin of axioms. While mathematicians certainly have the freedom to choose whatever axioms they wish, the ones that have proven most fruitful both within mathematics and for applications are ones that are abstractions of properties of the real numbers or of physical systems. Thus, the selection of axioms is not completely arbitrary; their usefulness is dependent on entities outside of mathematics. That is, mathematics is indeed not autonomous, and this is a limitation.

One of mathematics' greatest capabilities is that it is a means of communication. In Chapter 7, we discussed the role of mathematics in culture and saw that mathematics has enhanced communication in many ways — for example, it enabled seventeenth- and eighteenth-century scientists to communicate across cultural boundaries. Today, through vehicles as familiar as graphs, charts, and statistics, it enhances communication of many ideas to popular audiences. However, as a communication tool, mathematics has severe limitations. For example, to communicate an idea quickly, a metaphor is typically far more effective. Also, a well-chosen story can evoke empathy that a number or graph can never produce. Because of its precise measurements and its requirement that terms be defined unambiguously, mathematics can be very effective at enabling large groups of people to understand certain ideas in the same way. That is, it greatly reduces confusion among people who make the effort to understand an idea being presented mathematically. However, the price for this clarity is hard work — it is typically more difficult to understand mathematics than metaphors or stories. Thus, even though mathematics is an important communication tool, the scope of what it can communicate and to whom is quite limited.

Furthermore, mathematics is today widely used to address many issues in business, education, and government. Its preciseness is very appealing to political leaders who need to formulate policies that will stand up to public scrutiny and to government agencies that need to act consistently. However, as we saw in Chapter 7, the use of mathematics in such settings affects people and thus involves broader issues of values, norms, and purposes. Issues that arise in these settings often involve concepts that cannot be precisely defined or meaningfully quantified. Again, these are limitations of mathematics — values, norms, and purposes must be imported from outside mathematics; imprecisely defined concepts may still be important. Note that we are not saying these are failings of mathematics — mathematics can be very useful in such settings. Rather, we are saying that mathematics has limitations and must be accompanied by other sources of knowledge and other modes of thought.

During the "age of reason" — the eighteenth and nineteenth century — a group of artists and writers known as the romantics objected that rationalism neglected important dimensions of the human personality — intuition, aesthetics, affection, and so forth. We agree, and we suggest that the romantic objection applies to mathematics in general, not just to the historical movement known as rationalism. The capacity to do mathematics is a marvelous gift of God and a powerful tool. However, one of the consequences of Christian presuppositions is a perspective that such capacity is *a* tool, not *the* tool. It is one gift among many. The Christian mathematician can explore it with joy and gratitude, while also looking to other modes of expression such as drama, poetry, and visual imagery as of equal value. In his letter to the Philippians, the apostle Paul writes, "Finally, brothers, whatever is true, whatever is noble, whatever is right, whatever is pure, whatever is lovely, whatever is admirable — if anything is excellent or praiseworthy — think about such things."[15] Mathematics can indeed be a part of fulfilling this advice, particularly in identifying what is true. However, Paul also is referring to dimensions of value, justice, genuineness, aesthetics, and praiseworthiness that in many ways transcend mathematics. A Christian mathematician needs to be careful that in developing expertise in rational analysis, she recognizes the value of other matters.

In summary, then, we have here yet another way that a Christian might read the experience of mathematics differently from an unbeliever. She sees it in a much broader setting of God's purposes for human culture. This enables her to enjoy its capabilities but also to acknowledge its limitations. It also asks her to ask some questions not typically addressed in the mathematics community such as: What is the place of mathematics in the broader context of human thought? What are its unique capabilities and contributions? What are the important human tasks to which it cannot contribute?

15. Philippians 4:8, NIV.

Final Remarks

There are many more issues we have not addressed. Here are a few:

- The relationship between language and reality is complex and mysterious. For example, logic seems to correspond both to cause-and-effect relationships in nature and to permission schemas in human relationships. But the exact nature of this correspondence is still poorly understood. Also, as developmental psychologists such as Jean Piaget have extensively studied, symbolization is a human mental operation that develops very early in childhood and is fundamental to all human cognition. While the particular symbols people use vary from one culture to another, the process of symbolization itself seems to develop independently of culture. Considerable work needs to be done in understanding this process. Ideas drawn from Christian thought may assist in addressing these questions.
- "Truth" is an important word, but it is used differently in mathematics, science, ordinary language, and scripture. In mathematics, it refers to formal propositions that have been warranted by mathematical proof. In science, it refers to theories that have been empirically verified. In ordinary language, it refers to statements that correspond with the way things actually are. Scripture uses "truth" in various ways — the way things are, as indicating ultimate reality (when Jesus says, "I am the way, the truth, and the life"), and as correspondence to an ideal (as in "Here is a true Israelite"). There are different ways of coming to know each of these different kinds of truth. A great deal of harm has arisen from people demanding that spiritual truth be warranted by methods suitable for science or mathematics. These different concepts of truth need to be explicated more clearly and the means suitable to each expressed in a form that non-scholars can easily assimilate.
- Some good work was done in the 1990s on the ethical issues involved in applying statistics and mathematical models in situations that directly affect human beings. Does Christian thought have anything to add to this conversation?
- We have suggested purposes to form a basis for a set of control beliefs for mathematics, purposes that are based on principles such as stewardship, service, and the joy of discovery rather than on isolation and rationalism. Can the details of this view be fleshed out?

- Besides what we have discussed in this book, what additional contributions can mathematics make to our understanding of theology? For example, the concepts of paradox and infinity are very important in each discipline. Can mathematical thinking on these questions provide any insights?

Along the lines of our last point, let us close our book by returning to Hilbert's two problems mentioned at the beginning of this chapter, for the history of their investigation is rife with irony and paradox.

One of the most bizarre connections between Hilbert's first and second problems came as a result of a theorem Kurt Gödel proved in 1940, and yet another theorem proved 23 years later by Paul Cohen (1934-), a Stanford logician. To understand what happened, we need to look at the context of Hilbert's first problem in some detail. Its formulation is due to a contemporary of Hilbert, the great German mathematician Georg Cantor.

Born in 1845, Cantor is one of the few examples of a mathematician whose Christian faith directly influenced the mathematics he produced. Out of obedience to what he said was a voice from God, he proved the well-known result that in some sense there are more irrational numbers than rational numbers, something that would have caused the Pythagoreans we studied in Chapter 2 to shriek in horror, as the idea that there could be any incommensurable quantities at all seemed self-contradictory to them. Cantor designated the symbol $\aleph_0$ to represent collections of numbers whose size equaled that of all the rational numbers, and c to represent collections of numbers whose size equaled that of all the irrational numbers. The continuum hypothesis simply states that there is no collection of numbers whose size is *between* $\aleph_0$ and c, and the challenge of its proof was Hilbert's second problem.

Cantor was sure this result was true and spent years trying to verify it. In August of 1884 he finally had a proof, and wrote to a friend announcing his result. Three months later he wrote again saying that not only was his first proof flawed, but that he now had a proof that the continuum hypothesis was false. Again, he discovered an error in this proof.

By the time Hilbert gave his Paris address, the continuum hypothesis was not yet settled. As we mentioned, the first breakthrough in attacking this problem came from the work of Kurt Gödel in 1940. He was able to show that the continuum hypothesis is completely consistent with the rest of the axioms of arithmetic. This was quite a discovery, and gave mathematicians hope that a proof of the continuum hypothesis would soon be coming. In 1963, however, Paul Cohen proved that the denial of the continuum hypothesis is also completely consistent with the rest of the axioms of arithmetic.

This result is surprising. It seems on one level that the answer to the question, "Is the continuum hypothesis true?" ought to be either yes or no. It turns out, though, that the correct answer is, "Take your pick. If you want to assume the truth of the continuum hypothesis, fine, if not, also fine. In either case, you get a completely consistent system."

Note the irony. Hilbert hoped in conjunction with his second problem to find a way to create a formula that a person could use to determine whether any mathematical proposition were true or false. Kurt Gödel's famous undecidability theorem of 1931 showed that such a formula is impossible to attain, and in 1940 he even helped show that Hilbert's very first problem was an example of something that could not be verified by means of any such formula. That is, Cantor's continuum hypothesis demonstrated that arithmetic is incomplete, calling to mind Gödel's result that a sufficiently powerful consistent mathematical system can never be a complete system.

In formulating and working with this hypothesis, Cantor was forced to wrestle with ideas that are at least paradoxical. That there can be levels of infinity is surely bordering on the impossible, at least to the mind of a person not trained in mathematical thinking. Even expert mathematicians may think there is a paradox involved in the result that between any two rational numbers is an irrational number, and between any two irrational numbers is a rational number, yet the sets of rational and irrational numbers cannot be put into a one-to-one correspondence.

If things can get this convoluted in a simple, logically precise, carefully defined system such as arithmetic, imagine how strange the things that are true might be in other fields as well. We learn, for example, that electrons behave like a wave, also like a particle, and that there is a fundamental uncertainty as to where subatomic particles can be located. We can go on at length listing oddities from many disciplines; we will close our remarks, though, with some theological conundrums.

Christians believe that Jesus is fully God and fully man, that God is all loving and all powerful, but evil exists in the world, that the scriptures say we are to work out our own salvation, but that it is God who is working in us. We see apparently conflicting descriptions of certain events or statements given in the gospels and epistles, such as the apparently conflicting statements concerning human freedom and determinism with respect to salvation. Many Calvinists opt for a strong determinism, citing supporting scriptural passages such as,

> So then He has mercy on whom He desires, and He hardens whom He desires. You will say to me then, "Why does He still find fault? For who resists His will?" On the contrary, who are you, O man, who answers back to God? The thing molded will not say to the molder, "Why did you make me like this," will it? Or does not the potter have a right over the clay, to make from the same lump one vessel for honorable use, and another for common use? What if God, although willing to demonstrate His wrath and to make His power known, endured with much patience vessels of wrath prepared for destruction? And He did so in order that He might make known the riches of His glory upon vessels of mercy, which He prepared beforehand for glory.[16]

It can be convincingly argued that any view short of this compromises the absolute sovereignty of God. On the other hand, many will tolerate no such view, citing supporting scriptural passages such as the following:

> First of all, then, I urge that entreaties and prayers, petitions and thanksgivings, be made on behalf of all men, for kings and all who are in authority, in order that we may lead a tranquil and quiet life in all godliness and dignity. This is good and acceptable in the sight of God our Savior, who desires all men to be saved and to come to the knowledge of the truth.[17]

They argue that it makes no sense for God to pre-assign people to eternal destinies if he ". . . desires *all* men to be saved. . . ." Or, consider the passage, "Let the one who wishes take the water of life without cost."[18] Again, they argue that any sense of the normal meaning of an invitation intrinsic to this verse is mocked if indeed God pre-assigns people to their eternal destinies. It can also be convincingly argued that any view short of one that allows for human choice compromises the absolute justice and love of God.

16. Romans 9:18-23, NASB.
17. 1 Timothy 2:1-4, NASB.
18. Revelation 22:17, NASB.

How do we put all this together? Well, there are many ways, but we suggest that a strong case can be made here for a compatibilist view. This would be the approach that affirms both positions. Considering the paradoxical mathematical statements we affirm, it seems that this is certainly not intellectual suicide. No matter what position we adopt, we are left with some tough questions: For the Calvinist, why did God choose person *x* and not person *y*, together with scriptural data supporting an Arminian view. For the Arminian, why did person *x* choose to believe and person *y* choose not to believe, again with scriptural data supporting a Calvinist view. For a compatibilist, how can free will and predestination both be true?

We are not arguing for an anti-rationalist point of view. That is, we do not mean to imply that the paradoxes about our faith are really contradictory, and that we mindlessly affirm conflicting views, as obedient automata. But what we do want to suggest is this: Insofar as working out positions on various issues demands the interplay of logical ideas, perhaps the dichotomy between consistency and completeness applies there as well. In other words, (speaking metaphorically, of course) perhaps Gödel's incompleteness theorem applies in some way to all areas that operate with a certain core of logical substrata. In the working out of our world and life view, of course we strive for both consistency and completeness. However, as we near one of these ideals, it appears at times we also near the opposite of the other. Does our Christian world and life view appear paradoxical? Perhaps it is because it needs reworking, and because it is somewhat inconsistent. Perhaps, on the other hand, we see paradoxical things in the scriptures precisely because they give us a fairly *complete* picture of God. Perhaps to our finite and fallen minds a complete picture of God appears somewhat inconsistent. As Pascal wrote in his *Pensées,*

> What sort of freak, then, is man! How novel, how monstrous, how chaotic, how paradoxical, how prodigious! Judge of all things, feeble earthworm, repository of truth, sink of doubt and error, glory and refuse of the universe! . . . Know then, proud man, what a paradox you are to yourself. Be humble, impotent reason! Be silent, feeble nature! . . . Listen to God. . . . Is it not as clear as day that man's condition is dual? The point is that if man had never been corrupted, he would, in his innocence, confidently enjoy both truth and felicity, and, if man had never been anything but corrupt, he would have no idea either of truth or bliss. But unhappy as we are (and we should be less so if there were no

> element of greatness in our condition) we have an idea of happiness but we cannot attain it. We perceive an image of the truth and possess nothing but falsehood, being equally incapable of absolute ignorance and certain knowledge; so obvious is it that we once enjoyed a degree of perfection from which we have unhappily fallen.

Of course, if we accept this notion, it comes at a price. For one thing, it means that we cannot be so quick to brand views that are at variance with ours as contradictory. As we've indicated throughout this book, it demands a certain amount of epistemological humility in dialoguing with others. Furthermore, although we strive for both consistency and completeness, if we take an honest look at where we are at any moment, we'll probably admit that we don't have either. We'll have to confess that we are really somewhere in between — between consistency and completeness. We, along with all Christian thinkers, live with that tension. On the other hand, we continually try to resolve it. We hope the ideas we have addressed in this book have contributed to that end.

Select Bibliography

Archimedes, *Works, with a Supplement "The Method" of Archimedes Recently Discovered by Heiberg.* Ed. Thomas Little Heath. Cambridge University Press, 1897.

Aspray, William, and Kitcher, Philip, eds. *History and Philosophy of Modern Mathematics.* Minnesota Studies in the Philosophy of Science, vol. 11. University of Minnesota Press, 1988.

Barbour, Ian, *Issues in Science and Religion.* SCM Press, 1966.

Barnes, Jonathan, *Early Greek Philosophy.* Penguin, 1987.

Barrow, John, and Tipler, Frank. *The Anthropic Cosmological Principle.* Oxford University Press, 1986.

Barwise, J., and Etchemendy, J. *The Liar: An Essay on Truth and Circularity.* Oxford University Press, 1987.

Barwise, J., and Perry, J. *Situations and Attitudes.* MIT Press, 1983.

Bealer, G. *Quality and Concept.* Oxford University Press, 1982.

Behe, Michael. *Darwin's Black Box.* Free Press, 1996.

Berggren, J. Lennart. "Islamic Acquisition of the Foreign Sciences: A Cultural Perspective." *American Journal of Islamic Social Sciences* 9 (1992): 310-24.

Berliner, D., and Calfee, R., eds. *Handbook of Educational Psychology.* Macmillan, 1996.

Blackmore, John T. *Ernst Mach: His Work, Life, and Influence.* University of California Press, 1972.

Bohm, David. *The Undivided Universe: An Ontological Interpretation of Quantum Theory.* Routledge, 1993.

Bradley, W. James, and Schaefer, Kurt C. *The Uses and Misuses of Data and Models: The Mathematization of the Human Sciences.* Sage Publications, 1998.

Bradley, Raymond, and Swartz, Norman. *Possible Worlds: An Introduction to Logic and Its Philosophy.* Hackett, 1979.

Burtt, Edwin Arthur. *The Metaphysical Foundations of Physical Science.* Second edition. Doubleday Anchor Books, 1954.

Cantor, G. *Gesammelte Abhandlungen.* Springer, 1932.

Caspar, Max. *Kepler.* Revised edition. Dover Publications, 1993.

Cassirer, Ernst. *The Philosophy of the Enlightenment.* Princeton University Press, 1951.

Chaitin, Gregory J. "On the Length of Programs for Computing Finite Binary Sequences." *Journal of the Association for Computing Machinery* 13 (1966): 547-69.

Chalmers, David J. *The Conscious Mind: In Search of a Fundamental Theory.* Oxford University Press, 1996.

Chihara, C. *Constructibility and Mathematical Existence.* Oxford University Press, 1990.

Chisholm, R. *The First Person.* Minneapolis: University of Minnesota Press, 1981.

Clouser, Roy A. *The Myth of Religious Neutrality: An Essay on the Hidden Role of Religious Beliefs in Theories.* University of Notre Dame Press, 1991.

Cohen, H. Floris. *The Scientific Revolution: A Historiographical Inquiry.* University of Chicago Press, 1994.

Copenhaver, Brian P., and Schmitt, Charles B. *Renaissance Philosophy: A History of Western Philosophy.* Vol. 3. Oxford University Press, 1992.

Copleston, Frederick, S.J. *A History of Philosophy,* Vol. 4. Burns, Oates and Washbourne, Ltd., 1958.

Copleston, Frederick C. *A History of Medieval Philosophy.* Harper and Row, 1972.

Coulson, C. A. *Science and Religion: A Changing Relationship.* Cambridge University Press, 1955.

Crombie, Alistair Cameron. *Styles of Scientific Thinking in the European Tradition: The History of Argument and Explanation Especially in the Mathematical and Biomedical Sciences and Arts.* 3 vols. Duckworth, 1994.

Crosby, Alfred W. *The Measure of Reality: Quantification and Western Society, 1250-1600.* Cambridge University Press, 1997.

Damasio, Antonio R. *Descartes' Error: Emotion, Reason, and the Human Brain.* Avon Books, 1994.

Daston, Lorraine J. *Classical Probability in the Enlightenment.* Princeton University Press, 1988.

Davies, P. *The Mind of God: The Scientific Basis for a Rational World.* Simon and Schuster, 1992.

Davis, M. "Hilbert's Tenth Problem Is Unsolvable." *American Mathematical Monthly* 80 (1973): 233-69.

Davis, Philip J., and Hersh, Reuben. *Descartes' Dream: The World According to Mathematics.* Harcourt Brace Jovanovich., 1986.

Davis, R., Maher, C., and Noddings, N., eds. *Constructivist Views on the Teaching and Learning of Mathematics.* National Council of Teachers of Mathematics, 1990.

Dear, Peter. *Discipline and Experience: The Mathematical Way in the Scientific Revolution.* University of Chicago Press, 1995.

Dembski, William. *The Design Inference.* Cambridge University Press, 1998.

Detlefsen, Michael, "On Interpreting Gödel's Second Theorem." *Journal of Philosophical Logic* 8, no. 3 (1979): 297-313.

Dowty, D., Wall, R., and Peters, S. *Introduction to Montague Semantics.* Reidel, 1981

Drake, Stillman. *Discoveries and Opinions of Galileo.* Doubleday Anchor Book, 1957.

Drake, Stillman. *Galileo at Work: His Scientific Biography.* University of Chicago Press, 1978.

Dreyfus, H. L. *What Computers Can't Do.* Harper and Row, 1979.

Dyson, F. J. *Mathematics in the Modern World.* W. H. Freeman, 1968.

Eigen, Manfred. *Steps Towards Life: A Perspective on Evolution.* Trans. P. Woolley. Oxford University Press, 1992.

Ellul, J. *The Technological Bluff.* Eerdmans, 1990.

Ellul, J. *The Technological Society.* Knopf, 1964.

Ellul, J. *The Technological System.* Continuum, 1980.

Ernest, Paul. *Social Constructivism as a Philosophy of Mathematics.* SUNY Press, 1998.

Euwe, M., and Meiden, W. *Chess Master Vs. Chess Master.* David McKay Company, Inc., 1977.

Evans, Stephen. *Wisdom and Humanness in Psychology: Prospects for a Christian Approach.* Baker Book House, 1989.

Frege, Gottlob. *Translations from the Philosophical Writings of Gottlob Frege.* Third edition. Ed. Peter Geach and Max Black. Basil Blackwell, 1985.

Fong, G. T., Krantz, D. H., and Nisbett, R. E. "The Effects of Statistical Training on Thinking about Everyday Problems." *Cognitive Psychology* 18 (1986): 253-92.

Foucault, M. *Power/Knowledge: Selected Interviews and Other Writings.* Pantheon, 1980.

Fowler, David. *The Mathematics of Plato's Academy: A New Reconstruction.* Clarendon Press, 1987.

Fraenkel, A. *Abstract Set Theory.* North-Holland, 1961.

Frängsmyr, Tore, Heilbron, John L., and Rider, Robin E., eds. *The Quantifying Spirit in the Eighteenth Century.* University of California Press, 1990.

Frieden, Roy. *Physics from Fisher Information: A Unification.* Cambridge University Press, 1998.

Garey, M. R., and Johnson, D. S. *Computers and Intractability: A Guide to the Theory of NP-Completeness.* Freeman, 1979.

Gaukroger, Stephen. *Descartes: An Intellectual Biography.* Oxford University Press, 1995.

Gillispie, Charles Coulston. *The Edge of Objectivity: An Essay in the History of Scientific Ideas.* Princeton University Press, 1960.

Griggs, R. A., and Cox, J. R. "Permission Schemas and the Selection Task." *The Quarterly Journal of Experimental Psychology* 46 (1993): 637-51.

Hacking, Ian. *The Emergence of Probability.* Cambridge University Press, 1975.

Hallett, M. *Cantorian Set Theory and the Limitation of Size.* Oxford University Press, 1984.

Hamming, R. W. "Mathematics on a Distant Planet." *American Mathematical Monthly* 105 (1998): 640-50.

Hankins, Thomas L. *Science and the Enlightenment.* Cambridge University Press, 1985.

Hardy, G. H. *A Mathematician's Apology.* Cambridge University Press, 1940.

Haugeland, John, ed. *Mind Design.* MIT Press, 1981.

Heath, Thomas Little. *A History of Greek Mathematics.* 2 vols. Clarendon Press, 1921; reprinted Dover, 1981.

Hersh, Reuben. *What Is Mathematics, Really?* Oxford University Press, 1997.

Heyting, Arend. *Intuitionism: An Introduction.* Third revised edition. North Holland, 1971.

Heyting, Arend, ed. *Constructivity in Mathematics.* North-Holland, 1959.

Hiebert, James, ed. *Conceptual and Procedural Knowledge: The Case of Mathematics.* Lawrence Erlbaum Associates Publishers, 1986.

Hilbert, D. "Mathematische Probleme." *Proceedings of the International Congress of Mathematicians in Paris 1900,* Göttingen, 1900, pp. 253-97. Eng. trans. *Bulletin of the American Mathematical Society* 2 (1901-2): 437-39.

Hofstadter, D. R. *Gödel, Escher, Bach: An Eternal Golden Braid.* Basic Books, Inc., 1979.

Høyrup, Jens. "The Formation of 'Islamic Mathematics': Sources and Conditions." *Science in Context* 1, no. 2, pp. 281-329.

Høyrup, Jens. *In Measure, Number, and Weight: Studies in Mathematics and Culture.* SUNY Press, 1994.

Johnson-Laird, P. N., Legrenzi, P., and Legrenzi, M. S. "Reasoning and a Sense of Reality." *British Journal of Psychology* 63 (1972): 395-400.

Joseph, George Gheverghese. *The Crest of the Peacock: Non-European Roots of Mathematics.* Penguin Books, 1991.

Katz, Victor. *A History of Mathematics: An Introduction.* Second edition. Addison-Wesley, 1998.

Kauffman, S. A. *The Origins of Order: Self-Organization and Selection in Evolution.* Oxford University Press, 1993.

Kirkpatrick, S., Gelatt, C., and Vecchi, M., "Optimisation by Simulated Annealing," *Science* 220 (May 1983): 671-80.

Kitcher, P. *The Nature of Mathematical Knowledge.* Oxford University Press, 1983.

Kline, Morris. *Mathematics in Western Culture.* Oxford University Press, 1964.

Kneale, William and Martha. *The Development of Logic.* Oxford University Press, 1988.

Knorr, Wilbur. *The Evolution of the Euclidean Elements.* Reidel, 1975.

Knuth, D. E. *Fundamental Algorithms.* Addison-Wesley, 1969.

Köhler, W. *Gestalt Psychology.* The New American Library, 1947.

Krajicek, Jan, and Pudlak, Pavel. "Propositional Proof Systems: The Consistency of First Order Theories and the Complexity of Computations." *Journal of Symbolic Logic,* 54, no. 3 (1989): 1063-79.

Kramer, J. J., and Conoley, J. C., eds. *Buros' Mental Measurement Handbook.* University of Nebraska Press, 1992.

Kranakis, Evangelos. *Primality and Cryptography.* Wiley-Teubner, 1986.

Kristeller, Paul Oskar. *Renaissance Thought: The Classic, Scholastic, and Humanist Strains.* Harper and Row, 1955.

Kretzmann, Norman, Kenny, Anthony, and Pinborg, Jan, eds. *The Cambridge History of Later Medieval Philosophy.* University of Cambridge Press, 1982.

Kuhn, Thomas. *The Essential Tension: Selected Studies in Scientific Tradition and Change.* University of Chicago Press, 1977.

Kurzweil, Ray. *The Age of Spiritual Machines.* Viking, 1999.

Landauer, Rolf. "Information Is Physical." *Physics Today,* May 1991.

Lave, J. *Cognition in Practice: Mind, Mathematics, and Culture in Everyday Life.* Cambridge University Press, 1988.

Lenoir, Timothy. *The Strategy of Life: Teleology and Mechanics in Nineteenth-Century German Biology.* Reidel, 1982.

Lindberg, David C. *The Beginnings of Western Science: The European Scientific Tradition in Philosophical, Religious, and Institutional Context, 600 BC to AD 1450.* University of Chicago Press, 1992.

Lindberg, David C. "Science as Handmaiden: Roger Bacon and the Patristic Tradition." ISIS 78, no. 294 (Dec. 1987): 518-36.

Lloyd, G. E. R. *Adversaries and Authorities.* Cambridge University Press, 1996.

Lloyd, G. E. R. "Techniques and Dialectic: Method in Greek and Chinese Mathematics and Medicine." In *Method in Ancient Philosophy*, ed. Jyl Gentzler. Clarendon Press, 1998, pp. 354-70.

Lundy, M., and Mees, A. "Convergence of an Annealing Algorithm," *Mathematical Programming* 34 (1986): 111-24.

Lyotard, Jean-François. *The Postmodern Condition.* University of Minnesota Press, 1984.

Maddy, Penelope. *Realism in Mathematics.* Oxford University Press, 1990.

Mancosu, Paolo. *Philosophy of Mathematics and Mathematical Practice in the Seventeenth Century.* Oxford University Press, 1999.

Marsden, G. M. *The Outrageous Idea of Christian Scholarship.* Oxford University Press, 1997.

Martzloff, Jean-Claude. *A History of Chinese Mathematics.* Springer Verlag, 1997.

Matiyasevich, Y. *Hilbert's Tenth Problem.* MIT Press, 1993.

McCorduck, P. *Machines Who Think.* W. H. Freeman, 1979.

McKirahan, Richard D. "Aristotle's Subordinated Sciences." *British Journal for the History of Science* 11, no. 39 (1978): 197-220.

Menzel, C. "Theism, Platonism, and the Metaphysics of Mathematics." *Faith and Philosophy* 4 (1987): 365-82.

Morris, Thomas V. *Anselmian Explorations.* University of Notre Dame Press, 1987.

Morris, Thomas V., ed. *Divine and Human Action.* Cornell University Press, 1988.

National Council of Teachers of Mathematics, Commission on Standards for School Mathematics. *Principles and Standards for School Mathematics.* Reston, VA, 2000.

Needham, Joseph. *Science and Civilization in China.* Vol. 3. Cambridge University Press, 1959.

Newell, A., and Simon, H. A. *Human Problem Solving.* Prentice-Hall, 1972.

Nisbett, R. E., Fong, G. T., Lehman, D. R., and Chang, P. W. "Teaching Reasoning." *Science* 238 (1987): 625-31.

Nisbett, R. E., Krantz, D. H., Jepson, C., and Kunda, Z. "The Use of Statistical Heuristics in Everyday Inductive Reasoning." *Psychological Review* 90 (1983): 339-63.

Novick, L. R., and Holyoak, K. J. "Mathematical Problem Solving by Analogy." *Journal of Experimental Psychology: Learning, Memory, and Cognition* 17 (1991): 398-415.

Nye, Mary Jo, Richards, Joan L., and Stuewer, Roger H., eds. *The Invention of Physical Science: Intersections of Mathematics, Theology and Natural Philosophy since the Seventeenth Century.* Boston Studies in the Philosophy of Science. Vol. 139. Kluwer Academic Publishers, 1992.

Olson, Richard. *Science Deified and Science Defied: The Historical Significance of Science in Western Culture,* vol. 2: *From the Early Modern Age Through the Early Romantic Era, ca. 1640 to ca. 1820.* University of California Press, 1995.

Peacocke, A. *Theology for a Scientific Age: Being and Becoming — Natural and Divine.* Basil Blackwell Ltd, 1990.

Pearcey, Nancy R., and Thaxton, Charles B. *The Soul of Science: Christian Faith and Natural Philosophy.* Crossway Books, 1994.

Perkins, D. N., and Salomon, G. "Are Cognitive Skills Context Bound?" *Educational Researcher* 18 (1989): 23.

Pegis, A. C., ed. *Introduction to St. Thomas Aquinas.* Modern Library, 1948.

Pintrich, R. R., Brown, D. R., and Weinstein, C. E., eds. *Student Motivation, Cognition, and Learning.* Lawrence Erlbaum Associates, 1994.

Plantinga, A. *The Nature of Necessity.* Oxford University Press, 1974.

Polanyi, Michael. "Life Transcending Physics and Chemistry." *Chemical and Engineering News,* 21 Aug. 1967, pp. 54-66.

Polanyi, Michael. "Life's Irreducible Structure." *Science* 113 (1968): 1308-12.

Porter, Theodore M. *Trust in Numbers: The Pursuit of Objectivity in Science and Public Life.* Princeton University Press, 1995.

Reid, Constance. *Hilbert-Courant.* Springer, 1986.

Renyi, Alfred. *Dialogues on Mathematics.* Holden-Day, 1967.

Richardson, Alan. *The Miracle-Stories of the Gospels.* Harper & Brothers, 1942.

Rose, Paul Lawrence. *The Italian Renaissance of Mathematics: Studies on Humanists and Mathematicians from Petrarch to Galileo.* Librairie Droz, 1976.

Rucker, R. *Infinity and the Mind.* Birkhauser, 1982.

Schoenfeld, A. H. "Explorations of Students' Mathematical Beliefs and Behavior." *Journal for Research in Mathematics Education* 20 (1989): 228-355.

Schoenfield, J. *Mathematical Logic.* Addison-Wesley, 1967.

Shapin, Steven. *The Scientific Revolution.* University of Chicago Press, 1996.

Shea, William R. *The Magic of Numbers and Motion: The Scientific Career of René Descartes.* Science History Publications, 1991.

Shea, William R., ed. *Nature Mathematized: Historical and Philosophical Case Studies in Classical Modern Natural Philosophy.* D. Reidel Publishing Company, 1983.

Simon, H. "Studying Human Intelligence by Creating Artificial Intelligence." *American Scientist* 69 (May-June 1981): 300-309.

Snow, Charles Percy. *The Two Cultures and the Scientific Revolution.* Cambridge University Press, 1959.

Steffe, L., and Gale, J., eds. *Constructivism in Education.* Lawrence Erlbaum Associates Publishers, 1995.

Steiner, Mark. *The Applicability of Mathematics as a Philosophical Problem.* Harvard University Press, 1998.

Sternberg, R. J., and Ben-Zeev, T., eds. *The Nature of Mathematical Thinking.* Lawrence Erlbaum Associates, 1996.

Sternberg. R. J., and Smith, E. E., eds. *The Psychology of Human Thought.* Cambridge University Press, 1988.

Stevenson, H. W., and Stigler, J. W. *The Learning Gap.* Summit Books, 1992.

Stigler, J. W., Lee, S., and Stevenson, H. W. *Mathematical Knowledge of Japanese, Chinese, and American Elementary School Children.* National Council of Teachers of Mathematics, 1990.

Strauss, Danie. "A Historical Analysis of the Role of Beliefs in the Three Foundational Crises in Mathematics." In *Facets of Faith of Science,* vol. 2. University Press of America, 1996.

Swetz, Frank J. *Capitalism and Arithmetic: The New Math of the Fifteenth Century.* Open Court, 1987.

Tillich, Paul. *Dynamics of Faith.* Harper & Row, 1954.

Tymoczko, Thomas, ed. *New Directions in the Philosophy of Mathematics.* Revised and expanded edition. Princeton University Press, 1998.

van Heijenoort J., ed. *From Frege to Gödel.* Harvard University Press, 1967.

von Mises, Richard. *Probability, Statistics, and Truth.* Second edition. Dover, 1957.

Vosniadov, S., and Ortony, A., eds. *Similarity and Analogical Reasoning.* Cambridge University Press, 1989.

Waldrop, M. *Complexity: The Emerging Science at the Edge of Order and Chaos.* Simon & Schuster, 1992.

Wallace, William A. *Galileo's Logic of Discovery and Proof: The Background, Content, and Use of His Appropriated Treatises on Aristotle's Posterior Analytics.* Boston Studies in the Philosophy of Science. Vol. 137. Kluwer Academic Publishers, 1992.

Wallace, William A., ed. *Reinterpreting Galileo: Studies in Philosophy and History of Philosophy.* Vol. 15. Catholic University of America Press, 1986.

Wang, H. *From Mathematics to Philosophy.* Routledge & Kegan Paul, 1974.

Wells, Ronald A. *History through the Eyes of Faith: Western Civilization and the Kingdom of God.* Harper and Row, 1989.

Weizenbaum, J. *Computer Power and Human Reason.* W. H. Freeman, 1976.

Westfall, Richard S. *The Life of Isaac Newton.* Cambridge University Press, 1993.

Westfall, Richard S. *Never At Rest: A Biography of Isaac Newton.* Cambridge University Press, 1980.

Weyl, Hermann. *Philosophy of Mathematics and Natural Science.* Revised and augmented English edition, based on a translation by Olaf Helmer. Princeton University Press, 1949.

Wilder, R. L. *Mathematics as a Cultural System.* Pergamon Press, 1981.
Wiles, Andrew J. *Annals of Mathematics* 141, no. 3 (May 1995).
Wilson, Catherine. *The Invisible World: Early Modern Philosophy and the Invention of the Microscope.* Princeton University Press, 1995.
Winograd, T. *Understanding Natural Language.* Academic Press, 1972.
Wittgenstein, Ludwig. *Remarks on the Foundations of Mathematics.* Revised edition. Ed. G. H. von Wright, R. Rhees, and G. E. M. Anscombe. Trans. G. E. M. Anscombe. MIT Press, 1983.
Wolterstorff, Nicholas. *Reason Within the Bounds of Religion.* Second edition. Eerdmans, 1984.
Wouters, Amo. "Viability Explanation." *Biology and Philosophy* 10 (1995): 435-57.
Yanm, Lǐ, and Shíràn, Dù. *Chinese Mathematics: A Concise History.* Clarendon Press, 1987.
Yates, Frances A. *Giordano Bruno and the Hermetic Tradition.* University of Chicago Press, 1964.
Yates, Frances A. *The Rosicrucian Enlightenment.* Routledge, 1972.
Yockey, Hubert. *Information Theory and Molecular Biology.* Cambridge University Press, 1992.
Zalta, E. *Abstract Objects.* D. Reidel, 1983.

Index